Springer Undergraduate Mathematics Series

Advisory Board

Other books in this series

Gareth A. Jones and J. Mary Jones

Elementary Number Theory

 Springer

Gareth A. Jones, MA, DPhil
School of Mathematics, University of Southampton, Highfield, Southampton, SO17 1BJ, UK

J. Mary Jones, MA, DPhil
The Open University, Walton Hall, Milton Keynes, MK7 6AA, UK

Cover illustration elements reproduced by kind permission of:
Aptech Systems, Inc., Publishers of the GAUSS Mathematical and Statistical System, 23804 S.E. Kent-Kangley Road, Maple Valley, WA 98038, USA. Tel: (206) 432 - 7855 Fax (206) 432 - 7832 email: info@aptech.com URL: www.aptech.com
American Statistical Association: Chance Vol 8 No 1, 1995 article by KS and KW Heiner 'Tree Rings of the Northern Shawangunks' page 32 fig 2
Springer-Verlag: Mathematica in Education and Research Vol 4 Issue 3 1995 article by Roman E Maeder, Beatrice Amrhein and Oliver Gloor 'Illustrated Mathematics: Visualization of Mathematical Objects' page 9 fig 11, originally published as a CD ROM 'Illustrated Mathematics' by TELOS: ISBN 0-387-14222-3, German edition by Birkhauser: ISBN 3-7643-5100-4.
Mathematica in Education and Research Vol 5 Issue 3 1995 article by Richard J Gaylord and Kazume Nishidate 'Traffic Engineering with Cellular Automata' page 35 fig 2. Mathematica in Education and Research Vol 5 Issue 2 1996 article by Michael Trott 'The Implicitization of a Trefoil Knot' page 14.
Mathematica in Education and Research Vol 5 Issue 2 1996 article by Lee de Cola 'Coins, Trees, Bars and Bells: Simulation of the Binomial Process' page 19 fig 3. Mathematica in Education and Research Vol 5 Issue 2 1996 article by Richard Gaylord and Kazume Nishidate 'Contagious Spreading' page 33 fig 1. Mathematica in Education and Research Vol 5 Issue 2 1996 article by Joe Buhler and Stan Wagon 'Secrets of the Madelung Constant' page 50 fig 1.

British Library Cataloguing in Publication Data
Jones, Gareth A.
   Elementary number theory. - (Springer undergraduate mathematics series)
   1. Number theory
   I. Title II. Jones, J. Mary
   512.7'2
ISBN 3540761977

Library of Congress Cataloging-in-Publication Data
Jones, Gareth A.
   Elementary number theory / Gareth A. Jones and J. Mary Jones.
     p.   cm. -- (Springer undergraduate mathematics series)
   Includes bibliographical references and index.
   ISBN 3-540-76197-7 (pbk.: alk. paper)
   1. Number theory. I. Jones, J. Mary (Josephine Mary), 1946-
   II. Title. III. Series.
QA241. J62 1998   97-41193
512'.7—dc21   CIP

Springer Undergraduate Mathematics Series ISSN 1615-2085
ISBN 3-540-76197-7
Springer Science+Business Media
springeronline.com

© Springer-Verlag London Limited 1998
Printed in Great Britain
8th printing 2005

Typeset by Focal Image, London
Printed and bound at the Athenæum Press Ltd., Gateshead, Tyne & Wear
12/3830-7 Printed on acid-free paper SPIN 11383383

Preface

Our intention in writing this book is to give an elementary introduction to number theory which does not demand a great deal of mathematical background or maturity from the reader, and which can be read and understood with no extra assistance. Our first three chapters are based almost entirely on A-level mathematics, while the next five require little else beyond some elementary group theory. It is only in the last three chapters, where we treat more advanced topics, including recent developments, that we require greater mathematical background; here we use some basic ideas which students would expect to meet in the first year or so of a typical undergraduate course in mathematics. Throughout the book, we have attempted to explain our arguments as fully and as clearly as possible, with plenty of worked examples and with outline solutions for all the exercises.

There are several good reasons for choosing number theory as a subject. It has a long and interesting history, ranging from the earliest recorded times to the present day (see Chapter 11, for instance, on Fermat's Last Theorem), and its problems have attracted many of the greatest mathematicians; consequently the study of number theory is an excellent introduction to the development and achievements of mathematics (and, indeed, some of its failures). In particular, the explicit nature of many of its problems, concerning basic properties of integers, makes number theory a particularly suitable subject in which to present modern mathematics in elementary terms.

A second reason is that many students nowadays are unfamiliar with the notion of formal proof; this is best taught in a concrete setting, rather than as an abstract exercise in logic, but earlier choices of context, such as geometry and analysis, have suffered from the conceptual difficulty and abstract nature of their subject-matter, whereas number theory is about very familiar and easily manipulated objects, namely integers. We therefore see this book as a vehicle for explaining how mathematicians go about their business, finding experimental evidence, making conjectures, creating proofs and counterexamples, and so on.

A third reason is that many students prefer computation to abstraction, and number theory, with its discrete, precise nature, is an ideal topic in which to perform numerical experiments and calculations. Many of these can be done by hand, and throughout the book we have given examples and exercises of an algorithmic nature. Nowadays, almost every student has access to computing facilities far in excess of anything the great calculator Gauss could have imagined, and for a few of our exercises such electronic assistance is desirable or even essential. We have not linked our approach to any particular machine, programming language or computer algebra system, since even a fairly primitive pocket calculator or personal computer can greatly enhance one's ability to do number theory (and part of the fun lies in persuading it to do so).

A final reason for learning number theory is that, despite Hardy's (1940) famous but now out-dated claim, it is useful. Its best-known modern application is to the cryptographic systems which allow banks, commercial companies, military establishments, and so on to exchange information in securely-encoded form; many of these systems are based on such number-theoretic properties as the apparent difficulty of factorising very large integers (see Chapters 2 and 5). Physicists, engineers and computer scientists are also finding that number-theoretic concepts are playing an increasing role in their work. These applications were not the original motivation for the great developments in number theory, but their emergence can only add to the importance of the subject.

The first three chapters of this book are intended to be accessible to anyone with a little A-level mathematics. In particular, they are suitable for first-year university students and for the more advanced sixth-formers. Equivalence relations appear in Chapter 3, but otherwise no abstract mathematics is used. Proof by induction is used several times, and three versions of this (including strong induction and the well-ordering principle) are summarised in Appendix A. Chapters 4–8 are a little more algebraic in flavour, and require slightly greater mathematical maturity. Here, it is helpful if the reader has met some elementary group theory (subgroups, cyclic groups, direct products, isomorphisms), and knows what rings and fields are; these topics are summarised in Appendix B. Probabilities are also mentioned, though not in any essential way. These chapters are therefore suitable for second- or third-year students, and also for those first-year students sufficiently interested to want to read further. The last three chapters are more advanced, relying on ideas from other areas of mathematics such as analysis, calculus, geometry and algebra which students will almost certainly have met early in their undergraduate studies; these include convergence (summarised in Appendix C), power series, complex numbers and vector spaces. These chapters should therefore be suitable for students at second- or third-year level. The final chapter, which traces Fermat's Last Theorem from its ancient roots to its recent proof, is rather more descriptive and historical in style than the others, but we have tried to include sufficient technical detail to give the reader a flavour of this exciting topic.

The early parts of the book could be used as a first-year introduction to the concepts and methods of pure mathematics, while the rest could form the basis for a more specialised second- or third-year course in number theory. Indeed, many of the chapters are based on courses we have taught to first- and third-year mathematics students at the University of Southampton. The book is also suitable for other students, such as computer scientists and physicists, who want an elementary introduction which brings them up to date with recent developments in the subject.

The two essentials for starting number theory are confidence with traditional algebraic manipulation, and some conception of formal proof. Unfortunately, the recent expansion of university education in the UK has coincided with a decline in numbers taking Further Mathematics A-level, so mathematics students now arrive at university much less familiar with these topics than their predecessors were. In our first few chapters we have therefore taken a more leisurely approach than is traditional, using simple results in number theory to illustrate methods of proof, and emphasising algorithmic and computational aspects in parallel with theory. In later chapters, the pace is rather brisker, but even here we have attempted to present our arguments in as simple terms as possible in order to make them more widely accessible. In the case of some advanced results, this has forced us to concentrate on special cases, or to give only outline proofs, but we think this is a worthwhile sacrifice if it conveys to our readers some feeling of what high-level mathematics is like and how it is done – too many mathematics students graduate with only the vaguest idea of the great problems and achievements of their subject.

We would like to thank Peter Neumann for showing us how to discover and communicate mathematics, and many of our colleagues at Southampton, especially Ann and Keith Hirst and David Singerman, for their sound advice on teaching mathematics in general and number theory in particular. We are very grateful to Susan Hezlet and her colleagues at Springer for their advice and encouragement. It is also traditional to thank one's partner for patience and tolerance during the preparation of a book; instead, we shall simply thank our children for not playing their music any louder than was absolutely necessary.

Contents

Notes to the Reader

Mathematics is a difficult subject to read, and number theory is no exception, even if its subject matter is less abstract than some other topics. Do not be surprised, therefore, if it takes you several attempts before you completely understand an argument. It is often useful when reading mathematics to make notes and to do calculations as you go along; for instance, a general argument can often be clarified by seeing how it works in some specific cases.

Exercises are an important part of the learning process, and you are encouraged to attempt them while reading each section; we have generally placed them immediately after the topics on which they are based, to reinforce your understanding of those topics. Supplementary exercises, which are generally more demanding, are placed at the end of a chapter; they can refer to anything in that chapter, and possibly also to topics covered in earlier chapters. Answers or outline solutions for all the exercises are given at the end of the book; how-

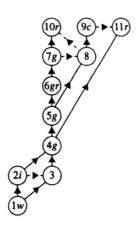

ever, there is a great deal more to be gained from trying the exercises first, before reading the solutions!

The diagram on page xiii shows the interdependence of chapters, with continuous and broken lines indicating strong and weak links. Thus, to understand Chapter 11 it is sufficient to have read Chapters 1–4, though it also helps to know a little of the material in Chapter 9. The letters i and w indicate that the principles of induction and well-ordering are used; these are summarised in Appendix A. Similarly g and r refer to material on groups and rings (Appendix B), and c to convergence (Appendix C).

1
Divisibility

We start with a number of fairly elementary results and techniques, mainly about greatest common divisors. You have probably met some of this material already, though it may not have been treated as formally as here. There are several good reasons for giving very precise definitions and proofs, even when there is general agreement about the validity of the mathematics involved. The first is that 'general agreement' is not the same as convincing proof: it is not unknown for majority opinion to be seriously mistaken about some point. A second reason is that, if we know exactly what assumptions are required in order to deduce certain conclusions, then we may be able to deduce similar conclusions in other areas where the same assumptions hold true. For example, this chapter is entirely devoted to the divisibility properties of *integers*, but it turns out that very similar definitions, methods and theorems are valid for certain other objects which can be added, subtracted and multiplied; some of these objects, such as polynomials, are very familiar, while others, such as Gaussian integers and quaternions, will be introduced in later chapters. These generalisations of the integers are also explored in algebra, under the heading of ring theory.

1.1 Divisors

Our starting-point is the *division algorithm*, which is as follows:

Theorem 1.1

If a and b are integers with $b > 0$, then there is a unique pair of integers q and r such that

$$a = qb + r \qquad \text{and} \qquad 0 \leq r < b.$$

Example 1.1

If $a = 9$ and $b = 4$ then we have $9 = 2 \times 4 + 1$ with $0 \leq 1 < 4$, so $q = 2$ and $r = 1$; if $a = -9$ and $b = 4$ then $q = -3$ and $r = 3$.

In Theorem 1.1, we call q the *quotient* and r the *remainder*. By dividing by b, so that

$$\frac{a}{b} = q + \frac{r}{b} \qquad \text{and} \qquad 0 \leq \frac{r}{b} < 1,$$

we see that q is the integer part $\lfloor a/b \rfloor$ of a/b, the greatest integer $i \leq a/b$. This makes it easy to calculate q, and then to find $r = a - qb$.

Proof

First we prove existence. Let

$$S = \{ a - nb \mid n \in \mathbb{Z} \} = \{ a, a \pm b, a \pm 2b, \dots \}.$$

This set of integers contains non-negative elements (take $n = -|a|$), so $S \cap \mathbb{N}$ is a non-empty subset of $\mathbb{N}$; by the well-ordering principle (see Appendix A), $S \cap \mathbb{N}$ has a least element, which has the form $r = a - qb \geq 0$ for some integer q. Thus $a = qb + r$ with $r \geq 0$. If $r \geq b$ then S contains a non-negative element $a - (q+1)b = r - b < r$; this contradicts the minimality of r, so we must have $r < b$.

To prove uniqueness, suppose that $a = qb + r = q'b + r'$ with $0 \leq r < b$ and $0 \leq r' < b$, so $r - r' = (q' - q)b$. If $q' \neq q$ then $|q' - q| \geq 1$, so $|r - r'| \geq |b| = b$, which is impossible since r and r' lie between 0 and $b-1$ inclusive. Hence $q' = q$ and so $r' = r$.   □

We can now deal with the case $b < 0$: since $-b > 0$, Theorem 1.1 implies that there exist integers q^* and r such that $a = q^*(-b) + r$ and $0 \leq r < -b$, so putting $q = -q^*$ we again have $a = qb + r$. Uniqueness is proved as before, so combining this with Theorem 1.1 we have:

Corollary 1.2

If a and b are integers with $b \neq 0$, then there is a unique pair of integers q and r such that

$$a = qb + r \qquad \text{and} \qquad 0 \leq r < |b|.$$

(Note that when $b < 0$ we have

$$\frac{a}{b} = q + \frac{r}{b} \qquad \text{and} \qquad 0 \geq \frac{r}{b} > -1,$$

so that in this case q is $\lceil a/b \rceil$, the least integer $i \geq a/b$.)

Example 1.2

As an application, we show that if n is a square then n leaves a remainder 0 or 1 when divided by 4. To prove this, let $n = a^2$. Theorem 1.1 (with $b = 4$) gives $a = 4q + r$ where $r = 0, 1, 2$ or 3, so that

$$n = (4q + r)^2 = 16q^2 + 8qr + r^2.$$

If $r = 0$ then $n = 4(4q^2 + 2qr) + 0$, if $r = 1$ then $n = 4(4q^2 + 2qr) + 1$, if $r = 2$ then $n = 4(4q^2 + 2qr + 1) + 0$, and if $r = 3$ then $n = 4(4q^2 + 2qr + 2) + 1$. In each case, the remainder is 0 or 1.

Exercise 1.1

Find a shorter proof for Example 1.2, based on putting $b = 2$ in Theorem 1.1.

Exercise 1.2

What are the possible remainders when a perfect square is divided by 3, or by 5, or by 6 ?

Definition

If a and b are any integers, and $a = qb$ for some integer q, then we say that b *divides* a, or b is a *factor* of a, or a is a *multiple* of b. For instance, the factors of 6 are $\pm 1, \pm 2, \pm 3$ and ± 6. When b divides a we write $b | a$, and we use the notation $b \nmid a$ when b does not divide a. To avoid common misconceptions, we note that every integer divides 0 (since $0 = 0.b$ for all b), 1 divides every integer, and every integer divides itself. We now record some simple but useful facts about divisibility, proving two of them, and leaving the rest for the reader.

Exercise 1.3

Prove that

(a) if $a|b$ and $b|c$ then $a|c$;

(b) if $a|b$ and $c|d$ then $ac|bd$;

(c) if $m \neq 0$, then $a|b$ if and only if $ma|mb$;

(d) if $d|a$ and $a \neq 0$ then $|d| \leq |a|$.

Theorem 1.3

(a) If c divides $a_1, \ldots, a_k$, then c divides $a_1 u_1 + \cdots + a_k u_k$ for all integers $u_1, \ldots, u_k$.

(b) $a|b$ and $b|a$ if and only if $a = \pm b$.

Proof

(a) If c divides a_i then $a_i = q_i c$ for some integers q_i $(i = 1, \ldots, k)$. Then $a_1 u_1 + \cdots + a_k u_k = q_1 c u_1 + \cdots + q_k c u_k = (q_1 u_1 + \cdots + q_k u_k)c$, and as $q_1 u_1 + \cdots + q_k u_k$ is an integer (since q_i and u_i are) we see that $c|(a_1 u_1 + \cdots + a_k u_k)$.

(b) If $a = \pm b$ then $b = qa$ and $a = q'b$ where $q = q' = \pm 1$, so $a|b$ and $b|a$. Conversely, let $a|b$ and $b|a$, so $b = qa$ and $a = q'b$ for some integers q and q'. If $b = 0$ then the second equation gives $a = 0$, so $a = \pm b$ as required. We can therefore assume that $b \neq 0$. Eliminating a from the two equations, we have $b = qq'b$; cancelling b (possible since $b \neq 0$) we have $qq' = 1$, so $q, q' = \pm 1$ (using Exercise 1.3(d)) and hence $a = \pm b$.   □

Exercise 1.4

If a divides b, and c divides d, must $a + c$ divide $b + d$?

The most useful form of Theorem 1.3(a) is the case $k = 2$, which we record in the following slightly simpler notation.

Corollary 1.4

If c divides a and b, then c divides $au + bv$ for all integers u and v.

Definition

If $d|a$ and $d|b$ we say that d is a *common divisor* (or *common factor*) of a and b; for instance, 1 is a common divisor of any pair of integers a and b. If a and b are not both 0, then Exercise 1.3(d) shows that no common divisor is greater than $\max(|a|, |b|)$, so that among all their common divisors there is a greatest one. This is the *greatest common divisor* (or *highest common factor*) of a and b; it is the unique integer d satisfying

(1) $d|a$ and $d|b$ (so that d is a common divisor),

(2) if $c|a$ and $c|b$ then $c \le d$ (so that no common divisor exceeds d).

However, the case $a = b = 0$ has to be excluded: every integer divides 0 and is therefore a common divisor of a and b, so there is no greatest common divisor in this case. When it exists, we denote the greatest common divisor of a and b by $\gcd(a, b)$, or simply (a, b). This definition extends in the obvious way to the greatest common divisor of any set of integers (not all 0).

One way of finding the greatest common divisor of a and b is simply to list all the divisors of a and all the divisors of b, and to choose the largest integer appearing in both lists. It is clearly sufficient to list positive divisors: if $a = 12$ and $b = -18$, for example, then by writing the positive divisors of 12 as $1, 2, 3, 4, 6, 12$, and those of -18 as $1, 2, 3, 6, 9, 18$, we immediately see that the greatest common divisor is 6. This method can be very tedious when a or b are large, but fortunately there is a more efficient method of calculating greatest common divisors, namely *Euclid's algorithm* (published in Book VII of Euclid's *Elements* around 300 BC). This is based on the following simple observation.

Lemma 1.5

If $a = qb + r$ then $\gcd(a, b) = \gcd(b, r)$.

Proof

By Corollary 1.4, any common divisor of b and r also divides $qb + r = a$; similarly, since $r = a - qb$, it follows that any common divisor of a and b also divides r. Thus the two pairs a, b and b, r have the same common divisors, so they have the same greatest common divisor.   □

Euclid's algorithm uses this repeatedly to simplify the calculation of greatest common divisors by reducing the size of the given integers without changing their greatest common divisor. Suppose we are given two integers a and b (not both 0), and we wish to find $d = \gcd(a, b)$. If $a = 0$ then $d = |b|$, and if $b = 0$ then $d = |a|$, so ignoring these trivial cases we may assume that a and b are both non-zero. Since

$$\gcd(a, b) = \gcd(-a, b) = \gcd(a, -b) = \gcd(-a, -b),$$

we may assume that a and b are both positive. Since $\gcd(a, b) = \gcd(b, a)$ we may assume that $a \geq b$, and by ignoring the trivial case $\gcd(a, a) = a$ we may assume that $a > b$, so

$$a > b > 0.$$

We now use the division algorithm (Theorem 1.1) to divide b into a, and write

$$a = q_1 b + r_1 \qquad \text{with} \qquad 0 \leq r_1 < b.$$

If $r_1 = 0$ then $b | a$, so $d = b$ and we halt. If $r_1 \neq 0$ then we divide r_1 into b and write

$$b = q_2 r_1 + r_2 \qquad \text{with} \qquad 0 \leq r_2 < r_1.$$

Now Lemma 1.5 gives $\gcd(a, b) = \gcd(b, r_1)$, so if $r_2 = 0$ then $d = r_1$ and we halt. If $r_2 \neq 0$ we write

$$r_1 = q_3 r_2 + r_3 \qquad \text{with} \qquad 0 \leq r_3 < r_2,$$

and we continue in this way; since $b > r_1 > r_2 > \ldots \geq 0$, we must eventually get a remainder $r_n = 0$ (after at most b steps) at which point we stop. The last two steps will have the form

$$r_{n-3} = q_{n-1} r_{n-2} + r_{n-1} \qquad \text{with} \qquad 0 < r_{n-1} < r_{n-2},$$

$$r_{n-2} = q_n r_{n-1} + r_n \qquad \text{with} \qquad r_n = 0.$$

Theorem 1.6

In the above calculation we have $d = r_{n-1}$ (the last non-zero remainder).

Proof

By applying Lemma 1.5 to the successive equations for $a, b, r_1, \ldots, r_{n-3}$ we see that

$$d = \gcd(a, b) = \gcd(b, r_1) = \gcd(r_1, r_2) = \cdots = \gcd(r_{n-2}, r_{n-1}).$$

The last equation $r_{n-2} = q_n r_{n-1}$ shows that $r_{n-1} | r_{n-2}$, so $\gcd(r_{n-2}, r_{n-1}) = r_{n-1}$ and hence $d = r_{n-1}$. $\qquad\square$

Example 1.3

To calculate $d = \gcd(1492, 1066)$ we write

$$
\begin{aligned}
1492 &= 1.1066 + 426 \\
1066 &= 2.426 + 214 \\
426 &= 1.214 + 212 \\
214 &= 1.212 + 2 \\
212 &= 106.2 + 0.
\end{aligned}
$$

The last non-zero remainder is 2, so $d = 2$.

In many cases, the value of d can be identified before a zero remainder is reached: since $d = \gcd(a, b) = \gcd(b, r_1) = \gcd(r_1, r_2) = \ldots$, one can stop as soon as one recognises the greatest common divisor of a pair of consecutive terms in the sequence $a, b, r_1, r_2, \ldots$. In Example 1.3, for instance, the remainders 214 and 212 clearly have greatest common divisor 2, so $d = 2$.

Exercise 1.5

Calculate $\gcd(1485, 1745)$.

Supplementary Exercises 1.17–1.24 consider the efficiency of Euclid's algorithm; see also Knuth (1968) for a detailed analysis. Stein's (1967) algorithm is similar, but more suitable for computer implementation: it avoids the time-consuming operation of division, and by concentrating on powers of 2 it exploits the binary arithmetic used in computers.

1.2 Bezout's identity

The following result uses Euclid's algorithm to give a simple expression for $d = \gcd(a, b)$ in terms of a and b:

Theorem 1.7

If a and b are integers (not both 0), then there exist integers u and v such that

$$
\gcd(a, b) = au + bv.
$$

(This equation is sometimes known as *Bezout's identity*. We will see later that the values of u and v are not uniquely determined by a and b.)

Proof

We use the equations which arise when we apply Euclid's algorithm to calculate $d = \gcd(a, b)$ as the last non-zero remainder r_{n-1}. The penultimate equation, in the form

$$r_{n-1} = r_{n-3} - q_{n-1}r_{n-2},$$

expresses d as a multiple of r_{n-3} plus a multiple of r_{n-2}. We then use the previous equation, in the form

$$r_{n-2} = r_{n-4} - q_{n-2}r_{n-3},$$

to eliminate r_{n-2} and express d as a multiple of r_{n-4} plus a multiple of r_{n-3}. We gradually work backwards through the equations in the algorithm, eliminating $r_{n-3}, r_{n-4}, \ldots$ in succession, until eventually we have expressed d as a multiple of a plus a multiple of b, that is, $d = au + bv$ for some integers u and v.   □

Example 1.4

In Example 1.3 we used Euclid's algorithm to calculate d, where $a = 1492$ and $b = 1066$. Using those equations again, we have

$$
\begin{aligned}
d &= 2 \\
&= 214 - 1.212 \\
&= 214 - 1.(426 - 1.214) \\
&= -1.426 + 2.214 \\
&= -1.426 + 2.(1066 - 2.426) \\
&= 2.1066 - 5.426 \\
&= 2.1066 - 5(1492 - 1.1066) \\
&= -5.1492 + 7.1066,
\end{aligned}
$$

so we can take $u = -5$ and $v = 7$. The next exercise shows that the values we have found for u and v are not unique. (Later, in Theorem 1.13, we will see how to determine all possible values for u and v.)

Exercise 1.6

Find a pair of integers $u' \neq -5$ and $v' \neq 7$ such that $\gcd(1492, 1066) = 1492u' + 1066v'$.

Exercise 1.7

Express $\gcd(1485, 1745)$ in the form $1485u + 1745v$.

Exercise 1.8

Show that $c|a$ and $c|b$ if and only if $c|\gcd(a,b)$.

Having seen how to calculate the greatest common divisor of two integers, it is a straightforward matter to extend this to any finite set of integers (not all 0). The method, which involves repeated use of Euclid's algorithm, is based on the following exercise.

Exercise 1.9

Prove that $\gcd(a_1, \ldots, a_k) = \gcd(\gcd(a_1, a_2), a_3, \ldots, a_k)$.

This reduces the problem of calculating the greatest common divisor d of k integers to two smaller problems: we calculate $d_2 = \gcd(a_1, a_2)$ and then $d = \gcd(d_2, a_3, \ldots, a_k)$, involving two and $k-1$ integers respectively. This second problem can be further reduced by calculating $d_3 = \gcd(d_2, a_3)$ and then $d = \gcd(d_3, a_4, \ldots, a_k)$, involving two and $k-2$ integers. Continuing, we eventually reduce the problem to a sequence of $k-1$ calculations involving pairs of integers, each of which can be performed by Euclid's algorithm: we find $d_2 = \gcd(a_1, a_2), d_i = \gcd(d_{i-1}, a_i)$ for $i = 3, \ldots, k$, and put $d = d_k$.·

Example 1.5

To calculate $d = \gcd(36, 24, 54, 27)$ we find $d_2 = \gcd(36, 24) = 12$, then $d_3 = \gcd(12, 54) = 6$, and finally $d = d_4 = \gcd(6, 27) = 3$.

Exercise 1.10

Calculate $\gcd(1092, 1155, 2002)$ and $\gcd(910, 780, 286, 195)$.

Exercise 1.11

Show that if $a_1, \ldots, a_k$ are non-zero integers, then their greatest common divisor has the form $a_1 u_1 + \cdots + a_k u_k$ for some integers $u_1, \ldots, u_k$. Find such an expression where $k = 3$ and $a_1 = 1092, a_2 = 1155, a_3 = 2002$.

Theorem 1.7 states that $\gcd(a, b)$ can be written as a multiple of a plus a multiple of b; using this we shall describe the set of all integers which can be written in this form.

Theorem 1.8

Let a and b be integers (not both 0) with greatest common divisor d. Then an integer c has the form $ax + by$ for some $x, y \in \mathbb{Z}$ if and only if c is a multiple of d. In particular, d is the least positive integer of the form $ax + by$ $(x, y \in \mathbb{Z})$.

Proof

If $c = ax + by$ where $x, y \in \mathbb{Z}$, then since d divides a and b, Corollary 1.4 implies that d divides c. Conversely, if $c = de$ for some integer e, then by writing $d = au + bv$ (as in Theorem 1.7) we get $c = aue + bve = ax + by$, where $x = ue$ and $y = ve$ are both integers. Thus the integers of the form $ax + by$ $(x, y \in \mathbb{Z})$ are the multiples of d, and the least positive integer of this form is the least positive multiple of d, namely d itself. $\qquad\square$

Example 1.6

We saw in Example 1.3 that if $a = 1492$ and $b = 1066$ then $d = 2$, so the integers of the form $c = 1492x + 1066y$ are the multiples of 2. Example 1.4 gives $2 = 1492.(-5) + 1066.7$, so multiplying through by e we can express any even integer $2e$ in the form $1492x + 1066y$: for instance, $-4 = 1492.10 + 1066.(-14)$.

Definition

Two integers a and b are *coprime* (or *relatively prime*) if $\gcd(a, b) = 1$. For example, 10 and 21 are coprime, but 10 and 12 are not. More generally, a set $a_1, a_2, \ldots$ of integers are *coprime* if $\gcd(a_1, a_2, \ldots) = 1$, and they are *mutually coprime* if $\gcd(a_i, a_j) = 1$ whenever $i \neq j$. If they are mutually coprime then they are coprime (since $\gcd(a_1, a_2, \ldots) | \gcd(a_i, a_j)$), but the converse is false: the integers $6, 10$ and 15 are coprime but are not mutually coprime.

Corollary 1.9

Two integers a and b are coprime if and only if there exist integers x and y such that

$$ax + by = 1.$$

Proof

Let $\gcd(a, b) = d$. If we put $c = 1$ in Theorem 1.8, we see that $ax + by = 1$ for some $x, y \in \mathbb{Z}$ if and only if $d | 1$, that is, $d = 1$. $\qquad\square$

For example, $10.(-2) + 21.1 = 1$, confirming that 10 and 21 are coprime.

Corollary 1.10

If $\gcd(a, b) = d$ then

$$\gcd(ma, mb) = md$$

for every integer $m > 0$, and

$$\gcd\left(\frac{a}{d}, \frac{b}{d}\right) = 1.$$

Proof

By Theorem 1.8, $\gcd(ma, mb)$ is the smallest positive value of $max + mby = m(ax + by)$, where $x, y \in \mathbb{Z}$, while d is the smallest positive value of $ax + by$, so $\gcd(ma, mb) = md$. Writing $d = au + bv$ and then dividing by d, we have

$$\frac{a}{d}.u + \frac{b}{d}.v = 1,$$

so Corollary 1.9 implies that the intergers a/d and b/d are coprime.   □

Corollary 1.11

Let a and b be coprime integers.

(a) If $a|c$ and $b|c$ then $ab|c$.

(b) If $a|bc$ then $a|c$.

Proof

(a) We have $ax + by = 1$, $c = ae$ and $c = bf$ for some integers x, y, e and f. Then $c = cax + cby = (bf)ax + (ae)by = ab(fx + ey)$, so $ab|c$.

(b) As in (a), $c = cax + cby$. Since $a|bc$ and $a|a$, Corollary 1.4 implies that $a|(cax + cby) = c$.   □

Exercise 1.12

Show that both parts of Corollary 1.11 can fail if a and b are not coprime.

1.3 Least common multiples

Definition

If a and b are integers, then a *common multiple* of a and b is an integer c such that $a|c$ and $b|c$. If a and b are both non-zero, then they have positive common multiples (such as $|ab|$), so by the well-ordering principle they have a *least common multiple* or, more precisely, a least *positive* common multiple; this is the unique positive integer l satisfying

(1) $a|l$ and $b|l$ (so l is a common multiple), and

(2) if $a|c$ and $b|c$, with $c > 0$, then $l \leq c$ (so no positive common multiple is less than l).

We usually denote l by $\mathrm{lcm}(a, b)$, or simply $[a, b]$. For example $\mathrm{lcm}(15, 10) = 30$, since the positive multiples of 15 are $15, 30, 45, \ldots$ while those of 10 are $10, 20, 30, \ldots$. The properties of the least common multiple can be deduced from those of the greatest common divisor, by means of the following result.

Theorem 1.12

Let a and b be positive integers, with $d = \gcd(a, b)$ and $l = \mathrm{lcm}(a, b)$. Then

$$dl = ab.$$

(Since $\gcd(a, b) = \gcd(|a|, |b|)$ and $\mathrm{lcm}(a, b) = \mathrm{lcm}(|a|, |b|)$, it is no great restriction to assume $a, b > 0$.)

Proof

Let $e = a/d$ and $f = b/d$, and consider

$$\frac{ab}{d} = \frac{de.df}{d} = def.$$

Clearly this is positive, so we can show that it is equal to l by showing that it satisfies conditions (1) and (2) of the definition of $\mathrm{lcm}(a, b)$. First,

$$def = (de)f = af \quad \text{and} \quad def = (df)e = be;$$

thus $a|def$ and $b|def$, so (1) is satisfied. Second, suppose that $a|c$ and $b|c$, with $c > 0$; we need to show that $def \leq c$. By Theorem 1.7 there exist integers u and v such that $d = au + bv$. Now

$$\frac{c}{def} = \frac{cd}{(de)(df)} = \frac{cd}{ab} = \frac{c(au + bv)}{ab} = \left(\frac{c}{b}\right)u + \left(\frac{c}{a}\right)v$$

is an integer, since a and b are factors of c; thus $def|c$ and hence (by Exercise 1.3(d)) we have $def \leq c$, as required. □

Example 1.7

If $a = 15$ and $b = 10$, then $d = 5$ and $l = 30$; thus $dl = 150 = ab$, agreeing with Theorem 1.12.

We can use Theorem 1.12 to find $l = \operatorname{lcm}(a, b)$ efficiently by first using Euclid's algorithm to find $d = \gcd(a, b)$, and then calculating $l = ab/d$.

Example 1.8

Since $\gcd(1492, 1066) = 2$ we have $\operatorname{lcm}(1492, 1066) = (1492 \times 1066)/2 = 795236$.

Exercise 1.13

Calculate $\operatorname{lcm}(1485, 1745)$.

Exercise 1.14

Show that c is a common multiple of a and b if and only if it is a multiple of $l = \operatorname{lcm}(a, b)$.

1.4 Linear Diophantine equations

In this book we will consider a number of *Diophantine equations* (named after the 3rd-century mathematician Diophantos of Alexandria): these are equations in one or more variables, for which we seek integer-valued solutions. One of the simplest of these is the *linear Diophantine equation* $ax + by = c$; we can use some of the preceding ideas to find all integer solutions x, y of this equation. The following result was known to the Indian mathematician Brahmagupta, around AD 628:

Theorem 1.13

Let a, b and c be integers, with a and b not both 0, and let $d = \gcd(a, b)$. Then the equation

$$ax + by = c$$

has an integer solution x, y if and only if c is a multiple of d, in which case there are infinitely many solutions. These are the pairs

$$x = x_0 + \frac{bn}{d}, \quad y = y_0 - \frac{an}{d} \quad (n \in \mathbb{Z}),$$

where x_0, y_0 is any particular solution.

Proof

The fact that there is a solution if and only if $d|c$ is merely a restatement of Theorem 1.8. For the second part of the theorem, let x_0, y_0 be a particular solution, so

$$ax_0 + by_0 = c.$$

If we put

$$x = x_0 + \frac{bn}{d}, \quad y = y_0 - \frac{an}{d}$$

where n is any integer, then

$$ax + by = a\left(x_0 + \frac{bn}{d}\right) + b\left(y_0 - \frac{an}{d}\right) = ax_0 + by_0 = c,$$

so x, y is also a solution. (Note that x and y are integers since d divides b and a respectively.) This gives us infinitely many solutions, for different integers n. To show that these are the only solutions, let x, y be any integer solution, so $ax + by = c$. Since $ax + by = c = ax_0 + by_0$ we have

$$a(x - x_0) + b(y - y_0) = 0,$$

so dividing by d we get

$$\frac{a}{d}(x - x_0) = -\frac{b}{d}(y - y_0). \tag{1.1}$$

Now a and b are not both 0, and we can suppose that $b \neq 0$ (if not, interchange the roles of a and b in what follows). Since b/d divides each side of (1.1), and is coprime to a/d by Corollary 1.10, it divides $x - x_0$ by Corollary 1.11(b). Thus $x - x_0 = bn/d$ for some integer n, so

$$x = x_0 + \frac{bn}{d}.$$

Substituting back for $x - x_0$ in (1.1) we get

$$-\frac{b}{d}(y - y_0) = \frac{a}{d}(x - x_0) = \frac{a}{d} \cdot \frac{bn}{d},$$

so dividing by b/d (which is non-zero) we have

$$y = y_0 - \frac{an}{d}.$$

$\square$

Thus we can find the solutions of any linear Diophantine equation $ax + by = c$ by the following method:

(1) Calculate $d = \gcd(a, b)$, either directly or by Euclid's algorithm.

(2) Check whether d divides c: if it does not, there are no solutions, so stop here; if it does, write $c = de$.

(3) If $d|c$, use the method of proof of Theorem 1.7 to find integers u and v such that $au + bv = d$; then $x_0 = ue, y_0 = ve$ is a particular solution of $ax + by = c$.

(4) Now use Theorem 1.13 to find the general solution x, y of the equation.

Example 1.9

Let the equation be

$$1492x + 1066y = -4,$$

so $a = 1492$, $b = 1066$ and $c = -4$. In step (1), we use Example 1.3 to see that $d = 2$. In step (2) we check that d divides c: in fact, $c = -2d$, so $e = -2$. In step (3) we use Example 1.4 to write $d = -5.1492 + 7.1066$; thus $u = -5$ and $v = 7$, so $x_0 = (-5).(-2) = 10$ and $y_0 = 7.(-2) = -14$ give a particular solution of the equation. By Theorem 1.13, the general solution has the form

$$x = 10 + \frac{1066n}{2} = 10 + 533n, \quad y = -14 - \frac{1492n}{2} = -14 - 746n \quad (n \in \mathbb{Z}).$$

Exercise 1.15

Find the general solution of the Diophantine equation $1485x + 1745y = 15$.

It is sometimes useful to interpret the linear Diophantine equation $ax + by = c$ geometrically. If we allow x and y to take any real values, then the graph of this equation is a straight line L in the xy-plane. The points (x, y) in the plane with integer coordinates x and y are the *integer lattice-points*, the vertices of a tessellation (tiling) of the plane by unit squares. Pairs of integers x and y satisfying the equation correspond to integer lattice-points (x, y) on L; thus Theorem 1.13 asserts that L passes through such a lattice-point if and only if $d|c$, in which case it passes through infinitely many of them, with the given values of x and y.

Exercise 1.16

If $a_1, \ldots, a_k$ and c are integers, when does the Diophantine equation $a_1 x_1 + \cdots + a_k x_k = c$ have integer solutions $x_1, \ldots, x_k$?

1.5 Supplementary exercises

Exercise 1.17

Let us define the *height* $h(a)$ of an integer $a \geq 2$ to be the greatest n such that Euclid's algorithm requires n steps to compute $\gcd(a, b)$ for some positive $b < a$ (that is, $\gcd(a, b) = r_{n-1}$). Show that $h(a) = 1$ if and only if $a = 2$, and find $h(a)$ for all $a \leq 8$.

Exercise 1.18

The *Fibonacci numbers* $f_n = 1, 1, 2, 3, 5, \ldots$ are defined by $f_1 = f_2 = 1$, and $f_{n+2} = f_{n+1} + f_n$ for all $n \geq 1$. Show that $0 \leq f_n < f_{n+1}$ for all $n \geq 2$. What happens if Euclid's algorithm is applied when a and b are a pair of consecutive Fibonacci numbers f_{n+2} and f_{n+1}? Show that $h(f_{n+2}) \geq n$.

Exercise 1.19

Suppose that $a > b > 0$, that Euclid's algorithm computes $\gcd(a, b)$ in n steps, and that a is the smallest integer with this property (that is, if $a' > b' > 0$ and $\gcd(a', b')$ requires n steps, then $a' \geq a$); show that a and b are consecutive Fibonacci numbers $a = f_{n+2}$ and $b = f_{n+1}$ (Lamé's Theorem, 1845).

Exercise 1.20

Show that $h(f_{n+2}) = n$, and f_{n+2} is the smallest integer of this height.

Exercise 1.21

Show that $f_n = (\phi^n - \psi^n)/\sqrt{5}$, where ϕ, ψ are the positive and negative roots of $\lambda^2 = \lambda + 1$. Deduce that $f_n = \{\phi^n/\sqrt{5}\}$, where $\{x\}$ denotes the integer closest to x. Hence obtain the approximate upper bound

$$\log_\phi(a\sqrt{5}) - 2 = \log_\phi(a) + \frac{1}{2}\log_\phi(5) - 2 \approx 4.785\log_{10}(a) - 0.328$$

for the number of steps required to compute $\gcd(a, b)$ by Euclid's algorithm, where $a \geq b > 0$.

Exercise 1.22

Show that if a and b are integers with $b \neq 0$, then there is a unique pair of integers q and r such that $a = qb + r$ and $-|b|/2 < r \leq |b|/2$. Use this result instead of Corollary 1.2 to devise an alternative algorithm to Euclid's for calculating greatest common divisors (the *least remainders algorithm*).

Exercise 1.23

Use the least remainders algorithm to compute $\gcd(1066, 1492)$ and $\gcd(1485, 1745)$, and compare the numbers of steps required by this algorithm with those required by Euclid's.

Exercise 1.24

What happens if the least remainders algorithm is applied to a pair of consecutive Fibonacci numbers?

Exercise 1.25

Show that if a and b are coprime positive integers, then every integer $c \geq ab$ has the form $ax + by$ where x and y are non-negative integers. Show that the integer $ab - a - b$ does not have this form.

2

Prime Numbers

The first main result in this chapter is the Fundamental Theorem of Arithmetic (Theorem 2.3), which asserts that each integer $n > 1$ can be written, in an essentially unique way, as a product of prime-powers. This allows many number-theoretic problems to be reduced to questions about prime numbers, so we devote this chapter to the properties of this important class of integers. The second major result is the theorem of Euclid (Theorem 2.6) that there are infinitely many prime numbers; this result is so fundamental that, during the course of this book, we will give several totally different proofs of it to illustrate different techniques in number theory. Although there are infinitely many prime numbers, they occur rather irregularly among the integers, and we have included a number of results which enable us to predict where primes will appear or how frequently they appear; some of these results, such as the Prime Number Theorem, are quite difficult, and are therefore stated without proof.

2.1 Prime numbers and prime-power factorisations

Definition

An integer $p > 1$ is said to be *prime* if the only positive divisors of p are 1 and p itself.

Note that 1 is not prime. The smallest prime is 2, and all the other primes (such as $3, 5, 7, 11, \ldots$) are odd. A list of the primes $p < 1000$ is given in Appendix D. An integer $n > 1$ which is not prime (such as $4, 6, 8, 9, \ldots$) is said to be *composite*; such an integer has the form $n = ab$ where $1 < a < n$ and $1 < b < n$.

Lemma 2.1

Let p be prime, and let a and b be any integers. Then

(a) either p divides a, or a and p are coprime;

(b) if p divides ab, then p divides a or p divides b.

Proof

(a) By its definition, $\gcd(a, p)$ is a positive divisor of p, so it must be 1 or p since p is prime. If $\gcd(a, p) = p$, then since $\gcd(a, p)$ divides a we have $p|a$; if $\gcd(a, p) = 1$ then a and p are coprime.

(b) Let $p|ab$. If p does not divide a then part (a) implies that $\gcd(a, p) = 1$. Now Bezout's identity (Theorem 1.7) gives $1 = au + pv$ for some integers u and v, so $b = aub + pvb$. By our assumption, p divides ab and hence divides aub; it clearly divides pvb, so it also divides b, as required.   □

Both parts of this result can fail if p is not prime: take $p = 4, a = 6$ and $b = 10$, for instance. Lemma 2.1(b) can be extended to products of any number of factors:

Corollary 2.2

If p is prime and p divides $a_1 \ldots a_k$, then p divides a_i for some i.

Proof

We use induction on k (see Appendix A). If $k = 1$ then the assumption is that $p|a_1$, so the conclusion is automatically true (with $i = 1$). Now assume that $k > 1$ and that the result is proved for all products of $k - 1$ factors a_i. If we put $a = a_1 \ldots a_{k-1}$ and $b = a_k$, then $a_1 \ldots a_k = ab$ and so $p|ab$. By Lemma 2.1(b), it follows that $p|a$ or $p|b$. In the first case we have $p|a_1 \ldots a_{k-1}$, so the induction hypothesis implies that $p|a_i$ for some $i = 1, \ldots, k - 1$; in the second case we have $p|a_k$. Thus in either case $p|a_i$ for some i, as required.   □

Exercise 2.1

Prove that if p is prime and $p|a^k$, then $p|a$, and hence $p^k|a^k$; is this still valid if p is composite?

As an application of Lemma 2.1(b) (which we will not require until Chapter 11), we consider polynomials with integer coefficients. Such a polynomial $f(x)$ is *reducible* if $f(x) = g(x)h(x)$, where $g(x)$ and $h(x)$ are non-constant polynomials with integer coefficients; otherwise, $f(x)$ is *irreducible*. Eisenstein's criterion states that if $f(x) = a_0 + a_1x + \cdots + a_nx^n$, where each $a_i \in \mathbb{Z}$, if p is a prime such that p divides $a_0, a_1, \ldots, a_{n-1}$ but not a_n, and if p^2 does not divide a_0, then $f(x)$ is irreducible. To prove this, suppose that $f(x)$ is reducible, say $f(x) = g(x)h(x)$ with $g(x) = b_0 + b_1x + \cdots + b_sx^s$, $h(x) = c_0 + c_1x + \cdots + c_tx^t$, and $s, t \geq 1$. Since $a_0 = b_0c_0$ is divisible by p but not p^2, precisely one of b_0, c_0 is divisible by p; transposing $g(x)$ and $h(x)$ if necessary, we may assume that p divides b_0 but not c_0. Now p cannot divide b_s, for otherwise it would divide $a_n = b_sc_t$; hence there exists $i \leq s$ such that p divides $b_0, b_1, \ldots, b_{i-1}$ but not b_i. Now $a_i = b_0c_i + b_1c_{i-1} + \cdots + b_{i-1}c_1 + b_ic_0$, with p dividing both a_i (since $i \leq s = n - t < n$) and $b_0c_i + \cdots + b_{i-1}c_1$, so p divides b_ic_0. Then Lemma 2.1(b) implies that p divides b_i or c_0, which is a contradiction, so $f(x)$ must be irreducible.

Example 2.1

The polynomial $f(x) = x^3 - 4x + 2$ is irreducible, since it satisfies Eisenstein's criterion with $p = 2$.

Example 2.2

Consider the p-th *cyclotomic polynomial*

$$\Phi_p(x) = 1 + x + x^2 + \cdots + x^{p-1},$$

where p is prime. (For an application of this polynomial, and an explanation of its name, see Chapter 11, Section 9.) To show that $\Phi_p(x)$ is irreducible, we cannot apply Eisenstein's criterion directly; however, it is sufficient to show that the polynomial $f(x) = \Phi_p(x + 1)$ is irreducible, since any factorisation $g(x)h(x)$ of $\Phi_p(x)$ would imply a similar factorisation $g(x+1)h(x+1)$ of $f(x)$. Now $\Phi_p(x) = (x^p - 1)/(x - 1)$, so replacing x with $x + 1$ we get

$$f(x) = \frac{(x+1)^p - 1}{x} = x^{p-1} + \binom{p}{p-1}x^{p-2} + \cdots + \binom{p}{2}x + \binom{p}{1}$$

by the Binomial Theorem. Since p is prime, the binomial coefficients $\binom{p}{i} = p!/i!(p-i)!$ are divisible by p for $i = 1, \ldots, p-1$; moreover, p does not divide the leading coefficient ($= 1$) of $f(x)$, and p^2 does not divide the constant term $\binom{p}{1} = p$. Thus $f(x)$ is irreducible by Eisenstein's criterion, and hence so is $\Phi_p(x)$.

The next result, known as the *Fundamental Theorem of Arithmetic*, explains why prime numbers are so important: they are the basic building blocks out of which all integers can be constructed.

Theorem 2.3

Each integer $n > 1$ has a prime-power factorisation

$$n = p_1^{e_1} \ldots p_k^{e_k},$$

where $p_1, \ldots, p_k$ are distinct primes and $e_1, \ldots e_k$ are positive integers; this factorisation is unique, apart from permutations of the factors.

(For instance, 200 has prime-power factorisation $2^3.5^2$, or alternatively $5^2.2^3$ if we permute the factors, but it has no other prime-power factorisations.)

Proof

First we use the principle of strong induction (see Appendix A) to prove the existence of prime-power factorisations. Since we are assuming that $n > 1$, the induction starts with $n = 2$. As usual, this case is easy: the required factorisation is simply $n = 2^1$. Now assume that $n > 2$ and that every integer strictly between 1 and n has a prime-power factorisation. If n is prime then $n = n^1$ is the required factorisation of n, so we can assume that n is composite, say $n = ab$ where $1 < a, b < n$. By the induction hypothesis, both a and b have prime-power factorisations, so by substituting these into the equation $n = ab$ and then collecting together powers of each prime p_i we get a prime-power factorisation of n.

Now we prove uniqueness. Suppose that n has prime-power factorisations

$$n = p_1^{e_1} \ldots p_k^{e_k} = q_1^{f_1} \ldots q_l^{f_l},$$

where $p_1, \ldots, p_k$ and $q_1, \ldots, q_l$ are two sets of distinct primes, and the exponents e_i and f_j are all positive. The first factorisation shows that $p_1 | n$, so Corollary 2.2 (applied to the second factorisation) implies that $p_1 | q_j$ for some $j = 1, \ldots, l$. By permuting (or renumbering) the prime-powers in the second factorisation we may assume that $j = 1$, so $p_1 | q_1$. Since q_1 is prime, it follows that $p_1 = q_1$, so cancelling this prime from the two factorisations we get

$$p_1^{e_1-1} p_2^{e_2} \ldots p_k^{e_k} = q_1^{f_1-1} q_2^{f_2} \ldots q_l^{f_l}.$$

We keep repeating this argument, matching primes in the two factorisations and then cancelling them, until we run out of primes in one of the factorisations. If one factorisation runs out before the other, then at that stage our reduced factorisations express 1 as a product of primes p_i or q_j, which is impossible since $p_i, q_j > 1$. It follows that both factorisations run out of primes simultaneously, so we must have cancelled the e_i copies of each p_i with the same number (f_i) of copies of q_i; thus $k = l$, each $p_i = q_i$ (after permuting factors), and each $e_i = f_i$, so we have proved uniqueness.   □

Theorem 2.3 allows us to use prime-power factorisations to calculate products, quotients, powers, greatest common divisors and least common multiples. Suppose that integers a and b have factorisations

$$a = p_1^{e_1} \ldots p_k^{e_k} \quad \text{and} \quad b = p_1^{f_1} \ldots p_k^{f_k}$$

(where we have $e_i, f_i \geq 0$ to allow for the possibility that some primes p_i may divide one but not both of a and b). Then we have

$$
\begin{aligned}
ab &= p_1^{e_1+f_1} \ldots p_k^{e_k+f_k}, \\
a/b &= p_1^{e_1-f_1} \ldots p_k^{e_k-f_k} \quad \text{(if } b|a), \\
a^m &= p_1^{me_1} \ldots p_k^{me_k}, \\
\gcd(a, b) &= p_1^{\min(e_1,f_1)} \ldots p_k^{\min(e_k,f_k)}, \\
\operatorname{lcm}(a, b) &= p_1^{\max(e_1,f_1)} \ldots p_k^{\max(e_k,f_k)},
\end{aligned}
$$

where $\min(e, f)$ and $\max(e, f)$ are the minimum and maximum of e and f. Unfortunately, finding the factorisation of a large integer can take a very long time!

The following notation is often useful: if p is prime, we write $p^e \| n$ to indicate that p^e is the highest power of p dividing n, that is, p^e divides n but p^{e+1} does not. For instance, $2^3 \| 200$, $5^2 \| 200$, and $p^0 \| 200$ for all primes $p \neq 2, 5$. The preceding results show that if $p^e \| a$ and $p^f \| b$ then $p^{e+f} \| ab$, $p^{e-f} \| a/b$ (if $b|a$), $p^{me} \| a^m$, etc.

The following result looks rather obvious and innocuous, but in later chapters we shall see that it can be extremely useful, especially in the case $m = 2$:

Lemma 2.4

If $a_1, \ldots, a_r$ are mutually coprime positive integers, and $a_1 \ldots a_r$ is an m-th power for some integer $m \geq 2$, then each a_i is an m-th power.

Proof

It follows from the above formula for a^m that a positive integer is an m-th power if and only if the exponent of each prime in its prime-power factorisation is divisible by m. If $a = a_1 \ldots a_r$, where the factors a_i are mutually coprime, then each prime power p^e appearing in the factorisation of any a_i also appears as the full power of p in the factorisation of a; since a is an m-th power, e is divisible by m, so a_i is an m-th power.   □

Of course, it is essential to assume that $a_1, \ldots, a_r$ are mutually coprime here: for instance, neither 24 nor 54 are perfect squares, but their product $24 \times 54 = 1296 = 36^2$ is a perfect square. (A *perfect square* is an integer of the form $m = n^2$, where n is an integer.)

Exercise 2.2

Find the prime-power factorisations of 132, of 400, and of 1995. Hence find $\gcd(132, 400)$, $\gcd(132, 1995)$, $\gcd(400, 1995)$ and $\gcd(132, 400, 1995)$

Exercise 2.3

Are the following statements true or false, where a and b are positive integers and p is prime? In each case, give a proof or a counterexample:

(a) If $\gcd(a, p^2) = p$ then $\gcd(a^2, p^2) = p^2$.

(b) If $\gcd(a, p^2) = p$ and $\gcd(b, p^2) = p^2$ then $\gcd(ab, p^4) = p^3$.

(c) If $\gcd(a, p^2) = p$ and $\gcd(b, p^2) = p$ then $\gcd(ab, p^4) = p^2$.

(d) If $\gcd(a, p^2) = p$ then $\gcd(a + p, p^2) = p$.

We can use prime-power factorisations to generalise the classic result (known to the Pythagoreans in the 5th century BC) that $\sqrt{2}$ is irrational. A *rational number* is a real number of the form a/b, where a and b are integers and $b \neq 0$; all other real numbers are *irrational*.

Corollary 2.5

If a positive integer m is not a perfect square, then $\sqrt{m}$ is irrational.

Proof

It is sufficient to prove the contrapositive, that if $\sqrt{m}$ is rational then m is a perfect square. Suppose that $\sqrt{m} = a/b$ where a and b are positive integers. Then

$$m = a^2/b^2 \, .$$

If a and b have prime-power factorisations

$$a = p_1^{e_1} \ldots p_k^{e_k} \quad \text{and} \quad b = p_1^{f_1} \ldots p_k^{f_k}$$

as above, then

$$m = p_1^{2e_1 - 2f_1} \ldots p_k^{2e_k - 2f_k}$$

must be the factorisation of m. Notice that every prime p_i appears an even number of times in this factorisation, and $e_i - f_i \geq 0$ for each i, so

$$m = \left(p_1^{e_1 - f_1} \ldots p_k^{e_k - f_k} \right)^2$$

is a perfect square. □

Exercise 2.4

If m and n are positive integers, under what condition is $m^{1/n}$ rational?

2.2 Distribution of primes

Euclid's Theorem, that there are infinitely many primes, is one of the oldest and most attractive in mathematics. In this book we will give several proofs of this result, very different in style, to illustrate some important techniques in number theory. (It is useful, rather than wasteful, to have several proofs of the same result, since one may be able to adapt these proofs to give different generalisations.) Our first proof (the earliest and simplest) is in Book IX of Euclid's *Elements*:

Theorem 2.6

There are infinitely many primes.

Proof

The proof is by contradiction: we assume that there are only finitely many primes, and then we obtain a contradiction from this, so it follows that there must be infinitely many primes.

Suppose then that the only primes are $p_1, p_2, \ldots, p_k$. Let

$$m = p_1 p_2 \ldots p_k + 1.$$

Since m is an integer greater than 1, the Fundamental Theorem of Arithmetic (Theorem 2.3) implies that it is divisible by some prime p (this includes the possibility that $m = p$). By our assumption, this prime p must be one of the primes $p_1, p_2, \ldots, p_k$, so p divides their product $p_1 p_2 \ldots p_k$. Since p divides both m and $p_1 p_2 \ldots p_k$ it divides $m - p_1 p_2 \ldots p_k = 1$, which is impossible. We deduce that our initial assumption was false, so there must be infinitely many primes. $\square$

We can use this proof to obtain a little more information about how frequently prime numbers occur. Let p_n denote the n-th prime (in increasing order), so that $p_1 = 2$, $p_2 = 3$, $p_3 = 5$, and so on.

Corollary 2.7

The n-th prime p_n satisfies $p_n \leq 2^{2^{n-1}}$ for all $n \geq 1$.

(This estimate is very weak, since in general p_n is significantly smaller than $2^{2^{n-1}}$: for instance $2^{2^3} = 256$, whereas p_4 is only 7. We will meet some better estimates soon.)

Proof

We use strong induction on n. The result is true for $n = 1$, since $p_1 = 2 = 2^{2^0}$. Now assume that the result is true for each $n = 1, 2, \ldots, k$. As in the proof of Theorem 2.6, $p_1 p_2 \ldots p_k + 1$ must be divisible by some prime p; this prime cannot be one of $p_1, p_2, \ldots, p_k$, for then it would divide 1, which is impossible. Now this new prime p must be at least as large as the $(k+1)$-th prime p_{k+1}, so

$$
\begin{aligned}
p_{k+1} \leq p \leq p_1 p_2 \ldots p_k + 1 \leq 2^{2^0} . 2^{2^1} \ldots 2^{2^{k-1}} + 1 &= 2^{1+2+4+\cdots+2^{k-1}} + 1 \\
&= 2^{2^k - 1} + 1 \\
&= \frac{1}{2} . 2^{2^k} + 1 \leq 2^{2^k}.
\end{aligned}
$$

(Here we have used the induction hypothesis, that $p_i \leq 2^{2^{i-1}}$ for $i \leq k$, together with the sum of the finite geometric series $1 + 2 + 4 + \cdots + 2^{k-1} = 2^k - 1$.) This proves the inequality for $n = k + 1$, so by induction it is true for all $n \geq 1$. □

For any real number $x > 0$, let $\pi(x)$ denote the number of primes $p \leq x$; thus $\pi(1) = 0$, $\pi(2) = \pi(2\frac{1}{2}) = 1$, and $\pi(10) = 4$, for example. Let $\lg x = \log_2 x$ denote the logarithm of x to the base 2, defined by $y = \lg x$ if $x = 2^y$ (so $\lg 8 = 3$ and $\lg(\frac{1}{2}) = -1$, for instance).

Corollary 2.8

$\pi(x) \geq \lfloor \lg(\lg x) \rfloor + 1$.

Proof

$\lfloor \lg(\lg x) \rfloor + 1$ is the largest integer n such that $2^{2^{n-1}} \leq x$. By Corollary 2.7 there are at least n primes $p_1, p_2, \ldots, p_n \leq 2^{2^{n-1}}$. These primes are all less than or equal to x, so $\pi(x) \geq n = \lfloor \lg(\lg x) \rfloor + 1$.   □

As before, this result is very weak, and $\pi(x)$ is generally much larger than $\lfloor \lg(\lg x) \rfloor + 1$; for instance, if $x = 10^9$ then $\lfloor \lg(\lg x) \rfloor + 1 = 5$, whereas the number of primes $p \leq 10^9$ is not 5 but approximately 5×10^7. By compiling extensive lists of primes, Gauss conjectured in 1793 that $\pi(x)$ is approximated by the function

$$\mathrm{li}\, x = \int_2^x \frac{dt}{\ln t},$$

or equivalently by $x/\ln x$ (see Exercise 2.5), in the sense that

$$\frac{\pi(x)}{x/\ln x} \to 1 \quad \text{as} \quad x \to \infty.$$

(Here $\ln x = \log_e x$ is the natural logarithm $\int_1^x t^{-1} dt$ of x.) This result, known as the *Prime Number Theorem*, was eventually proved by Hadamard and de la Vallée Poussin in 1896. Its proof is beyond the scope of this book; see Hardy and Wright (1979) or Rose (1988), for example. One can interpret the Prime Number Theorem as showing that the proportion $\pi(x)/\lfloor x \rfloor$ of primes among the positive integers $i \leq x$ is approximately $1/\ln x$ for large x. Since $1/\ln x \to 0$ as $x \to \infty$, this shows that the primes occur less frequently among larger integers than among smaller integers. For instance there are 168 primes between 1 and 1000, then 135 primes between 1001 and 2000, then 127 between 2001 and 3000, and so on.

Exercise 2.5

One version of l'Hôpital's rule states that if $f'(x)/g'(x) \to l$ as $x \to \infty$, then $f(x)/g(x) \to l$ also. Use this to show that

$$\lim_{x \to \infty} \frac{\text{li}(x)}{x/\ln x} = 1,$$

so that the two approximations for $\pi(x)$ given above are equivalent to each other.

One can use the method of proof of Theorem 2.6 to show that certain sets of integers contain infinitely many primes, as in the next theorem. Every odd integer n must have a remainder 1 or 3 when divided by 4, so it must have the form $4q+1$ or $4q+3$ for some integer q. Since $(4s+1)(4t+1) = 4(4st+s+t)+1$, the product of two integers of the form $4q + 1$ also has this form, and hence (by induction) so has the product of any number of integers of this form.

Theorem 2.9

There are infinitely many primes of the form $4q + 3$.

Proof

The proof is by contradiction. Suppose that there are only finitely many primes of this form, say $p_1, \ldots, p_k$. Let $m = 4p_1 \ldots p_k - 1$, so m also has the form $4q+3$ (with $q = p_1 \ldots p_k - 1$). Since m is odd, so is each prime p dividing m, so p has the form $4q + 1$ or $4q + 3$ for some q. If each such p has the form $4q + 1$, then m (being a product of such integers) must also have this form, which is false. Hence m must be divisible by at least one prime p of the form $4q + 3$. By our assumption, $p = p_i$ for some i, so p divides $4p_1 \ldots p_k - m = 1$, which is impossible. This contradiction proves the result. $\square$

There are also infinitely many primes of the form $4q+1$; however, the proof is a little more subtle, so we will return to this result later, in Corollary 7.8. (Where does the method of proof of Theorem 2.9 break down in this case?)

Exercise 2.6

Prove that every prime $p \neq 3$ has the form $3q + 1$ or $3q + 2$ for some integer q; prove that there are infinitely many primes of the form $3q + 2$.

These results are all special cases of a general theorem proved by Dirichlet in 1837, concerning primes in arithmetic progressions:

Theorem 2.10

If a and b are coprime integers then there are infinitely many primes of the form $aq + b$.

The proof uses rather advanced techniques, so we will omit it; it is given in several books, such as Apostol (1976). Notice that the theorem fails if a and b have greatest common divisor $d > 1$, since then every integer of the form $aq + b$ is divisible by d and so at most one of them can be prime.

Despite the above results proving the existence of infinite sets of primes, it is difficult to give explicit examples of such infinite sets, since primes seem to occur so irregularly among the integers. For instance, Exercise 2.7 shows that the gaps between successive primes can be arbitrarily large. At the opposite extreme, apart from the gap 1 between the primes 2 and 3, the smallest possible gap is 2 between pairs of so-called *twin primes* p and $p + 2$. There are enough examples of twin primes, such as 3 and 5, or 41 and 43, and so on, to give rise to the conjecture that infinitely many such pairs exist, but nobody has yet been able to prove this.

Exercise 2.7

Find five consecutive composite integers. Show that for each integer $k \geq 1$, there is a sequence of k consecutive composite integers.

Another open question concerning prime numbers is *Goldbach's Conjecture*, that every even integer $n \geq 4$ is the sum of two primes: thus $4 = 2 + 2$, $6 = 3 + 3$, $8 = 3 + 5$, and so on. The evidence for this is quite strong, but the best general result we have in this direction is a theorem of Chen Jing-Run (1973) that every sufficiently large even integer has the form $n = p + q$ where p is prime and q is the product of at most two primes. Similarly, Vinogradov proved in 1937 that every sufficiently large odd integer is the sum of three primes, so it immediately follows that every sufficiently large even integer is the sum of at most four primes.

2.3 Fermat and Mersenne primes

In order to find specific examples of primes, it seems reasonable to look at integers of the form $2^m \pm 1$, since many small primes, such as $3, 5, 7, 17, 31, \ldots$, have this form.

Lemma 2.11

If $2^m + 1$ is prime then $m = 2^n$ for some integer $n \geq 0$.

Proof

We prove the contrapositive, that if m is not a power of 2 then $2^m + 1$ is not prime. If m is not a power of 2, then m has the form $2^n q$ for some odd $q > 1$. Now the polynomial $f(t) = t^q + 1$ has a root $t = -1$, so it is divisible by $t + 1$; this is a proper factor since $q > 1$, so putting $t = x^{2^n}$ we see that the polynomial $g(x) = f(x^{2^n}) = x^m + 1$ has a proper factor $x^{2^n} + 1$. Taking $x = 2$ we see that $2^{2^n} + 1$ is a proper factor of the integer $g(2) = 2^m + 1$, which cannot therefore be prime. $\qquad\square$

Numbers of the form $F_n = 2^{2^n} + 1$ are called *Fermat numbers*, and those which are prime are called *Fermat primes*. Fermat conjectured that F_n is prime for every $n \geq 0$. For $n = 0, \ldots, 4$ the numbers $F_n = 3, 5, 17, 257, 65537$ are indeed prime, but in 1732 Euler showed that the next Fermat number

$$F_5 = 2^{2^5} + 1 = 4294967297 = 641 \times 6700417$$

is composite. The Fermat numbers have been studied intensively, often with the aid of computers, but no further Fermat primes have been found. It is conceivable that there are further Fermat primes (perhaps infinitely many) which we have not yet found, but the evidence is not very convincing. These primes are important in geometry: in 1801 Gauss showed that a regular polygon with k sides can be constructed by ruler-and-compass methods if and only if $k = 2^e p_1 \ldots p_r$ where $p_1, \ldots, p_r$ are distinct Fermat primes.

Exercise 2.8

Use the equation $641 = 2^4 + 5^4 = 5 \times 2^7 + 1$ to show that $2^{32} = 641q - 1$ for some integer q, so that F_5 is divisible by 641.

Even if not many of the Fermat numbers F_n turn out to be prime, the following result shows that their factors include an infinite set of primes:

Lemma 2.12

Distinct Fermat numbers F_n are mutually coprime.

Proof

Let $d = \gcd(F_n, F_{n+k})$ be the greatest common divisor of two Fermat numbers F_n and F_{n+k}, where $k > 0$. The polynomial $x^{2^k} - 1$ has a root $x = -1$, so it is divisible by $x + 1$. Putting $x = 2^{2^n}$ we see that F_n divides $F_{n+k} - 2$, so d divides 2 and hence d is 1 or 2. Since all Fermat numbers are odd, $d = 1$. □

This provides another proof of Theorem 2.6, since it follows from Lemma 2.12 that any infinite set of Fermat numbers must have infinitely many distinct prime factors.

Exercise 2.9

Show that if $a \geq 2$ and $a^m + 1$ is prime (for instance $37 = 6^2 + 1$), then a is even and m is a power of 2.

Theorem 2.13

If $m > 1$ and $a^m - 1$ is prime, then $a = 2$ and m is prime.

Exercise 2.10

Prove Theorem 2.13.

Integers of the form $2^p - 1$, where p is prime, are called *Mersenne numbers*, after Mersenne who studied them in 1644; those which are prime are called *Mersenne primes*. For the primes $p = 2, 3, 5, 7$, the Mersenne numbers

$$M_p = 3, 7, 31, 127$$

are indeed prime, but $M_{11} = 2047 = 23 \times 89$, so M_p is not prime for every prime p. At the time of writing, 35 Mersenne primes have been found, the latest two being $M_{1257787}$ and $M_{1398269}$ (discovered in 1996 by David Slowinski and Joel Armengaud respectively, with the aid of powerful computers)[*]. As in the case of the Fermat primes, it is not known whether there are finitely or infinitely many Mersenne primes. There is a result similar to Lemma 2.12, that distinct Mersenne numbers are mutually coprime, but it is more convenient to prove this in Chapter 6, as an application of groups (Corollary 6.3). We will meet the Mersenne primes again in Section 8.2, in connection with perfect numbers.

---

[*] Gordon Spence and Roland Clarkson have since discovered $M_{2976221}$ and $M_{3021377}$.

2.4 Primality-testing and factorisation

There are two practical problems which arise from the theory we have considered in this chapter:

(1) How do we determine whether a given integer n is prime?

(2) How do we find the prime-power factorisation of a given integer n?

In relation to the first problem, known as *primality-testing*, we have:

Lemma 2.14

An integer $n > 1$ is composite if and only if it is divisible by some prime $p \leq \sqrt{n}$.

Proof

If n is divisible by such a prime p, then since $1 < p \leq \sqrt{n} < n$ it follows that n is composite. Conversely, if n is composite then $n = ab$ where $1 < a < n$ and $1 < b < n$; at least one of a and b is less than or equal to $\sqrt{n}$ (if not, $ab > n$), and this factor will be divisible by a prime $p \leq \sqrt{n}$, which then divides n. $\square$

For example, we can see that 97 is prime by checking that it is divisible by none of the primes $p \leq \sqrt{97}$, namely 2, 3, 5 and 7. This method requires us to test whether an integer n is divisible by various primes p. For certain small primes p there are simple ways of doing this, based on properties of the decimal number system. In decimal notation we write a positive integer n in the form $a_k a_{k-1} \ldots a_1 a_0$, meaning that

$$n = a_k 10^k + a_{k-1} 10^{k-1} + \cdots + a_1 10 + a_0$$

where $a_0, \ldots, a_k$ are integers with $0 \leq a_i \leq 9$ for all i, and $a_k \neq 0$. From this, we see that n is divisible by 2 if and only if a_0 is divisible by 2, that is, $a_0 = 0, 2, 4, 6$ or 8; similarly, n is divisible by 5 if and only if $a_0 = 0$ or 5. With a little more ingenuity, we can also get tests for divisibility by 3 and 11. If we expand $10^i = (9+1)^i$ by the Binomial Theorem we get an integer of the form $9q + 1$; by doing this for each $i = 1, \ldots, k$ we see that

$$n = 9m + a_k + a_{k-1} + \cdots + a_1 + a_0$$

for some integer m, so n is divisible by 3 if and only if the sum

$$n' = a_k + a_{k-1} + \cdots + a_1 + a_0$$

of its digits is divisible by 3. For instance, if $n = 21497$ then $n' = 2 + 1 + 4 + 9 + 7 = 23$; this is not divisible by 3, so neither is n. (In general, if it is not obvious whether n' is divisible by 3 we can consider *its* digit-sum $n'' = (n')'$, and repeat this process as often as required.) Similarly, by putting $10^i = (11 - 1)^i = 11q + (-1)^i$ we see that

$$n = 11m + (-1)^k a_k + (-1)^{k-1} a_{k-1} + \cdots - a_1 + a_0$$

for some integer m, so n is divisible by 11 if and only if the alternating sum

$$n^* = (-1)^k a_k + (-1)^{k-1} a_{k-1} + \cdots - a_1 + a_0$$

of its digits is divisible by 11. Thus $n = 21497$ has $n^* = 2 - 1 + 4 - 9 + 7 = 3$, so it is not divisible by 11. For primes $p \neq 2, 3, 5$ and 11, one simply has to divide p into n and see whether or not the remainder is 0.

Exercise 2.11

Is 8703585473 divisible by 3? Is it divisible by 11?

Exercise 2.12

Are 157, 221, 641 or 1103 prime?

This method of primality-testing is effective for fairly small integers n, since there are not too many primes p to consider, but when n becomes large it is very time-consuming: by the Prime Number Theorem, the number of primes $p \leq \sqrt{n}$ is given by

$$\pi(\sqrt{n}) \approx \frac{\sqrt{n}}{\ln(\sqrt{n})} = \frac{2\sqrt{n}}{\ln n}.$$

In cryptography (the study of secret codes), one regularly uses integers with several hundred decimal digits; if $n \approx 10^{100}$, for example, then this method would involve testing about 8×10^{47} primes p, and even the fastest available supercomputers would take far longer than the current estimate for the age of the universe (about 15 billion years) to complete this task! Fortunately there are alternative algorithms (using some very sophisticated number theory) which will determine primality for very large integers much more efficiently. Some of the fastest of these are probabilistic algorithms, such as the Solovay–Strassen test, which will always detect a prime integer n, but which may incorrectly declare a composite number n as being prime; this may appear to be a disastrous fault, but in fact the probability of such an incorrect outcome is so low (far lower than the probability of a computational error due to a machine fault) that for most practical purposes these tests are very reliable. For detailed accounts of primality-testing and cryptography, see Koblitz (1994) and Kranakis (1986).

The *Sieve of Eratosthenes* is a systematic way of compiling a list of all the primes up to a given integer N. First we list the integers 2, 3, ... , N in increasing order. Then we underline 2 (which is prime) and cross out all the proper multiples 4, 6, 8, ... of 2 in the list (since these are composite). The first integer which is neither underlined nor crossed out is 3: this is prime, so we underline it and then cross out all its proper multiples 6, 9, 12, ... At the next stage we underline 5 and cross out 10, 15, 20, ... We continue like this until every integer in the list is either underlined or crossed out. At each stage, the first integer which is neither underlined nor crossed out must be prime, for otherwise it would have been crossed out, as a proper multiple of an earlier prime; thus only primes are underlined, and conversely, each prime in the list is eventually underlined at some stage, so when the process terminates the underlined numbers are precisely the primes $p \leq N$. (We can actually stop earlier, when the proper multiples of all the primes $p \leq \sqrt{N}$ have been crossed out, since Lemma 2.14 implies that every remaining integer in the list must be prime.)

Exercise 2.13

Use the Sieve of Eratosthenes to find all the primes $p \leq 100$.

Exercise 2.14

Evaluate the Mersenne number $M_{13} = 2^{13} - 1$. Is it prime?

Our second practical problem, *factorisation*, is apparently much harder than primality-testing. (It cannot be any easier, since the prime-power factorisation of an integer immediately tells us whether or not it is prime.) In theory we could factorise any integer n by testing it for divisibility by the primes 2, 3, 5, ... until a prime factor p is found; we then replace n with n/p and continue this process until a prime factor of n/p is found; eventually, we obtain all the prime factors of n, with their multiplicities. This algorithm is quite effective for small integers, but when n is large we meet the same problem as in primality-testing, that there are just too many possible prime factors to consider. There are, of course, more subtle approaches to factorisation, but at present the fastest known algorithms and computers cannot, in practice, factorise integers several hundred digits long (though nobody has yet proved that an efficient factorisation algorithm will never be found). A very effective cryptographic system (known as the RSA public key system, after its inventors Rivest, Shamir and Adleman, 1978) is based on the fact that it is relatively easy to calculate the product $n = pq$ of two very large primes p and q, while it is extremely difficult to reverse this process and obtain the factors p and q from n. We will examine this system in more detail in Chapter 5.

Exercise 2.15

Factorise 247 and 6887.

Exercise 2.16

Use a computer or a programmable calculator to factorise 3992003. (By hand, this could take several years!)

2.5 Supplementary exercises

Exercise 2.17

For which primes p is $p^2 + 2$ also prime?

Exercise 2.18

Show that if $p > 1$ and p divides $(p-1)! + 1$, then p is prime.

Exercise 2.19

Extend Theorem 2.3 so that it describes the factorisations of all positive *rational* numbers.

Exercise 2.20

Show that if $n, q \geq 1$ then the number of multiples of q among $1, 2, \ldots, n$ is $\lfloor n/q \rfloor$. Hence show that if p is prime and $p^e \| n!$, then $e = \lfloor n/p \rfloor + \lfloor n/p^2 \rfloor + \lfloor n/p^3 \rfloor + \cdots$.

Exercise 2.21

What is the relationship between the number of 0s at the end of the decimal expansion of an integer n, and the prime-power factorisation of n? Find the corresponding result for the base b expansion of n (where we write $n = \sum_{i=0}^{k} a_i b^i$ with $0 \leq a_i < b$).

Exercise 2.22

Show that $F_0 F_1 \ldots F_{n-1} = F_n - 2$ for all $n \geq 1$.

Exercise 2.23

Evaluate the Mersenne number M_{17}, and determine whether it is prime.

Exercise 2.24

Show that if p is prime and $n < p \leq 2n$, then $p | \binom{2n}{n}$. Deduce that $n^{\pi(2n)-\pi(n)} < 2^{2n}$, and hence $(\pi(2n) - \pi(n))/n < 2/\lg n$. (Since $2/\lg n \to 0$ as $n \to \infty$, this shows that the density of primes between n and $2n$ decreases towards 0, a weak form of the Prime Number Theorem.)

3

Congruences

In this chapter, we will study modular arithmetic, that is, the arithmetic of congruence classes, where we simplify number-theoretic problems by replacing each integer with its remainder when divided by some fixed positive integer n. This has the effect of replacing the infinite number system $\mathbb{Z}$ with a number system $\mathbb{Z}_n$ which contains only n elements. We find that we can add, subtract and multiply the elements of $\mathbb{Z}_n$ (just as in $\mathbb{Z}$), though there are some difficulties with division. Thus $\mathbb{Z}_n$ inherits many of the properties of $\mathbb{Z}$, but being finite it is often easier to work with. After a thorough study of linear congruences (the analogues in $\mathbb{Z}_n$ of the equation $ax = b$), we will consider simultaneous linear congruences, where the Chinese Remainder Theorem and its generalisations play a major role.

3.1 Modular arithmetic

Many problems involving large integers (such as some of those in the last chapter) can be simplified by a technique called *modular arithmetic*, where we use congruences in place of equations. The basic idea is to choose a particular integer n (depending on the problem), called the *modulus*, and replace every integer with its remainder when divided by n. In general, this remainder is smaller, and hence easier to deal with. Before going into the general theory, let us look at two simple examples.

Example 3.1

What day of the week will it be 100 days from now? We could solve this by getting out a diary and counting 100 days ahead, but a simpler method is to use the fact that the days of the week recur in cycles of length 7. Now $100 = 14 \times 7 + 2$, so the day will be the same as it is two days ahead, and this is easy to determine. Here we chose $n = 7$, and replaced 100 with its remainder on division by 7, namely 2.

Example 3.2

Is 22051946 a perfect square? We could solve this by computing $\sqrt{22051946}$ and determining whether it is an integer, or alternatively by squaring various integers and seeing whether 22051946 occurs, but there is a much simpler way of seeing that this number cannot be a perfect square. In Chapter 1 (Example 1.2) we showed that a perfect square must leave a remainder 0 or 1 when divided by 4. By looking at its last two digits, we see that

$$22051946 = 220519 \times 100 + 46 = 220519 \times 25 \times 4 + 46$$

leaves the same remainder as 46, and since $46 = 11 \times 4 + 2$ this remainder is 2. It follows that 22051946 is not a perfect square. (Of course, if the remainder had been 0 or 1, it would not follow that the number was a square: we would have to use some other method to find out.) In this case we chose $n = 4$, and replaced 22051946 first with 46, and then with 2.

Exercise 3.1

Show that the last decimal digit of a perfect square cannot be $2, 3, 7$ or 8. Is 3190491 a perfect square?

Definition

Let n be a positive integer, and let a and b be any integers. We say that a is *congruent* to b mod (n), or a is a *residue* of b mod (n), written

$$a \equiv b \mod (n),$$

if a and b leave the same remainder when divided by n. (Other notations for this include $a \equiv b \pmod{n}$ and $a \equiv_n b$; we will often use simply $a \equiv b$ if the value of n is understood.) To be more precise, we use the division algorithm (Theorem 1.1) to put $a = qn + r$ with $0 \leq r < n$, and $b = q'n + r'$ with $0 \leq r' < n$, and then we say that $a \equiv b \mod (n)$ if and only if $r = r'$. For instance, $100 \equiv 2 \bmod (7)$ in Example 3.1, and $22051946 \equiv 46 \equiv 2 \bmod (4)$ in Example 3.2. We will use the notation $a \not\equiv b \bmod (n)$ to denote that a and b are not congruent mod (n), that is, that they leave different remainders when divided by n. Our first result gives a useful alternative definition of congruence mod (n):

Lemma 3.1

For any fixed $n \geq 1$ we have $a \equiv b \bmod (n)$ if and only if $n \mid (a - b)$.

Proof

Putting $a = qn + r$ and $b = q'n + r'$ as above, we have $a - b = (q - q')n + (r - r')$ with $-n < r - r' < n$. If $a \equiv b \bmod (n)$ then $r = r'$, so $r - r' = 0$ and $a - b = (q - q')n$, which is divisible by n. Conversely, if n divides $a - b$ then it divides $(a - b) - (q - q')n = r - r'$; now the only integer strictly between $-n$ and n which is divisible by n is 0, so $r - r' = 0$, giving $r = r'$ and hence $a \equiv b \bmod (n)$. □

Our next result records some trivial but useful observations about congruences:

Lemma 3.2

For any fixed $n \geq 1$ we have

(a) $a \equiv a$ for all integers a;

(b) if $a \equiv b$ then $b \equiv a$;

(c) if $a \equiv b$ and $b \equiv c$ then $a \equiv c$.

Proof

(a) We have $n \mid (a - a)$ for all a.

(b) If $n \mid (a - b)$ then $n \mid (b - a)$.

(c) If $n \mid (a - b)$ and $n \mid (b - c)$ then $n \mid (a - b) + (b - c) = a - c$. □

These three properties are the reflexivity, symmetry and transitivity axioms for an equivalence relation, so Lemma 3.2 proves that for each fixed n, congruence mod (n) is an equivalence relation on $\mathbb{Z}$. It follows that $\mathbb{Z}$ is partitioned into disjoint equivalence classes; these are the *congruence classes*

$$[a] = \{b \in \mathbb{Z} \mid a \equiv b \bmod (n)\}$$
$$= \{\ldots, a - 2n, a - n, a, a + n, a + 2n, \ldots\}$$

for $a \in \mathbb{Z}$. (If we want to emphasise the particular value of n being used, we can use the notation $[a]_n$ here.) Each class corresponds to one of the n possible remainders $r = 0, 1, \ldots, n - 1$ on division by n, so there are n different congruence classes. They are

$$[0] = \{\ldots, 2n, -n, 0, n, 2n, \ldots\},$$
$$[1] = \{\ldots, 1 - 2n, 1 - n, 1, 1 + n, 1 + 2n, \ldots\},$$
$$\cdots$$
$$[n - 1] = \{\ldots, -n - 1, -1, n - 1, 2n - 1, 3n - 1, \ldots\}.$$

There are no further classes distinct from these: for example

$$[n] = \{\ldots, -n, 0, n, 2n, 3n, \ldots\} = [0].$$

More generally, we have

$$[a] = [b] \quad \text{if and only if} \quad a \equiv b \bmod (n).$$

When $n = 1$ all integers are congruent to each other, so there is a single congruence class, coinciding with $\mathbb{Z}$. When $n = 2$ the two classes $[0] = [0]_2$ and $[1] = [1]_2$ consist of the even and odd integers respectively. We can regard Theorem 2.9 as asserting that there are infinitely many primes $p \equiv 3 \bmod (4)$, that is, the class $[3]_4$ contains infinitely many primes.

For a given $n \geq 1$, we denote the set of n equivalence classes mod (n) by $\mathbb{Z}_n$, known as the set of integers mod (n). Our next aim is to show how to do arithmetic with these congruence classes, so that $\mathbb{Z}_n$ becomes a number system with properties very similar to those of $\mathbb{Z}$. We do this by using the operations of addition, subtraction and multiplication in $\mathbb{Z}$ to define the corresponding operations on the congruence classes in $\mathbb{Z}_n$. If $[a]$ and $[b]$ are elements of $\mathbb{Z}_n$ (that is, congruence classes mod (n)), we define their sum, difference and product to be the classes

$$[a] + [b] = [a + b],$$
$$[a] - [b] = [a - b],$$
$$[a][b] = [ab]$$

containing the integers $a + b, a - b$ and ab respectively. (We will leave the question of *division* of congruence classes until later; the difficulty is that if a and b are integers then a/b need not be an integer, in which case there is no congruence class $[a/b]$ for us to use.)

Before going further, we need to show that these three operations are well-defined, in the sense that the right-hand sides of the three equations defining them depend only on the classes $[a]$ and $[b]$, and not on the particular elements a and b we have chosen from those classes. More specifically, we must show that if $[a] = [a']$ and $[b] = [b']$, then $[a + b] = [a' + b']$, $[a - b] = [a' - b']$ and $[ab] = [a'b']$. These follow immediately from the following result:

Lemma 3.3

For a given $n \geq 1$, if $a' \equiv a$ and $b' \equiv b$ then $a' + b' \equiv a + b$, $a' - b' \equiv a - b$ and $a'b' \equiv ab$.

Proof

If $a' \equiv a$ then $a' = a + kn$ for some integer k, and similarly we have $b' = b + ln$ for some integer l; then $a' \pm b' = (a \pm b) + (k \pm l)n \equiv a \pm b$, and $a'b' = ab + (al + bk + kln)n \equiv ab$.   □

It follows that addition, subtraction and multiplication of pairs of classes in $\mathbb{Z}_n$ are all well-defined. In particular, by repeated use of the addition and multiplication parts of this lemma we can define arbitrary finite sums, products and powers of classes in $\mathbb{Z}_n$ by

$$
\begin{aligned}
[a_1] + [a_2] + \cdots + [a_k] &= [a_1 + a_1 + \cdots + a_k]\,, \\
[a_1][a_2] \cdots [a_k] &= [a_1 a_1 \cdots a_k]\,, \\
[a]^k &= [a^k]
\end{aligned}
$$

for any integer $k \geq 2$.

To emphasise why we have to be so careful about checking that the operations of arithmetic in $\mathbb{Z}_n$ are well-defined, let us look at what happens if we try to define exponentiation of classes in $\mathbb{Z}_n$ in the obvious way. We could define

$$
[a]^{[b]} = [a^b]\,,
$$

restricting b to non-negative values to ensure that a^b is an integer. If we take $n = 3$, for instance, this gives

$$
[2]^{[1]} = [2^1] = [2]\,;
$$

unfortunately, $[1] = [4]$ in $\mathbb{Z}_3$, and our definition also gives

$$
[2]^{[4]} = [2^4] = [16] = [1] \neq [2]\,;
$$

thus we can get different congruence classes for $[a]^{[b]}$ by choosing different elements b and b' in the same class $[b]$, namely $b = 1$ and $b' = 4$. This is because

$a' \equiv a$ and $b' \equiv b$ do *not* imply $(a')^{b'} \equiv a^b$, so exponentiation of congruence classes is not well-defined. We therefore confine arithmetic in $\mathbb{Z}_n$ to operations which are well-defined, like addition, subtraction, multiplication and powers; we shall see later that a restricted form of division can also be defined.

A set of n integers, containing one representative from each of the n congruence classes in $\mathbb{Z}_n$, is called a *complete set of residues* mod (n). A sensible choice of such a set can ease calculations considerably. One obvious choice is provided by the division algorithm (Theorem 1.1): we can divide any integer a by n to give $a = qn + r$ for some unique r satisfying $0 \le r < n$; thus each class $[a] \in \mathbb{Z}_n$ contains a unique $r = 0, 1, \ldots, n - 1$, so these n integers form a complete set of residues, called the *least non-negative residues* mod (n). For many purposes these are the most convenient residues to use, but sometimes it is better to replace Theorem 1.1 with Exercise 1.22 of Chapter 1, which gives a remainder r satisfying $-n/2 < r \le n/2$. These remainders are the *least absolute residues* mod (n), those with least absolute value; when n is odd they are $0, \pm 1, \pm 2, \ldots, \pm(n - 1)/2$, and when n is even they are $0, \pm 1, \pm 2, \ldots, \pm(n - 2)/2, n/2$. The following calculations illustrate these complete sets of residues.

Example 3.3

Let us calculate the least non-negative residue of 28×33 mod 35. Using least absolute residues mod (35), we have $28 \equiv -7$ and $33 \equiv -2$, so Lemma 3.3 implies that $28 \times 33 \equiv (-7) \times (-2) \equiv 14$. Since $0 \le 14 < 35$ it follows that 14 is the required least non-negative residue.

Example 3.4

Let us calculate the least absolute residue of 15×59 mod 75. We have $15 \times 59 \equiv 15 \times (-16)$, and a simple way to evaluate this is to do the multiplication in several stages, reducing the product mod (75) each time. Thus

$$15 \times (-16) = 15 \times (-4) \times 4 = (-60) \times 4 \equiv 15 \times 4 = 60 \equiv -15,$$

and since $-75/2 < -15 \le 75/2$ the required residue is -15.

Example 3.5

Let us calculate the least non-negative residue of 3^8 mod (13). Again, we do this in several stages, reducing mod (13) whenever possible:

$$3^2 = 9 \equiv -4,$$

so that

$$3^4 = (3^2)^2 \equiv (-4)^2 = 16 \equiv 3,$$

and hence

$$3^8 = (3^4)^2 \equiv 3^2 = 9;$$

the required residue is therefore 9.

Exercise 3.2

Without using a calculator, find:

(a) the least non-negative residue of 34×17 mod (29);

(b) the least absolute residue of 19×14 mod (23);

(c) the remainder when 5^{10} is divided by 19;

(d) the final decimal digit of $1! + 2! + 3! + \cdots + 10!$.

Since n divides m if and only if $m \equiv 0$ mod (n), it follows that problems about divisibility are equivalent to problems about congruences, and these can sometimes be easier to solve. Here is a typical illustration of this:

Example 3.6

Let us prove that $a(a + 1)(2a + 1)$ is divisible by 6 for every integer a. By taking least absolute residues mod (6), we see that $a \equiv 0, \pm 1, \pm 2$ or 3. If $a \equiv 0$ then $a(a + 1)(2a + 1) \equiv 0.1.1 \equiv 0$, if $a \equiv 1$ then $a(a + 1)(2a + 1) \equiv 1.2.3 = 6 \equiv 0$, and similar calculations (which you should try for yourself) show that $a(a + 1)(2a + 1) \equiv 0$ in the other four cases, so $6 | a(a + 1)(2a + 1)$ for all a.

Exercise 3.3

Find a quicker proof of this, based on the observation that $6 | m$ if and only if $2 | m$ and $3 | m$.

Exercise 3.3 uses the following more general principle, in which a single congruence mod (n) is replaced with a set of simultaneous congruences mod (p^e) for the various prime powers p^e dividing n (and these are often easier to deal with than the original congruence):

Theorem 3.4

Let n have prime-power factorisation

$$n = p_1^{e_1} \cdots p_k^{e_k},$$

where $p_1, \ldots, p_k$ are distinct primes. Then for any integers a and b we have $a \equiv b \mod (n)$ if and only if $a \equiv b \mod (p_i^{e_i})$ for each $i = 1, \ldots, k$.

This is quite easy to prove directly (try it for yourself, using Corollary 1.11), but we will deduce it later in this chapter as a corollary of the Chinese Remainder Theorem, which deals with simultaneous congruences in a more general setting.

Having seen how to add, subtract and multiply congruence classes, we can now combine these operations to form polynomials.

Lemma 3.5

Let $f(x)$ be a polynomial with integer coefficients, and let $n \geq 1$. If $a \equiv b \mod (n)$ then $f(a) \equiv f(b) \mod (n)$.

Proof

Write $f(x) = c_0 + c_1 x + \cdots + c_k x^k$, where each $c_i \in \mathbb{Z}$. If $a \equiv b \mod (n)$, then repeated use of Lemma 3.3 implies that $a^i \equiv b^i$ for all $i \geq 0$, so $c_i a^i \equiv c_i b^i$ for all i, and hence $f(a) = \sum c_i a^i \equiv \sum c_i b^i = f(b)$. $\square$

For an illustration of this, look at Example 3.6, where we took $f(x) = x(x + 1)(2x + 1) = 2x^3 + 3x^2 + x$ and $n = 6$; we then used the fact that if $a \equiv 0, \pm 1, \pm 2$ or 3 then $f(a) \equiv f(0), f(\pm 1), f(\pm 2)$ or $f(3)$ respectively, all of which are easily seen to be congruent to $0 \mod (6)$.

Suppose that a polynomial $f(x)$, with integer coefficients, has an integer root $x = a \in \mathbb{Z}$, so that $f(a) = 0$. It follows then that $f(a) \equiv 0 \mod (n)$ for all integers $n \geq 1$. We can often use the contrapositive of this to show that certain polynomials $f(x)$ have no integer roots: if there exists an integer $n \geq 1$ such that the congruence $f(x) \equiv 0 \mod (n)$ has no solutions x, then the equation $f(x) = 0$ can have no solutions x. If n is small we can check whether $f(x) \equiv 0 \mod (n)$ has any solutions simply by evaluating $f(x_1), \ldots, f(x_n)$ where $x_1, \ldots, x_n$ form a complete set of residues $\mod (n)$: each $x \in \mathbb{Z}$ is congruent to some x_i, so Lemma 3.5 implies that $f(x) \equiv f(x_i)$, and we simply determine whether any of $f(x_1), \ldots, f(x_n)$ is divisible by n.

Example 3.7

Let us prove that the polynomial $f(x) = x^5 - x^2 + x - 3$ has no integer roots. To do this, we take $n = 4$ (a choice which we will explain later), and consider the congruence

$$f(x) = x^5 - x^2 + x - 3 \equiv 0 \mod (4).$$

Using the least absolute residues $0, \pm 1, 2$ as a complete set of residues mod (4), we find that

$$f(0) = -3, \quad f(1) = -2, \quad f(-1) = -6 \quad \text{and} \quad f(2) = 27.$$

None of these values is divisible by 4, so the congruence $f(x) \equiv 0 \mod (4)$ has no solutions and hence the polynomial $f(x)$ has no integer roots.

You may be wondering why we took $n = 4$ in this example; the reason, which you can easily check, is that for each $n < 4$ the congruence $f(x) \equiv 0 \mod (n)$ *does* have a solution $x \in \mathbb{Z}$, even though the equation $f(x) = 0$ does not; thus 4 is the smallest value of n for which this method works. In general, the correct choice of n is often a matter of insight, experience, or simply trial and error: if one value of n fails to prove that a polynomial has no integer roots, do not give up (or even worse, do not assume that there must be a root); try a few more values, and if they also fail, this suggests that perhaps there really is an integer root.

Exercise 3.4

Prove that the following polynomials have no integer roots:

(a) $x^3 - x + 1$;

(b) $x^3 + x^2 - x + 1$;

(c) $x^3 + x^2 - x + 3$.

Example 3.8

Unfortunately, the method used in Example 3.7 is not always strong enough to prove the non-existence of integer roots. For instance, the polynomial

$$f(x) = (x^2 - 13)(x^2 - 17)(x^2 - 221)$$

clearly has no integer roots: indeed, since $13, 17$ and 221 ($= 13 \times 17$) are not perfect squares, the roots $\pm\sqrt{13}, \pm\sqrt{17}$ and $\pm\sqrt{221}$ of $f(x)$ are all irrational by Corollary 2.5. However, in Chapter 7 (Example 7.16) we will show that for every integer $n \geq 1$ there is a solution of $f(x) \equiv 0 \mod (n)$, so in this case there is no suitable choice of n for our method of congruences.

As a second application of Lemma 3.5, let us consider prime values of polynomials. The polynomial

$$f(x) = x^2 + x + 41$$

has the remarkable property that $f(x)$ is prime for each integer $x = -40, -39,$ $\ldots, 38, 39$ (though not for $x = -41$ or $x = 40$). Motivated by this, one might ask whether there is a polynomial $f(x)$, with integer coefficients, such that $f(x)$ is prime for *every* integer x. Apart from the trivial examples of constant polynomials $f(x) = p$ (p prime), there are none:

Theorem 3.6

There is no non-constant polynomial $f(x)$, with integer coefficients, such that $f(x)$ is prime for all integers x.

Proof

Suppose that $f(x)$ is prime for all integers x, and is not constant. If we choose any integer a, then $f(a)$ is a prime p. For each $b \equiv a$ mod (p), Lemma 3.5 implies that $f(b) \equiv f(a)$ mod (p), so $f(b) \equiv 0$ mod (p) and hence p divides $f(b)$. By our hypothesis, $f(b)$ is prime, so $f(b) = p$. There are infinitely many integers $b \equiv a$ mod (p), so the polynomial $g(x) = f(x) - p$ has infinitely many roots. However, this is impossible: having degree $d \geq 1$, $g(x)$ can have at most d roots, so such a polynomial $f(x)$ cannot exist.   □

Theorem 2.10 shows that if a and b are coprime then the linear polynomial $f(x) = ax + b$ has infinitely many prime values, but it is not known whether any polynomial $f(x)$ of degree $d \geq 2$, such as $x^2 + 1$, can have this property. There are polynomials $f(x_1, \ldots, x_m)$ of *several* variables, whose positive values coincide with the set of primes as $x_1, \ldots, x_m$ range over the positive integers, but unfortunately the known examples are rather complicated.

3.2 Linear congruences

We now return to the question of division of congruence classes, postponed from earlier in this chapter. In order to assign a meaning to a quotient $[b]/[a]$ of two congruence classes $[a], [b] \in \mathbb{Z}_n$, we need to consider the solutions of the linear congruence $ax \equiv b$ mod (n). Note that if x is a solution, and if $x' \equiv x$, then $ax' \equiv ax \equiv b$ and so x' is also a solution; thus the solutions (if they exist) form a union of congruence classes. Now $ax \equiv b$ mod (n) if and only if $ax - b$

is a multiple of n, so x is a solution of this linear congruence if and only if there is integer y such that x and y satisfy the linear Diophantine equation $ax + ny = b$. We studied this equation (with slightly different notation) in Chapter 1, so translating Theorem 1.13 into the language of congruences we have

Theorem 3.7

If $d = \gcd(a, n)$, then the linear congruence

$$ax \equiv b \bmod (n)$$

has a solution if and only if d divides b. If d does divide b, and if x_0 is any solution, then the general solution is given by

$$x = x_0 + \frac{nt}{d}$$

where $t \in \mathbb{Z}$; in particular, the solutions form exactly d congruence classes mod (n), with representatives

$$x = x_0, \; x_0 + \frac{n}{d}, \; x_0 + \frac{2n}{d}, \; \ldots, \; x_0 + \frac{(d-1)n}{d} \, .$$

(In fact, the equation $x = x_0 + t(n/d)$ shows that the solutions form a *single* congruence class $[x_0]$ mod (n/d), but since the problem is phrased in terms of congruences mod (n), it is traditional (and often more useful) to phrase the answer in the same way.)

Proof

Apart from a slight change of notation (n and b replacing b and c), the only part of this which is not a direct translation of Theorem 1.13 is the statement about congruence classes. To prove this, note that

$$x_0 + \frac{nt}{d} \equiv x_0 + \frac{nt'}{d} \bmod (n)$$

if and only if n divides $n(t - t')/d$, that is, if and only if d divides $t - t'$, so the congruence classes of solutions mod (n) are obtained by letting t range over a complete set of residues mod (d), such as $0, 1, \ldots, d - 1$.   □

Example 3.9

Consider the congruence

$$10x \equiv 3 \bmod (12) \, .$$

Here $a = 10$, $b = 3$ and $n = 12$, so $d = \gcd(10, 12) = 2$; this does not divide 3, so there are no solutions. (This can be seen directly: the elements of the congruence class $[3]$ in $\mathbb{Z}_{12}$ are all odd, whereas any elements of $[10][x]$ must be even.)

Example 3.10

Now consider the congruence

$$10x \equiv 6 \mod (12).$$

As before we have $d = 2$, and now this does divide $b = 6$, so there are two classes of solutions. We can take $x_0 = 3$ as a particular solution, so the general solution has the form

$$x = x_0 + \frac{nt}{d} = 3 + \frac{12t}{2} = 3 + 6t,$$

where $t \in \mathbb{Z}$. These solutions form two congruence classes $[3]$ and $[9]$ mod (12), with representatives $x_0 = 3$ and $x_0 + (n/d) = 9$; equivalently, they form a single congruence class $[3]$ mod (6).

Exercise 3.5

Find the general solution of the congruence $12x \equiv 9$ mod (15).

Corollary 3.8

If $\gcd(a, n) = 1$ then the solutions x of the linear congruence $ax \equiv b$ mod (n) form a single congruence class mod (n).

Proof

Put $d = 1$ in Theorem 3.7.   □

This means that if a and n are coprime then for each b there is a unique class $[x]$ such that $[a][x] = [b]$ in $\mathbb{Z}_n$; we can regard this class $[x]$ as the quotient class $[b]/[a]$ obtained by dividing $[b]$ by $[a]$ in $\mathbb{Z}_n$. If $d = \gcd(a, n) > 1$, however, there is either more than one such class $[x]$ (when d divides b), or there is no such class (when d does not divide b), so we cannot define a quotient class $[b]/[a]$ in this case.

Example 3.11

Consider the congruence

$$7x \equiv 3 \mod (12).$$

Here $a = 7$ and $n = 12$, and since these are coprime there is a single congruence class of solutions; this is the class $[x] = [9]$, since $7 \times 9 = 63 \equiv 3 \mod (12)$.

In Examples 3.9, 3.10 and 3.11, we had $n = 12$. When n is as small as this, it is feasible to find all solutions of a congruence $ax \equiv b \mod (n)$ by inspection: one can simply calculate ax for each of the n elements x of a complete set of residues mod (n), and see which of these products are congruent to b. When n is larger, however, a more efficient method is needed for solving linear congruences. We shall give an algorithm for this, based on Theorem 3.7, but first we need some preliminary results which help to simplify the problem.

Lemma 3.9

(a) Let m divide a, b and n, and let $a' = a/m$, $b' = b/m$ and $n' = n/m$; then

$$ax \equiv b \mod (n) \quad \text{if and only if} \quad a'x \equiv b' \mod (n').$$

(b) Let a and n be coprime, let m divide a and b, and let $a' = a/m$ and $b' = b/m$; then

$$ax \equiv b \mod (n) \quad \text{if and only if} \quad a'x \equiv b' \mod (n).$$

Proof

(a) We have $ax \equiv b \mod (n)$ if and only if $ax - b = qn$ for some integer q; dividing by m, we see that this is equivalent to $a'x - b' = qn'$, that is, to $a'x \equiv b' \mod (n')$.

(b) If $ax \equiv b \mod (n)$, then as in (a) we have $ax - b = qn$ and hence $a'x - b' = qn/m$; in particular, m divides qn. Now m divides a, which is coprime to n, so m is also coprime to n and hence m must divide q by Corollary 1.11(b). Thus $a'x - b' = (q/m)n$ is a multiple of n, so $a'x \equiv b' \mod (n)$. For the converse, if $a'x \equiv b' \mod (n)$ then $a'x - b' = q'n$ for some integer q', so multiplying through by m we have $ax - b = mq'n$ and hence $ax \equiv b \mod (n)$.   □

Note that in (a), where m divides a, b and n, we divide all three of these integers by m, whereas in (b), where m divides a and b, we divide just these two integers by m, leaving n unchanged.

Exercise 3.6

Show, by means of a counterexample, that Lemma 3.9(b) can fail if a and n are not coprime.

We now give an algorithm to solve the linear congruence $ax \equiv b \bmod (n)$. To help you understand each step, it may be useful to try this algorithm out on the congruence $10x \equiv 6 \bmod (14)$; when you have finished, look at Example 3.12 to see if your working agrees with ours.

Step 1. We calculate $d = \gcd(a, n)$ (as in Chapter 1), and see whether d divides b. If it does not, there are no solutions, so we stop. If it does, we go on to step 2.

Theorem 3.7 gives us the general solution, provided we can find a particular solution x_0, so from now on we concentrate on a method for finding x_0. The general strategy is to reduce $|a|$ until $a = \pm 1$, since in this case the solution $x_0 = \pm b$ is obvious.

Step 2. Since d divides a, b and n, Lemma 3.9(a) implies that we can replace the original congruence with

$$a'x \equiv b' \bmod (n'),$$

where $a' = a/d$, $b' = b/d$ and $n' = n/d$. By Corollary 1.10, a' and n' are coprime.

Step 3. We can therefore use Lemma 3.9(b) to divide this new congruence through by $m = \gcd(a', b')$, giving a congruence

$$a''x \equiv b'' \bmod (n')$$

where a'' $(= a'/m)$ is coprime to both b'' $(= b'/m)$ and n'. If $a'' = \pm 1$ then $x_0 = \pm b''$ is the required solution. Otherwise, we go on to step 4.

Step 4. Noting that

$$b'' \equiv b'' \pm n' \equiv b'' \pm 2n' \equiv \ldots \bmod (n'),$$

we may be able to replace b'' with some congruent number $b''' = b'' + kn'$ such that $\gcd(a'', b''') > 1$; by applying step 3 to the congruence $a''x \equiv b''' \bmod (n')$ we can again reduce $|a''|$. An alternative at this stage is to multiply through by some suitably chosen constant c, giving $ca''x \equiv cb'' \bmod (n')$; if c is chosen so that the least absolute residue a''' of ca'' satisfies $|a'''| < |a''|$, then we have reduced $|a''|$ to give a linear congruence $a'''x \equiv b''' \bmod (n')$ with $b''' = cb''$.

A combination of the methods in step 4 will eventually reduce a to ± 1, in which case the solution x_0 can be read off; then Theorem 3.7 gives the general solution.

Example 3.12

Consider the congruence

$$10x \equiv 6 \bmod (14).$$

Step 1 gives $\gcd(10, 14) = 2$, which divides 6, so solutions do exist. If x_0 is any solution, then the general solution is $x = x_0 + (14/2)t = x_0 + 7t$, where $t \in \mathbb{Z}$; these form the congruence classes $[x_0]$ and $[x_0 + 7]$ in $\mathbb{Z}_{14}$. To find x_0 we use step 2: we divide the original congruence through by $\gcd(10, 14) = 2$ to give

$$5x \equiv 3 \bmod (7).$$

Since $\gcd(5, 3) = 1$, step 3 has no effect, so we move on to step 4. Noting that $3 \equiv 10 \bmod (7)$, with 10 divisible by 5, we replace the congruence with

$$5x \equiv 10 \bmod (7)$$

and then divide by 5 (which is coprime to 7) to give

$$x \equiv 2 \bmod (7).$$

Thus $x_0 = 2$ is a solution, so the general solution has the form

$$x = 2 + 7t \quad (t \in \mathbb{Z}).$$

Example 3.13

Consider the congruence

$$4x \equiv 13 \bmod (47).$$

Step 1 gives $\gcd(4, 47) = 1$, which divides 13, so the congruence has solutions. If x_0 is any solution, then the general solution is $x = x_0 + 47t$ where $t \in \mathbb{Z}$, forming a single congruence class $[x_0]$ in $\mathbb{Z}_{47}$. Since $\gcd(4, 47) = 1$, step 2 has no effect, so we move on to step 3. Since $\gcd(4, 13) = 1$, step 3 also has no effect, so we go to step 4. We could now employ the method used in the previous example, but as an alternative we will illustrate the other technique described in step 4: noting that $4 \times 12 = 48 \equiv 1 \bmod (47)$, we multiply by 12 to give

$$48x \equiv 12 \times 13 \ \bmod (47),$$

that is,

$$x \equiv 3 \times 4 \times 13 \equiv 3 \times 52 \equiv 3 \times 5 \equiv 15 \ \bmod (47).$$

Thus we can take $x_0 = 15$, so the general solution is $x = 15 + 47t$.

Exercise 3.7

For each of the following congruences, decide whether a solution exists, and if it does exist, find the general solution:

(a) $3x \equiv 5 \bmod (7)$;

(b) $12x \equiv 15 \bmod (22)$;

(c) $19x \equiv 42 \bmod (50)$;

(d) $18x \equiv 42 \bmod (50)$.

3.3 Simultaneous linear congruences

We will now consider the solutions of simultaneous congruences. In the 1st century AD, the Chinese mathematician Sun-Tsu considered problems like 'find a number which leaves remainders $2, 3, 2$ when divided by $3, 5, 7$ respectively'. Equivalently, he wanted to find x such that the congruences

$$x \equiv 2 \ \bmod (3), \quad x \equiv 3 \ \bmod (5), \quad x \equiv 2 \ \bmod (7)$$

are simultaneously true. Note that if x_0 is any solution, then so is $x_0 + (3 \times 5 \times 7)t$ for any integer t, so the solutions form a union of congruence classes mod (105). We shall show that in this particular case the solutions form a *single* congruence class, but in other cases we may find that there are several classes or none: as an example, the simultaneous congruences

$$x \equiv 3 \ \bmod (9), \quad x \equiv 2 \ \bmod (6)$$

have no solutions, for if $x \equiv 3 \bmod (9)$ then 3 divides x, whereas if $x \equiv 2 \bmod (6)$ then 3 does not divide x. The difficulty here is that the moduli 9 and 6 have a factor 3 in common, so both congruences have implications about the congruence class of $x \bmod (3)$, and in this particular case these implications are mutually inconsistent. To avoid this type of difficulty, we will initially restrict our attention to cases like our first example, where the moduli are mutually coprime. Fortunately, the following result, known as the *Chinese Remainder Theorem*, gives a very satisfactory solution to this type of problem.

Theorem 3.10

Let $n_1, n_2, \ldots, n_k$ be positive integers, with $\gcd(n_i, n_j) = 1$ whenever $i \neq j$, and let $a_1, a_2, \ldots, a_k$ be any integers. Then the solutions of the simultaneous congruences

$$x \equiv a_1 \ \mathrm{mod} \ (n_1), \quad x \equiv a_2 \ \mathrm{mod} \ (n_2), \quad \ldots \quad x \equiv a_k \ \mathrm{mod} \ (n_k)$$

form a single congruence class mod (n), where $n = n_1 n_2 \ldots n_k$.

(This result has applications in many areas, including astronomy: if k events occur regularly, with periods $n_1, \ldots, n_k$, and with the i-th event happening at times $x = a_i, a_i + n_i, a_i + 2n_i, \ldots$, then the k events occur simultaneously at time x where $x \equiv a_i \ \mathrm{mod} \ (n_i)$ for all i; the theorem shows that if the periods n_i are mutually coprime then such a coincidence occurs with period n. Planetary conjunctions and eclipses are obvious examples of such regular events, and predicting these may have been the original motivation for the theorem.)

Proof

Let $c_i = n/n_i = n_1 \ldots n_{i-1} n_{i+1} \ldots n_k$ for each $i = 1, \ldots, k$. Since each of its factors n_j ($j \neq i$) is coprime to n_i, so is c_i. Corollary 3.8 therefore implies that for each i, the congruence $c_i x \equiv 1 \ \mathrm{mod} \ (n_i)$ has a single congruence class $[d_i]$ of solutions mod (n_i). We now claim that the integer

$$x_0 = a_1 c_1 d_1 + a_2 c_2 d_2 + \cdots + a_k c_k d_k$$

simultaneously satisfies the given congruences, that is, $x_0 \equiv a_i \ \mathrm{mod} \ (n_i)$ for each i. To see this, note that each c_j (other than c_i) is divisible by n_i, so $a_j c_j d_j \equiv 0$ and hence $x_0 \equiv a_i c_i d_i \ \mathrm{mod} \ (n_i)$; now $c_i d_i \equiv 1$, by choice of d_i, so $x_0 \equiv a_i$ as required. Thus x_0 is a solution of the simultaneous congruences, and it immediately follows that the entire congruence class $[x_0]$ of x_0 mod (n) consists of solutions.

To see that this class is unique, suppose that x is any solution; then $x \equiv a_i \equiv x_0 \ \mathrm{mod} \ (n_i)$ for all i, so each n_i divides $x - x_0$. Since $n_1, \ldots, n_k$ are mutually coprime, repeated use of Corollary 1.11(a) shows that their product n also divides $x - x_0$, so $x \equiv x_0 \ \mathrm{mod} \ (n)$. $\qquad\square$

Comments

1 The proof of Theorem 3.4, which we postponed until later, now follows immediately: given $n = p_1^{e_1} \ldots p_k^{e_k}$, we put $n_i = p_i^{e_i}$ for $i = 1, \ldots, k$, so $n_1, \ldots, n_k$ are mutually coprime with product n; the Chinese Remainder Theorem therefore implies that the solutions of the simultaneous congruences $x \equiv b$ mod (n_i) form a single congruence class mod (n); clearly b is a solution, so these congruences are equivalent to $x \equiv b$ mod (n).

2 Note that the proof of the Chinese Remainder Theorem does not merely show that there is a solution for the simultaneous congruences; it also gives us a formula for a particular solution x_0, and hence for the general solution $x = x_0 + nt \ (t \in \mathbb{Z})$.

Example 3.14

In our original problem

$$x \equiv 2 \ \text{mod} \ (3), \quad x \equiv 3 \ \text{mod} \ (5), \quad x \equiv 2 \ \text{mod} \ (7),$$

we have $n_1 = 3, n_2 = 5$ and $n_3 = 7$, so $n = 105, c_1 = 35, c_2 = 21$ and $c_3 = 15$. We first need to find a solution $x = d_1$ of $c_1 x \equiv 1$ mod (n_1), that is, $35x \equiv 1$ mod (3); this is equivalent to $-x \equiv 1$ mod (3), so we can take $x = d_1 = -1$ for example. Similarly, $c_2 x \equiv 1$ mod (n_2) gives $21x \equiv 1$ mod (5), that is, $x \equiv 1$ mod (5), so we can take $x = d_2 = 1$, while $c_3 x \equiv 1$ mod (n_3) gives $15x \equiv 1$ mod (7), that is, $x \equiv 1$ mod (7), so we can also take $x = d_3 = 1$. Of course, different choices of d_i are possible here, leading to different values of x_0, but they will all give the same congruence class of solutions mod (105). We now have

$$x_0 = a_1 c_1 d_1 + a_2 c_2 d_2 + a_3 c_3 d_3 = 2.35.(-1) + 3.21.1 + 2.15.1 = 23,$$

so the solutions form the congruence class $[23]$ mod (105), that is, the general solution is $x = 23 + 105t \ (t \in \mathbb{Z})$.

We can also use the Chinese Remainder Theorem as the basis for a second method for solving simultaneous linear congruences, which is less direct but often more efficient. We start by finding a solution $x = x_1$ of one of the congruences. It is usually best to start with the congruence involving the largest modulus, so in Example 3.14 we could start with $x \equiv 2$ mod (7), which has $x_1 = 2$ as an obvious solution. The remaining solutions of this congruence are found by adding or subtracting multiples of 7, and among these we can find an integer $x_2 = x_1 + 7t$ which also satisfies the second congruence $x \equiv 3$

mod (5): trying $x_1, x_1 \pm 7, x_1 \pm 14, \ldots$ in turn, we soon find $x_2 = 2 - 14 = -12$. This satisfies $x \equiv 2$ mod (7) and $x \equiv 3$ mod (5), and by the Chinese Remainder Theorem the general solution of this pair of congruences has the form $x_2 + 35t = -12 + 35t$ ($t \in \mathbb{Z}$). Trying $x_2, x_2 \pm 35, x_2 \pm 70, \ldots$ in turn, we soon find a solution $x_3 = -12 + 35t$ which also satisfies the third congruence $x \equiv 2$ mod (3), namely $x_3 = -12 + 35 = 23$. This satisfies all three congruences, so by the Chinese Remainder Theorem their general solution consists of the congruence class [23] mod (105).

Exercise 3.8

Solve the simultaneous congruences

$$x \equiv 1 \ \ \text{mod (4)}, \quad x \equiv 2 \ \ \text{mod (3)}, \quad x \equiv 3 \ \ \text{mod (5)}.$$

Exercise 3.9

Solve the simultaneous congruences

$$x \equiv 2 \ \ \text{mod (7)}, \quad x \equiv 7 \ \ \text{mod (9)}, \quad x \equiv 3 \ \ \text{mod (4)}.$$

The linear congruences in the Chinese Remainder Theorem are all of the form $x \equiv a_i$ mod (n_i). If we are given a set of simultaneous linear congruences, with one (or more) of them in the more general form $ax \equiv b$ mod (n_i), then we will first need to use the earlier algorithm to solve this congruence, expressing its general solution as a congruence class modulo some divisor of n_i; it will then be possible to apply the techniques based on the Chinese Remainder Theorem to solve the resulting simultaneous congruences.

Example 3.15

Consider the simultaneous congruences

$$7x \equiv 3 \ \ \text{mod (12)}, \quad 10x \equiv 6 \ \ \text{mod (14)}.$$

We saw in Examples 3.11 and 3.12 that the first of these congruences has the general solution $x = 9 + 12t$, and that the second has the general solution $x = 2 + 7t$. It follows that we can replace the original pair of congruences with the pair

$$x \equiv 9 \ \ \text{mod (12)}, \quad x \equiv 2 \ \ \text{mod (7)}.$$

Clearly, $x_0 = 9$ is a particular solution; since the moduli 12 and 7 are coprime, with product 84, the Chinese Remainder Theorem implies that the general solution has the form $9 + 84t$.

Exercise 3.10

Solve the simultaneous congruences

$$3x \equiv 6 \mod (12), \quad 2x \equiv 5 \mod (7), \quad 3x \equiv 1 \mod (5).$$

The Chinese Remainder Theorem can be used to convert a single congruence, with a large modulus, into several simultaneous congruences with smaller moduli, which may be easier to solve.

Example 3.16

Consider the linear congruence

$$13x \equiv 71 \mod (380).$$

Instead of using the algorithm described earlier for solving a single linear congruence, we can use the factorisation $380 = 2^2 \times 5 \times 19$, together with Theorem 3.4, to replace this congruence with the three simultaneous congruences

$$13x \equiv 71 \mod (4), \quad 13x \equiv 71 \mod (5), \quad 13x \equiv 71 \mod (19).$$

These immediately reduce to

$$x \equiv 3 \mod (4), \quad 3x \equiv 1 \mod (5), \quad 13x \equiv 14 \mod (19).$$

The first of these needs no further simplification, but we can apply the single congruence algorithm to simplify each of the other two. We write the second congruence as $3x \equiv 6 \mod (5)$, so dividing by 3 (which is coprime to 5) we get $x \equiv 2 \mod (5)$. Similarly, the third congruence can be written as $-6x \equiv 14 \mod (19)$, so dividing by -2 we get $3x \equiv -7 \equiv 12 \mod (19)$, and now dividing by 3 we have $x \equiv 4 \mod (19)$. Our original congruence is therefore equivalent to the simultaneous congruences

$$x \equiv 3 \mod (4), \quad x \equiv 2 \mod (5), \quad x \equiv 4 \mod (19).$$

Now these have mutually coprime moduli, so the Chinese Remainder Theorem applies, and we can use either of our two methods to find the general solution. Using the second method, we start with a solution $x_1 = 4$ of the third congruence; adding and subtracting multiples of 19, we find that $x_2 = 42$ also satisfies the second congruence, and then adding and subtracting multiples of $19 \times 5 = 95$ we find that 327 (or equivalently -53) also satisfies the first congruence. Thus the general solution has the form $x = 327 + 380t$ $(t \in \mathbb{Z})$.

Exercise 3.11

Solve the congruence $91x \equiv 419 \mod (440)$.

3.4 Simultaneous non-linear congruences

It is sometimes possible to solve simultaneous congruences by the Chinese Remainder Theorem, even when the congruences are not all linear.

Example 3.17

Consider the simultaneous congruences

$$x^2 \equiv 1 \mod (3) \quad \text{and} \quad x \equiv 2 \mod (4).$$

By inspection of the three congruence classes mod (3), we see that the first of these (which is not linear) is equivalent to $x \equiv 1$ or $2 \mod (3)$, so the pair of congruences are equivalent to

$$x \equiv 1 \text{ or } x \equiv 2 \mod (3), \quad \text{and} \quad x \equiv 2 \mod (4).$$

We have to be careful how we read the logical connectives 'and' and 'or' here: precisely one of the two congruences mod (3) is true, and the single congruence mod (4) is also true. Now $(p \vee q) \wedge r$ (meaning 'p or q, and r') is logically equivalent to $(p \wedge r) \vee (q \wedge r)$ (meaning 'p and r, or q and r'), so either

$$x \equiv 1 \mod (3) \quad \text{and} \quad x \equiv 2 \mod (4),$$

or

$$x \equiv 2 \mod (3) \quad \text{and} \quad x \equiv 2 \mod (4).$$

We now have two pairs of simultaneous linear congruences, and each pair can be solved by using the Chinese Remainder Theorem. The first pair has general solution $x \equiv -2 \mod (12)$, while the second pair has general solution $x \equiv 2 \mod (12)$, so our original pair of congruences has general solution $x \equiv \pm 2 \mod (12)$.

Exercise 3.12

Solve the simultaneous congruences

$$x^2 + 2x + 2 \equiv 0 \mod (5) \quad \text{and} \quad 7x \equiv 3 \mod (11).$$

The Chinese Remainder Theorem is useful for solving polynomial congruences when the modulus is composite.

Theorem 3.11

Let $n = n_1 \ldots n_k$ where the integers n_i are mutually coprime, and let $f(x)$ be a polynomial with integer coefficients. Suppose that for each $i = 1, \ldots, k$ there are N_i congruence classes $x \in \mathbb{Z}_{n_i}$ such that $f(x) \equiv 0 \bmod (n_i)$. Then there are $N = N_1 \ldots N_k$ classes $x \in \mathbb{Z}_n$ such that $f(x) \equiv 0 \bmod (n)$.

Proof

Since the moduli n_i are mutually coprime, we have $f(x) \equiv 0 \bmod (n)$ if and only if $f(x) \equiv 0 \bmod (n_i)$ for all i. Thus each class of solutions $x \in \mathbb{Z}_n$ of $f(x) \equiv 0 \bmod (n)$ determines a class of solutions $x = x_i \in \mathbb{Z}_{n_i}$ of $f(x_i) \equiv 0 \bmod (n_i)$ for each i. Conversely, if for each i we have a class of solutions $x_i \in \mathbb{Z}_{n_i}$ of $f(x_i) \equiv 0 \bmod (n_i)$, then by the Chinese Remainder Theorem there is a unique class $x \in \mathbb{Z}_n$ satisfying $x \equiv x_i \bmod (n_i)$ for all i, and this class satisfies $f(x) \equiv 0 \bmod (n)$. Thus there is a one-to-one correspondence between classes $x \in \mathbb{Z}_n$ satisfying $f(x) \equiv 0 \bmod (n)$, and k-tuples of classes $x_i \in \mathbb{Z}_{n_i}$ satisfying $f(x_i) \equiv 0 \bmod (n_i)$ for all i. For each i there are N_i choices for the class $x_i \in \mathbb{Z}_{n_i}$, so there are $N_1 \ldots N_k$ such k-tuples and hence this is the number of classes $x \in \mathbb{Z}_n$ satisfying $f(x) \equiv 0 \bmod (n)$. $\qquad\square$

Example 3.18

Putting $f(x) = x^2 - 1$, let us find the number N of classes $x \in \mathbb{Z}_n$ satisfying $x^2 \equiv 1 \bmod (n)$. We first count solutions of $x^2 \equiv 1 \bmod (p^e)$, where p is prime. If p is odd, then there are just two classes of solutions: clearly the classes $x \equiv \pm 1$ both satisfy $x^2 \equiv 1$, and conversely if $x^2 \equiv 1$ then p^e divides $x^2 - 1 = (x-1)(x+1)$ and hence (since $p > 2$) it divides $x - 1$ or $x + 1$, giving $x \equiv \pm 1$. If $p^e = 2$ or 4 then there are easily seen to be one or two classes of solutions, but if $p^e = 2^e \geq 8$ then a similar argument shows that there are four, given by $x \equiv \pm 1$ and $x \equiv 2^{e-1} \pm 1$: for any solution x, one of the factors $x \pm 1$ must be congruent to $2 \bmod (4)$, so the other factor must be divisible by 2^{e-1}. Now in general let n have prime-power factorisation $n_1 \ldots n_k$, where $n_i = p_i^{e_i}$ and each $e_i \geq 1$. We have just seen that for each odd p_i there are $N_i = 2$ classes in $\mathbb{Z}_{n_i}$ of solutions of $x^2 \equiv 1 \bmod (n_i)$, whereas if $p_i = 2$ we may have $N_i = 1, 2$ or 4, depending on e_i. By Theorem 3.11 there are $N = N_1 \ldots N_k$ classes in $\mathbb{Z}_n$ of solutions of $x^2 \equiv 1 \bmod (n)$, found by solving the simultaneous congruences $x^2 \equiv 1 \bmod (n_i)$. Substituting the values we have obtained for N_i, we therefore have

$$N = \begin{cases} 2^{k+1} & \text{if } n \equiv 0 \bmod (8), \\ 2^{k-1} & \text{if } n \equiv 2 \bmod (4), \\ 2^k & \text{otherwise,} \end{cases}$$

where k is the number of distinct primes dividing n. For instance, if $n = 60 = 2^2.3.5$ then $k = 3$ and there are $2^k = 8$ classes of solutions, namely $x \equiv \pm 1, \pm 11, \pm 19, \pm 29 \mod (60)$.

Exercise 3.13

How many classes of solutions are there for each of the following congruences?

(a) $x^2 - 1 \equiv 0 \mod (168)$.

(b) $x^2 + 1 \equiv 0 \mod (70)$.

(c) $x^2 + x + 1 \equiv 0 \mod (91)$.

(d) $x^3 + 1 \equiv 0 \mod (140)$.

3.5 An extension of the Chinese Remainder Theorem

Our final result, known to Yih-Hing in the 7th century AD, generalises the Chinese Remainder Theorem to the case where the moduli are not necessarily coprime. First we consider a simple illustration:

Example 3.19

We saw, in the comments preceding Theorem 3.10, that the simultaneous congruences
$$x \equiv 3 \mod (9) \quad \text{and} \quad x \equiv 2 \mod (6)$$
have no solution, so let us consider under what circumstances any pair of simultaneous congruences
$$x \equiv a_1 \mod (9) \quad \text{and} \quad x \equiv a_2 \mod (6)$$
have a solution. The greatest common divisor of the moduli 9 and 6 is 3, and the two congruences imply that
$$x \equiv a_1 \mod (3) \quad \text{and} \quad x \equiv a_2 \mod (3) \,,$$
so if a solution exists then $a_1 \equiv a_2 \mod (3)$, that is, 3 divides $a_1 - a_2$. Conversely, suppose that 3 divides $a_1 - a_2$, so $a_1 = a_2 + 3c$ for some integer c. Then the general solution of the first congruence $x \equiv a_1 \mod (9)$ has the form
$$x = a_1 + 9s = a_2 + 3c + 9s = a_2 + 3(c + 3s) \quad \text{where} \quad s \in \mathbb{Z} \,,$$

while the general solution of the second congruence $x \equiv a_2 \bmod (6)$ is

$$x = a_2 + 6t \quad \text{where} \quad t \in \mathbb{Z}.$$

This means that an integer $x = a_1 + 9s$ will satisfy both congruences provided $c + 3s = 2t$ for some t, that is, provided $s \equiv c \bmod (2)$. Thus the pair of congruences have a solution if and only if $3|(a_1 - a_2)$, in which case the general solution is

$$x = a_1 + 9(c + 2u) = a_1 + 9c + 18u \quad \text{where} \quad u \in \mathbb{Z},$$

forming a single congruence class $[a_1 + 9c] \bmod (18)$.

The final modulus, 18, is the least common multiple $[9, 6] = \text{lcm}(9, 6)$ of the moduli 9 and 6. A similar argument (which you should try for yourself) shows that in general, a pair of simultaneous congruences

$$x \equiv a_1 \bmod (n_1) \quad \text{and} \quad x \equiv a_2 \bmod (n_2)$$

have a solution if and only if $\gcd(n_1, n_2)$ divides $a_1 - a_2$, in which case the general solution is a single congruence class mod $\text{lcm}(n_1, n_2)$. Yih-Hing's result extends this to any finite set of linear congruences, showing that they have a solution if and only if each pair of them have a solution:

Theorem 3.12

Let $n_1, \ldots, n_k$ be positive integers, and let $a_1, \ldots, a_k$ be any integers. Then the simultaneous congruences

$$x \equiv a_1 \bmod (n_1), \quad \ldots, \quad x \equiv a_k \bmod (n_k)$$

have a solution x if and only if $\gcd(n_i, n_j)$ divides $a_i - a_j$ whenever $i \neq j$. When this condition is satisfied, the general solution forms a single congruence class mod (n), where n is the least common multiple of $n_1, \ldots, n_k$.

(Note that if the moduli n_i are mutually coprime then $\gcd(n_i, n_j) = 1$ for all $i \neq j$, so the condition $\gcd(n_i, n_j) | (a_i - a_j)$ is always satisfied; moreover, the least common multiple n of $n_1, \ldots, n_k$ is then their product $n_1 \ldots n_k$, so we obtain the Chinese Remainder Theorem as a special case of Theorem 3.12.)

Proof

If a solution x exists, then $x \equiv a_i \bmod (n_i)$ and hence $n_i | (x - a_i)$ for each i. For each pair $i \neq j$ let $n_{ij} = \gcd(n_i, n_j)$, so n_{ij} divides both n_i and n_j; it therefore divides $x - a_i$ and $x - a_j$, so it divides $(x - a_j) - (x - a_i) = a_i - a_j$, as required.

Let x_0 be any solution; then an integer x is a solution if and only if $x \equiv x_0$ mod (n_i) for each i, that is, $x - x_0$ is divisible by each n_i, or equivalently by their least common multiple $n = \mathrm{lcm}(n_1, \ldots, n_k)$. Thus the general solution consists of a single congruence class $[x_0]$ mod (n).

To complete the proof, we have to show that if n_{ij} divides $a_i - a_j$ for each pair $i \neq j$, then a solution exists. The strategy is to replace the given set of congruences with an equivalent set of congruences having mutually coprime moduli, and then to apply the Chinese Remainder Theorem to show that this new set has a solution. First we use Theorem 3.4 to replace each congruence $x \equiv a_i$ mod (n_i) with an equivalent finite set of congruences $x \equiv a_i$ mod (p^e), where p^e ranges over all the prime powers in the factorisation of n_i. This gives us a set of congruences, equivalent to the first set, in which all the moduli are prime powers. These moduli are not necessarily coprime, since some primes p may divide n_i for several i. For a given prime p, let us choose i so that n_i is divisible by the highest power of p, and let this power be p^e. If $p^f \mid n_j$, so that $f \leq e$, then p^f divides n_{ij} and hence (by our hypothesis) divides $a_i - a_j$; thus $a_i \equiv a_j$ mod (p^f), so the congruence $x \equiv a_i$ mod (p^e), if true, will imply $x \equiv a_i$ mod (p^f) and hence $x \equiv a_j$ mod (p^f). This means that we can discard all the congruences $x \equiv a_j$ mod (p^f) for this prime p from our set, with the exception of the single congruence $x \equiv a_i$ mod (p^e) involving the highest power of p, since this last congruence implies the others. If we do this for each prime p, we are then left with a finite set of congruences of the form $x \equiv a_i$ mod (p^e) involving distinct primes p; since these moduli p^e are mutually coprime, the Chinese Remainder Theorem implies that the congruences have a common solution, which is automatically a solution of the original set of congruences. $\qquad\square$

Example 3.20

Consider the congruences

$$x \equiv 11 \ \mathrm{mod}\ (36), \quad x \equiv 7 \ \mathrm{mod}\ (40), \quad x \equiv 32 \ \mathrm{mod}\ (75).$$

Here $n_1 = 36, n_2 = 40$ and $n_3 = 75$, so we have

$$n_{12} = \gcd(36, 40) = 4, \quad n_{13} = \gcd(36, 75) = 3 \quad \text{and} \quad n_{23} = \gcd(40, 75) = 5.$$

Since

$$a_1 - a_2 = 11 - 7 = 4, \quad a_1 - a_3 = 11 - 32 = -21 \quad \text{and} \quad a_2 - a_3 = 7 - 32 = -25,$$

the conditions $n_{ij} \mid (a_i - a_j)$ are all satisfied, so there are solutions, forming a single congruence class mod (n) where $n = \mathrm{lcm}(36, 40, 75) = 1800$. To find the general solution, we follow the procedure described in the last paragraph of the proof of Theorem 3.12. Factorising each n_i, we replace the first congruence with

$$x \equiv 11 \mod (2^2) \quad \text{and} \quad x \equiv 11 \mod (3^2),$$

the second with

$$x \equiv 7 \mod (2^3) \quad \text{and} \quad x \equiv 7 \mod (5),$$

and the third with

$$x \equiv 32 \mod (3) \quad \text{and} \quad x \equiv 32 \mod (5^2).$$

This gives us a set of six congruences, in which the moduli are powers of the primes $p = 2, 3$ and 5. From these, we select one congruence involving the highest power of each prime: for $p = 2$ we must choose $x \equiv 7 \mod (2^3)$ (which implies $x \equiv 11 \mod (2^2)$), for $p = 3$ we must choose $x \equiv 11 \mod (3^2)$ (which implies $x \equiv 32 \mod (3)$), and for $p = 5$ we must choose $x \equiv 32 \mod (5^2)$ (which implies $x \equiv 7 \mod (5)$). These three congruences, which can be simplified to

$$x \equiv 7 \mod (8), \quad x \equiv 2 \mod (9), \quad x \equiv 7 \mod (25),$$

have mutually coprime moduli, and you can check that our earlier methods, based on the Chinese Remainder Theorem, now give the general solution $x \equiv 407 \mod (1800)$.

Exercise 3.14

Determine which of the following sets of simultaneous congruences have solutions, and when they do, find the general solution:

(a) $x \equiv 1 \mod (6), \quad x \equiv 5 \mod (14), \quad x \equiv 4 \mod (21)$.

(b) $x \equiv 1 \mod (6), \quad x \equiv 5 \mod (14), \quad x \equiv -2 \mod (21)$.

(c) $x \equiv 13 \mod (40), \quad x \equiv 5 \mod (44), \quad x \equiv 38 \mod (275)$.

(d) $x^2 \equiv 9 \mod (10), \quad 7x \equiv 19 \mod (24), \quad 2x \equiv -1 \mod (45)$.

3.6 Supplementary exercises

Exercise 3.15

As a party trick, you ask a friend to choose an integer from 1 to 100, and to tell you its remainders on division by $3, 5$ and 7. How can you instantly identify the chosen number?

Exercise 3.16

Solve the following sets of simultaneous congruences:

(a) $x \equiv 1$ mod (4), $\quad x \equiv 2$ mod (3), $\quad x \equiv 3$ mod (5).

(b) $3x \equiv 6$ mod (12), $\quad 2x \equiv 5$ mod (7), $\quad 3x \equiv 1$ mod (5).

(c) $x^2 \equiv 3$ mod (6), $\quad x^3 \equiv 3$ mod (5).

Exercise 3.17

Find all the solutions of $x^3 + 3x - 8 \equiv 0$ mod (33).

Exercise 3.18

Seven thieves try to share a hoard of gold bars equally between themselves. Unfortunately, six bars are left over, and in the fight over them, one thief is killed. The remaining six thieves, still unable to share the bars equally since two are left over, again fight, and another is killed. When the remaining five share the bars, one bar is left over, and it is only after yet another thief is killed that an equal sharing is possible. What is the minimum number of bars which allows this to happen?

Exercise 3.19

An integer is *square-free* if it is a product of distinct primes. Show that for each integer $k \geq 1$ there is a set of k consecutive integers, none of which is square-free.

Exercise 3.20

Find complete sets of residues mod (7), all of whose elements are (a) odd, (b) even, (c) prime. Is there a complete set of residues mod (7) consisting of perfect squares?

Exercise 3.21

Show that if $n = n_1 \ldots n_k$ where $n_1, \ldots, n_k$ are mutually coprime integers, and R_i is a complete set of residues mod (n_i) for each i, then the integers $r = r_1 + r_2 n_1 + r_3 n_1 n_2 + \cdots + r_k n_1 n_2 \ldots n_{k-1}$ $(r_i \in R_i)$ form a complete set of residues mod (n).

4
Congruences with a Prime-power Modulus

As we saw in the last chapter, a single congruence mod (n) is equivalent to a set of simultaneous congruences modulo the prime powers p^e appearing in the factorisation of n. In this chapter we will therefore study congruences mod (p^e), where p is prime. We will first deal with the simplest case $e = 1$, and then, after a digression concerning primality-testing, we will consider the case $e > 1$. A good reason for starting with the prime case is that whereas modular addition, subtraction and multiplication behave much the same whether the modulus is prime or composite, division works much more smoothly when it is prime.

4.1 The arithmetic of $\mathbb{Z}_p$

We saw in Corollary 3.8 that a linear congruence $ax \equiv b \bmod (n)$ has a unique solution mod (n) if $\gcd(a, n) = 1$. Now if n is a prime p, then $\gcd(a, n) = \gcd(a, p)$ is either 1 or p; in the first case, we have a unique solution mod (p), while in the second case (where $p \mid a$), either every x is a solution (when $p \mid b$) or no x is a solution (when $p \nmid b$).

One can view this elementary result as saying that if the polynomial $ax - b$ has degree $d = 1$ over $\mathbb{Z}_p$ (that is, if $a \not\equiv 0 \bmod (p)$), then it has at most one root in $\mathbb{Z}_p$. Now in algebra we learn that a non-trivial polynomial of degree d, with real or complex coefficients, has at most d distinct roots in $\mathbb{R}$ or $\mathbb{C}$; it is reasonable to ask whether this is also true for the number system $\mathbb{Z}_p$, since we have just seen that it is true when $d = 1$. Our first main theorem, due to Lagrange, states that this is indeed the case.

65

Theorem 4.1

Let p be prime, and let $f(x) = a_d x^d + \cdots + a_1 x + a_0$ be a polynomial with integer coefficients, where $a_i \not\equiv 0 \bmod (p)$ for some i. Then the congruence $f(x) \equiv 0 \bmod (p)$ is satisfied by at most d congruence classes $[x] \in \mathbb{Z}_p$.

Comments

1 Note that this theorem allows the possibility that $a_d = 0$, so that $f(x)$ has degree less than d; if so, then by deleting $a_d x^d$ we see that there are strictly fewer than d classes $[x]$ satisfying $f(x) \equiv 0$. The same argument applies if we merely have $a_d \equiv 0 \bmod (p)$.

2 Even if $a_d \not\equiv 0$, $f(x)$ may still have fewer than d roots in $\mathbb{Z}_p$: for instance $f(x) = x^2 + 1$ has only one root in $\mathbb{Z}_2$, namely the class $[1]$, and it has no roots in $\mathbb{Z}_3$.

3 The condition that $a_i \not\equiv 0$ for some i ensures that $f(x)$ yields a non-trivial polynomial when we reduce it mod (p). If $a_i \equiv 0$ for all i then all p classes $[x] \in \mathbb{Z}_p$ satisfy $f(x) \equiv 0$, so the result will fail if $d < p$.

4 In the theorem, it is essential to assume that the modulus is prime: for example, the polynomial $f(x) = x^2 - 1$, of degree $d = 2$, has four roots in $\mathbb{Z}_8$, namely the classes $[1], [3], [5]$ and $[7]$. Indeed, Example 3.18, together with Theorem 2.6, shows that this polynomial can have an arbitrarily large number of roots in $\mathbb{Z}_n$ for composite n.

Proof

We use induction on d. If $d = 0$ then $f(x) = a_0$ with p not dividing a_0, so there are no solutions of $f(x) \equiv 0$, as required. For the inductive step, we now assume that $d \geq 1$, and that all polynomials $g(x) = b_{d-1} x^{d-1} + \cdots + b_0$ with some $b_i \not\equiv 0$ have at most $d - 1$ roots $[x] \in \mathbb{Z}_p$.

If the congruence $f(x) \equiv 0$ has no solutions, there is nothing left to prove, so suppose that $[a]$ is a solution; thus $f(a) \equiv 0$, so p divides $f(a)$. Now

$$f(x) - f(a) = \sum_{i=0}^{d} a_i x^i - \sum_{i=0}^{d} a_i a^i = \sum_{i=0}^{d} a_i (x^i - a^i) = \sum_{i=1}^{d} a_i (x^i - a^i).$$

For each $i = 1, \ldots, d$ we can put

$$x^i - a^i = (x - a)(x^{i-1} + a x^{i-2} + \cdots + a^{i-2} x + a^{i-1}),$$

so that by taking out the common factor $x - a$ we have

$$f(x) - f(a) = (x - a)g(x)$$

for some polynomial $g(x)$ with integer coefficients, of degree at most $d-1$. Now p cannot divide all the coefficients of $g(x)$: if it did, then since it also divides $f(a)$, it would have to divide all the coefficients of $f(x) = f(a) + (x - a)g(x)$, against our assumption. We may therefore apply the induction hypothesis to $g(x)$, so that at most $d-1$ classes $[x]$ satisfy $g(x) \equiv 0$. We now count classes $[x]$ satisfying $f(x) \equiv 0$: if any class $[x] = [b]$ satisfies $f(b) \equiv 0$, then p divides both $f(a)$ and $f(b)$, so it divides $f(b) - f(a) = (b - a)g(b)$; since p is prime, Lemma 2.1(b) implies that p divides $b - a$ or $g(b)$, so either $[b] = [a]$ or $g(b) \equiv 0$. There are at most $d-1$ classes $[b]$ satisfying $g(b) \equiv 0$, and hence at most $1+(d-1) = d$ satisfying $f(b) \equiv 0$, as required. $\qquad\square$

Exercise 4.1

Find the roots of the polynomial $f(x) = x^2 + 1$ in $\mathbb{Z}_p$ for each prime $p \leq 17$. Make a conjecture about how many roots $f(x)$ has in $\mathbb{Z}_p$ for each prime p.

A useful equivalent version of Lagrange's Theorem is the contrapositive:

Corollary 4.2

Let $f(x) = a_d x^d + \cdots + a_1 x + a_0$ be a polynomial with integer coefficients, and let p be prime. If $f(x)$ has more than d roots in $\mathbb{Z}_p$, then p divides each of its coefficients a_i.

Lagrange's Theorem tells us nothing new about polynomials $f(x)$ of degree $d \geq p$: there are only p classes in $\mathbb{Z}_p$, so it is trivial that at most d classes satisfy $f(x) \equiv 0$. The following result, useful in studying polynomials of high degree, is known as *Fermat's Little Theorem* (not to be confused with Fermat's Last Theorem, the subject of Chapter 11), though it was also known to Leibniz, and the first published proof was given by Euler.

Theorem 4.3

If p is prime and $a \not\equiv 0 \bmod (p)$, then $a^{p-1} \equiv 1 \bmod (p)$.

Proof

We give two proofs. Proof A is very short, relying on a little group theory (summarised in Appendix B), while Proof B is purely number-theoretic.

Proof A.   Since p is prime, the classes $[a] \neq [0]$ in $\mathbb{Z}_p$ are closed under taking products and inverses, so they form a group under multiplication, with identity element $[1]$. (This is the group U_p of units mod (p), which we will study more closely in Chapter 6.) The only non-trivial fact to check here is the existence of inverses: if $[a] \neq [0]$ then the congruence $ax \equiv 1$ has a unique solution $[x] \neq [0]$ in $\mathbb{Z}_p$, and this class is the inverse of $[a]$. This group of non-zero classes has order $p-1$, that is, it contains $p-1$ elements. Now a theorem of Lagrange (see Appendix B) implies that if g is any element of a group of finite order n, then g^n is the identity element in that group. Applying this result here, we see that each class $[a] \neq [0]$ satisfies $[a]^{p-1} = [1]$, so that $a^{p-1} \equiv 1$.

Proof B.   The integers $1, 2, \ldots, p-1$ form a complete set of non-zero residues mod (p). If $a \not\equiv 0$ mod (p) then $xa \equiv ya$ implies $x \equiv y$, by Corollary 3.8, so that the integers $a, 2a, \ldots, (p-1)a$ lie in distinct classes mod (p). None of these integers is divisible by p, so they also form a complete set of non-zero residues. It follows that $a, 2a, \ldots, (p-1)a$ are congruent to $1, 2, \ldots, p-1$ in some order. (For instance, if $p = 5$ and $a = 3$ then multiplying the residues $1, 2, 3, 4$ by 3 we get $3, 6, 9, 12$, which are respectively congruent to $3, 1, 4, 2$.) The products of these two sets of integers must therefore lie in the same class, that is,

$$1 \times 2 \times \cdots \times (p-1) \equiv a \times 2a \times \cdots \times (p-1)a \mod (p),$$

or equivalently

$$(p-1)! \equiv (p-1)! \, a^{p-1} \mod (p).$$

Since $(p-1)!$ is coprime to p, Corollary 3.8 allows us to divide through by $(p-1)!$ and deduce that $a^{p-1} \equiv 1$ mod (p).   □

Theorem 4.3 states that all the classes in $\mathbb{Z}_p$ except $[0]$ are roots of the polynomial $x^{p-1} - 1$. For a polynomial satisfied by *all* the classes in $\mathbb{Z}_p$, we simply multiply by x, to get $x^p - x$:

Corollary 4.4

If p is prime then $a^p \equiv a$ mod (p) for every integer a.

Proof

If $a \not\equiv 0$ then Theorem 4.3 gives $a^{p-1} \equiv 1$, so multiplying each side by a gives the result. If $a \equiv 0$ then $a^p \equiv 0$ also, so the result is again true.   □

These two results are very useful in dealing with large powers of integers.

Example 4.1

Let us find the least non-negative residue of 2^{68} mod (19). Since 19 is prime and 2 is not divisible by 19, we can apply Theorem 4.3 with $p = 19$ and $a = 2$, so that $2^{18} \equiv 1 \bmod (19)$. Now $68 = 18 \times 3 + 14$, so

$$2^{68} = (2^{18})^3 \times 2^{14} \equiv 1^3 \times 2^{14} \equiv 2^{14} \bmod (19).$$

Since $2^4 = 16 \equiv -3 \bmod (19)$, we can write $14 = 4 \times 3 + 2$ and deduce that

$$2^{14} = (2^4)^3 \times 2^2 \equiv (-3)^3 \times 2^2 \equiv -27 \times 4 \equiv -8 \times 4 \equiv -32 \equiv 6 \bmod (19),$$

so that $2^{68} \equiv 6 \bmod (19)$.

Example 4.2

We will show that $a^{25} - a$ is divisible by 30 for every integer a. Here Corollary 4.4 is more appropriate, since it refers to all integers a, rather than just those coprime to p. By factorising 30, we see that it is sufficient to prove that $a^{25} - a$ is divisible by each of the primes $p = 2, 3$ and 5. Let us deal with $p = 5$ first. Applying Corollary 4.4 twice, we have

$$a^{25} = (a^5)^5 \equiv a^5 \equiv a \bmod (5),$$

so 5 divides $a^{25} - a$ for all a. Similarly $a^3 \equiv a \bmod (3)$, so

$$a^{25} = (a^3)^8 a \equiv a^8 a = a^9 = (a^3)^3 \equiv a^3 \equiv a \bmod (3),$$

as required. For $p = 2$ a direct argument easily shows that $a^{25} - a$ is always even, but we can also continue with this method and use $a^2 \equiv a \bmod (2)$ to deduce (rather laboriously) that

$$
\begin{aligned}
a^{25} = (a^2)^{12} a &\equiv a^{12} a = (a^2)^6 a \equiv a^6 a = (a^2)^3 a \\
&\equiv a^3 a = a^4 = (a^2)^2 \\
&\equiv a^2 \equiv a \bmod (2).
\end{aligned}
$$

Exercise 4.2

Find the least non-negative residue of 3^{91} mod (23).

Corollary 4.4 shows that if $f(x)$ is any polynomial of degree $d \geq p$, then by repeatedly replacing any occurrence of x^p with x we can find a polynomial $g(x)$ of degree less than p with the property that $f(x) \equiv g(x)$ for all integers x. In other words, when considering polynomials mod (p), it is sufficient to restrict attention to those of degree $d < p$. Similarly, the coefficients can also be simplified by reducing them mod (p).

Example 4.3

Let us find all the roots of the congruence

$$f(x) = x^{17} + 6x^{14} + 2x^5 + 1 \equiv 0 \mod (5).$$

Here $p = 5$, so by replacing x^5 with x we can replace the leading term $x^{17} = (x^5)^3 x^2$ with $x^3 x^2 = x^5$, and hence with x. Similarly x^{14} is replaced with x^2, and x^5 with x, so giving the polynomial $x + 6x^2 + 2x + 1$. Reducing the coefficients mod (5) gives $x^2 + 3x + 1$. Thus $f(x) \equiv 0$ is equivalent to the much simpler congruence

$$g(x) = x^2 + 3x + 1 \equiv 0 \mod (5).$$

We will see later how to solve quadratic congruences, but here we can simply try all five classes $[x] \in \mathbb{Z}_5$, or else note that $g(x) \equiv (x - 1)^2$; either way, we find that $[x] = [1]$ is the only root of $g(x) \equiv 0$, so this class is the only root of $f(x) \equiv 0$.

As another application of Fermat's Little Theorem, we prove a result known as *Wilson's Theorem*, though it was first proved by Lagrange in 1770:

Corollary 4.5

An integer n is prime if and only if $(n - 1)! \equiv -1 \mod (n)$.

Proof

Suppose that n is a prime p. If $p = 2$ then $(p - 1)! = 1 \equiv -1 \mod (p)$, as required, so we may assume that p is odd. Define

$$f(x) = (1 - x)(2 - x)\ldots(p - 1 - x) + 1 - x^{p-1},$$

a polynomial with integer coefficients. This has degree $d < p - 1$, since when the product is expanded, the two terms in $f(x)$ involving x^{p-1} cancel. If $a = 1, 2, \ldots, p - 1$ then $f(a) \equiv 0 \mod (p)$: the product $(1 - a)(2 - a)\ldots(p - 1 - a)$ vanishes since it has a factor equal to 0, and $1 - a^{p-1} \equiv 0$ by Fermat's Little Theorem. Thus $f(x)$ has more than d roots mod (p), so by Corollary 4.2 its coefficients are all divisible by p. In particular, p divides the constant term $(p - 1)! + 1$, so $(p - 1)! \equiv -1$.

For the converse, suppose that $(n - 1)! \equiv -1 \bmod (n)$. We then have $(n - 1)! \equiv -1 \bmod (m)$ for any factor m of n. If $m < n$ then m appears as a factor of $(n - 1)!$, so $(n - 1)! \equiv 0 \bmod (m)$ and hence $-1 \equiv 0 \bmod (m)$. This implies that $m = 1$, so we conclude that n has no proper factors and is therefore prime. $\square$

Exercise 4.3

Prove that if p is an odd prime then the numerator of the rational number

$$r = 1 + \frac{1}{2} + \frac{1}{3} + \cdots + \frac{1}{p-1}$$

(in reduced form) is divisible by p; prove that if $p > 3$ then it is divisible by p^2 (Wolstenholme's Theorem).

We solved a quadratic congruence in Example 4.3, and we will deal with this subject thoroughly in Chapter 7; here we consider a simple but important example as an application of the theorems we have just proved.

Theorem 4.6

Let p be an odd prime. Then the quadratic congruence $x^2 + 1 \equiv 0 \bmod (p)$ has a solution if and only if $p \equiv 1 \bmod (4)$.

Proof

Suppose that p is an odd prime, and let $k = (p - 1)/2$. In the product

$$(p - 1)! = 1 \times 2 \times \cdots \times k \times (k + 1) \times \cdots \times (p - 2) \times (p - 1),$$

we have $p - 1 \equiv -1, p - 2 \equiv -2, \ldots, k + 1 = p - k \equiv -k \bmod (p)$, so by replacing each of the k factors $p - i$ with $-i$ for $i = 1, \ldots, k$ we see that

$$(p - 1)! \equiv (-1)^k . (k!)^2 \bmod (p).$$

Now Wilson's Theorem gives $(p - 1)! \equiv -1$, so $(-1)^k (k!)^2 \equiv -1$ and hence $(k!)^2 \equiv (-1)^{k+1}$. If $p \equiv 1 \bmod (4)$ then k is even, so $(k!)^2 \equiv -1$ and hence $x = k!$ is a solution of $x^2 + 1 \equiv 0 \bmod (p)$.

On the other hand, suppose that $p \equiv 3 \bmod (4)$, so that $k = (p-1)/2$ is odd. If x is any solution of $x^2 + 1 \equiv 0 \bmod (p)$, then x is coprime to p, so Fermat's Little Theorem gives $x^{p-1} \equiv 1 \bmod (p)$. Thus $1 \equiv (x^2)^k \equiv (-1)^k \equiv -1 \bmod (p)$, which is impossible since p is odd, so there can be no solution. $\square$

Example 4.4

Let $p = 13$, so $p \equiv 1 \bmod (4)$. Then $k = 6$, and $6! = 720 \equiv 5 \bmod (13)$, so $x = 5$ is a solution of $x^2 + 1 \equiv 0 \bmod (13)$, as is easily verified. The other solution is $-5 \equiv 8 \bmod (13)$.

Lagrange's Theorem implies that if p is any prime then there are at most two classes $[x] \in \mathbb{Z}_p$ of solutions of $x^2 + 1 \equiv 0 \bmod (p)$. When $p \equiv 1 \bmod (4)$ these are the two classes $\pm [k!]$, when $p \equiv 3 \bmod (4)$ there are no solutions, and when $p = 2$ there is a unique class $[1]$ of solutions.

4.2 Pseudoprimes and Carmichael numbers

In theory, Wilson's Theorem solves the primality-testing problem considered in Chapter 2. However, the difficulty of computing factorials makes it a very inefficient test, even for fairly small integers. In many cases we can do better by using Corollary 4.4, or rather its contrapositive, which asserts that if there is an integer a satisfying $a^n \not\equiv a \bmod (n)$, then n is composite. This test is much easier to apply, since in modular arithmetic, large powers can be calculated much more easily than factorials, as we shall soon show. This is particularly true if a computer, or even a calculator, is available. Although we will restrict attention to examples which are small enough to deal with by hand, it is a good exercise to write programs to extend the techniques to much larger integers.

The method is as follows. If we are given an integer n to test for primality, we choose an integer a and compute $a^n \bmod (n)$, reducing the numbers mod (n) whenever possible to simplify the calculations. Let us say that n passes the base a test if $a^n \equiv a \bmod (n)$, and fails the test if $a^n \not\equiv a \bmod (n)$; thus if n fails the test for any a then Corollary 4.4 implies that n must be composite, whereas if n passes the test then it might be prime or composite. For computational simplicity, it is sensible to start with $a = 2$ (clearly $a = 1$ is useless). If we find that $2^n \not\equiv 2 \bmod (n)$, then n has failed the base 2 test and must therefore be composite, so we can stop. For instance, $2^6 = 64 \not\equiv 2 \bmod (6)$, so 6 fails the test and is therefore composite. The Chinese knew this test, and they conjectured 25 centuries ago that the converse was also true, that if n passes the base 2 test then n is prime. This turns out to be false, but it took until 1819 for a counterexample to be discovered: there are composite integers n satisfying $2^n \equiv 2 \bmod (n)$, so they pass the base 2 test and yet they are not prime. We call such integers *pseudoprimes*: they look as if they are prime numbers, but they are in fact composite.

Example 4.5

Let us apply the base 2 test to the integer $n = 341$. Computing 2^{341} mod (341) is greatly simplified by noting that $2^{10} = 1024 \equiv 1$ mod (341), so

$$2^{341} = (2^{10})^{34}.2 \equiv 2 \mod (341),$$

and 341 has passed the test. However $341 = 11.31$, so it is not a prime but a pseudo-prime. (In fact, knowing this factorisation in advance, one could 'cheat' in the base 2 test to avoid large computations: since 11 and 31 are primes, Theorem 4.3 gives $2^{10} \equiv 1$ mod (11) and $2^{30} \equiv 1$ mod (31), which easily imply that $2^{341} - 2$ is divisible by both 11 and 31, and hence by 341.) By checking that all composite numbers $n < 341$ fail the base 2 test, one can show that 341 is the smallest pseudo-prime.

Exercise 4.4

Apply the base 2 test to the integers $n = 511$ and 509. What do you deduce about them? (Hint: $2^9 = 512$.)

Exercise 4.5

Show that the integer $n = 161038$, which has prime-power factorisation 2.73.1103, is a pseudo-prime. (It is, in fact, the smallest even pseudo-prime.)

Fortunately, pseudo-primes are quite rare, but nevertheless, there are infinitely many of them.

Theorem 4.7

There are infinitely many pseudo-primes.

Proof

We will show that if n is a pseudo-prime then so is $2^n - 1$. Since $2^n - 1 > n$ one can iterate this, starting with $n = 341$, to generate an infinite sequence of pseudo-primes.

If n is a pseudo-prime then n is composite, so Theorem 2.13 implies that $2^n - 1$ is composite. The proof of Theorem 2.13 was set as an exercise, so if you haven't done it, then here it is. We have $n = ab$, where $1 < a < n$ and

$1 < b < n$. In the polynomial identity

$$x^m - 1 = (x - 1)(x^{m-1} + x^{m-2} + \cdots + 1), \qquad (4.1)$$

which is valid for all $m \geq 1$, we put $x = 2^a$ and $m = b$, giving

$$2^n - 1 = 2^{ab} - 1 = (2^a - 1)(2^{a(b-1)} + 2^{a(b-2)} + \cdots + 1).$$

Since $1 < 2^a - 1 < 2^n - 1$, this shows that $2^n - 1$ is composite.

Now we need to prove that $2^{2^n - 1} \equiv 2 \bmod (2^n - 1)$. Since n is a pseudo-prime we have $2^n \equiv 2 \bmod (n)$, so $2^n = nk + 2$ for some integer $k \geq 1$. If we put $x = 2^n$ and $m = k$ in (4.1), we see that $2^n - 1$ divides $(2^n)^k - 1$; thus $2^{nk} \equiv 1 \bmod (2^n - 1)$, so $2^{2^n - 1} = 2^{nk+1} = 2^{nk}.2 \equiv 2 \bmod (2^n - 1)$, as required. $\qquad \square$

Exercise 4.6

Show that the Mersenne numbers $M_p = 2^p - 1$ (p prime) and the Fermat numbers $F_n = 2^{2^n} + 1$ ($n \geq 0$) all pass the base 2 test.

Let us return to our primality-testing method. If n fails the base 2 test then we can stop, knowing that n is composite; however, if n passes then it could be prime or pseudo-prime, and we do not know which. We therefore repeat the test with a different value of a. As with $a = 2$, failing the test shows that n is composite, whereas passing tells us nothing. In general, we test n repeatedly, each time using a different value of a. Note that if n has passed the tests for bases a and b (possibly equal), so that $a^n \equiv a$ and $b^n \equiv b \bmod (n)$, then $(ab)^n = a^n b^n \equiv ab \bmod (n)$, so n must also pass the base ab test; there is therefore no point in applying this test, and it is sensible to restrict the values of a to successive prime numbers. We call n a *pseudo-prime to the base a* if n is composite and satisfies $a^n \equiv a \bmod (n)$; thus a pseudo-prime to the base 2 is just a pseudo-prime, as defined earlier.

Example 4.6

Let us take $n = 341$ again. This passed the base 2 test, so let us try base 3 next. We will compute $3^{341} \bmod (341)$ by first computing it mod (11) and mod (31). Since $3^5 = 243 \equiv 1 \bmod (11)$ and $341 \equiv 1 \bmod (5)$, we have $3^{341} \equiv 3 \bmod (11)$. Theorem 4.3 gives $3^{30} \equiv 1 \bmod (31)$, and since $341 \equiv 11 \bmod (30)$, we therefore have $3^{341} \equiv 3^{11} \bmod (31)$; now $3^5 \equiv -5 \bmod (31)$, so $3^{341} \equiv 3.(-5)^2 = 75 \not\equiv 3 \bmod (31)$. Thus $3^{341} \not\equiv 3 \bmod (341)$, so 341 fails the base 3 test.

Exercise 4.7

Does 341 pass the base 5 or base 7 tests?

In our implementations of base a tests, we have so far avoided the problem of computing a^n mod (n) directly, either by using our knowledge of some smaller power of a (such as $2^{10} \equiv 1$ mod (341) in Example 4.5), or by using a factorisation of n to replace the modulus n with smaller moduli (such as 11 and 31 in Example 4.6). In general, neither of these short-cuts may be available to us, so how can we calculate a^n mod (n) efficiently when n is large? Simply calculating $a, a^2, a^3, \ldots, a^n$ mod (n) in turn will be very time-consuming, and a much better method is to use repeated squaring and multiplying, a technique which is also effective for computing n-th powers of other objects such as integers or matrices. The basic idea is that if $n = 2m$ is even then $x^n = (x^m)^2$, and if $n = 2m + 1$ is odd then $x^n = (x^m)^2 x$, so repeated use of this rule reduces the computation of n-th powers to a fairly small number of application of the functions

$$f : \mathbb{Z}_n \to \mathbb{Z}_n, \quad x \mapsto x^2 \quad \text{and} \quad g : \mathbb{Z}_n \to \mathbb{Z}_n, \quad x \mapsto x^2 a,$$

both of which are easy to evaluate.

Example 4.7

Let $n = 91$. This is odd, so $a^{91} = (a^{45})^2 a = g(a^{45})$. Similarly, 45 is odd, so $a^{45} = g(a^{22})$, giving $a^{91} = g(g(a^{22})) = (g \circ g)(a^{22})$. Since 22 is even, we have $a^{22} = (a^{11})^2 = f(a^{11})$, so $a^{91} = g(g(f(a^{11}))) = (g \circ g \circ f)(a^{11})$. You should check that by continuing we eventually reach

$$a^{91} = (g \circ g \circ f \circ g \circ g \circ f \circ g)(1), \tag{4.2}$$

which can be evaluated by starting with 1, and applying f twice and g five times, in the appropriate order. Since f involves one multiplication, and g involves two, the total number of multiplications required is 12, which is significantly less than the 90 required if we successively evaluate $a, a^2, a^3, \ldots, a^{91}$. (In fact, by halting the iteration a step earlier, with $a^{91} = (g \circ g \circ f \circ g \circ g \circ f)(a)$, one can reduce the number of multiplications to 10.) Since each multiplication is performed in $\mathbb{Z}_{91}$, the numbers involved never become excessively large: if we use least absolute moduli then we cannot meet any number larger than $45^2 = 2025$.

Exercise 4.8

Use this method to apply the base 2 test to 91. What does the test tell you about 91?

Given any integer n, we can easily construct the appropriate sequence of applications of f and g from the binary representation of n: this is a finite sequence of symbols 0 and 1, which we read from left to right, applying f or g whenever we meet 0 or 1 respectively. For instance,

$$91 = 1.2^6 + 0.2^5 + 1.2^4 + 1.2^3 + 0.2^2 + 1.2^1 + 1.2^0 = 1011011$$

in binary notation, so starting with the integer 1, we apply the functions g, f, g, g, f, g, g in that order; since we are writing functions on the left, we write this sequence in reverse to obtain (4.2). This argument implies that, for any n, the number of multiplications required to compute a^n is at most twice the number of digits in the binary expansion of n, that is, at most $2(1 + \lfloor \lg n \rfloor)$.

Exercise 4.9

Write the integer 133 in binary notation, and use this to apply the base 2 test to it; what do you deduce from this?

Returning to primality testing, if we eventually find a base a test which n fails, then we have proved that n is composite. If, on the other hand, n continues to pass successive tests, then we have proved nothing definite about n; however, it can be shown that the probability of n being prime rapidly approaches 1 as it passes more and more independent tests, so after a sufficient number of tests we can assert that n has a very high probability of being prime. While this is not definite enough for a rigorous proof of primality, for many practical purposes (such as cryptography) a high level of probability is quite adequate: the chance of n being composite, after passing sufficiently many tests, is significantly smaller than the chance of a machine or human error in computing with n. This is a typical example of a probabilistic algorithm, where we accept a slight degree of uncertainty about the outcome in order to obtain an answer in a reasonable amount of time. By contrast, the primality test based on Wilson's Theorem gives absolute certainty (if we can ensure accurate computation), at the cost of unreasonable computing time.

It is tempting to conjecture that if n is composite, then it will fail the base a test for some a, so the above algorithm will eventually detect this (possibly after a very large number of tests have been passed). Unfortunately, this is not the case: there are composite integers n which pass the base a test for every a, so they cannot be detected by this algorithm. These are the *Carmichael numbers*, composite integers n with the property that $a^n \equiv a \mod (n)$ for all integers a, so they satisfy the conclusion of Corollary 4.4 without being prime.

The smallest example of a Carmichael number is $n = 561 = 3.11.17$. This is clearly composite, so to show that it is a Carmichael number we need to show that $a^{561} \equiv a$ mod (561) for all integers a, and to do this it is sufficient to show that the congruence $a^{561} \equiv a$ is satisfied modulo 3, 11 and 17 for all a. Consider $a^{561} \equiv a$ mod (17) first. This is obvious if $a \equiv 0$ mod (17), so assume that $a \not\equiv 0$ mod (17). Since 17 is prime, Theorem 4.3 gives $a^{16} \equiv 1$ mod (17); since $561 \equiv 1$ mod (16), we therefore have $a^{561} \equiv a^1 = a$ mod (17). Similar calculations show that $a^{561} \equiv a$ mod (3) and $a^{561} \equiv a$ mod (11), so $a^{561} \equiv a$ mod (561) as required. As with pseudo-primes, showing that this is the smallest Carmichael number depends on the tedious but routine task of verifying that every smaller composite number fails the base a test for some a.

Exercise 4.10

Show that $a^{561} \equiv a$ mod (3) and $a^{561} \equiv a$ mod (11) for all integers a.

Carmichael numbers occur much less frequently than primes, and they are quite difficult to construct. In 1912, Carmichael conjectured that there are infinitely many of them, and this was proved in 1992 by Alford, Granville and Pomerance. The proof is difficult, but a crucial step is the following elementary and useful result:

Lemma 4.8

If n is square-free (a product of distinct primes) and if $p - 1$ divides $n - 1$ for each prime p dividing n, then n is either a prime or a Carmichael number.

Exercise 4.11

Prove Lemma 4.8.

In fact, the converse of Lemma 4.8 is also true, but we will postpone the proof of this until Theorem 6.15, since it needs ideas we have not yet considered.

Example 4.8

The number $n = 561 = 3.11.17$ is square-free and composite; since $n - 1 = 560$ is divisible by $p - 1 = 2, 10$ and 16, Lemma 4.8 implies that 561 is a Carmichael number.

Exercise 4.12

Show that 1729 and 2821 are Carmichael numbers.

Exercise 4.13

Find a Carmichael number of the form $7.23.p$, where p is prime.

4.3 Solving congruences mod (p^e)

Suppose that p is prime and $e \geq 1$. If x is a solution of a congruence $f(x) \equiv 0$ mod (p^e), then x satisfies $f(x) \equiv 0$ mod (p), so one way of finding all solutions of $f(x) \equiv 0$ mod (p^e) is first to solve the simpler congruence $f(x) \equiv 0$ mod (p), and then to see which solutions of this are also solutions of the more restrictive congruence $f(x) \equiv 0$ mod (p^e). In many cases an effective strategy is to increase the exponent of p one step at a time, solving $f(x) \equiv 0$ mod (p), then $f(x) \equiv 0$ mod (p^2), and so on until we reach the modulus p^e. Before considering the general theory, we will first study some simple examples.

Example 4.9

To solve the congruence
$$2x \equiv 3 \bmod (5^e),$$
we take $p = 5$ and $f(x) = 2x - 3$. By inspection, the only solution of $2x \equiv 3$ mod (5) is $x \equiv 4$ mod (5). Any solution of $2x \equiv 3$ mod (5^2) must satisfy $2x \equiv 3$ mod (5), and must therefore have the form $x \equiv 4 + 5k_1$ mod (5^2) for some integer k_1. Then $3 \equiv 2x \equiv 8 + 10k_1$ mod (5^2), so $10k_1 \equiv -5$ mod (5^2) and hence $2k_1 \equiv -1$ mod (5). This has solution $k_1 \equiv 2$ mod (5), so we obtain $x \equiv 4 + 5k_1 \equiv 14$ mod (5^2) as the general solution of $2x \equiv 3$ mod (5^2). We can now repeat this process to solve $2x \equiv 3$ mod (5^3). Putting $x \equiv 14 + 5^2 k_2$ mod (5^3) we see that $28 + 50k_2 \equiv 3$ mod (5^3), so $50k_2 \equiv -25$ mod (5^3) and hence $2k_2 \equiv -1$ mod (5), with solution $k_2 \equiv 2$ mod (5); thus $x \equiv 14 + 5^2 k_2 \equiv 64$ mod (5^3) is the general solution of $2x \equiv 3$ mod (5^3).

We can iterate this as often as we like, a typical step being as follows. Suppose that, for some i, the general solution of $2x \equiv 3$ mod (5^i) is $x \equiv x_i$ mod (5^i) for some x_i, so $2x_i - 3 = 5^i q_i$ for some integer q_i. (We took $x_1 = 4$ and $q_1 = 1$ in the above calculation, for instance.) We put $x \equiv x_i + 5^i k_i$ mod (5^{i+1}) for some unknown integer k_i, so $3 \equiv 2x \equiv 2x_i + 2.5^i k_i$ mod (5^{i+1}), or equivalently $2k_i \equiv -q_i$ mod (5), with solution $k_i \equiv 2q_i$ mod (5). Thus $x \equiv x_{i+1} \equiv x_i + 2.5^i q_i$ mod (5^{i+1}) is the general solution of $2x \equiv 3$ mod (5^{i+1}).

Exercise 4.14

Show that $k_i \equiv 2$ mod (5) for all i in Example 4.9.

We could have solved the congruence in Example 4.9 more directly by writing it as $2x \equiv 3 + 5^e \bmod (5^e)$; since $3 + 5^e$ is even, and 2 is coprime to 5^e, we get the general solution $x \equiv (3 + 5^e)/2 \bmod (5^e)$. Instead, we used the longer iterative method to give a simple illustration of how this method works. In the next example, where the congruence is non-linear, no such short-cut is available.

Example 4.10

Let us solve

$$f(x) = x^3 - x^2 + 4x + 1 \equiv 0 \bmod (5^e)$$

for $e = 1, 2$ and 3. By inspection, the only solutions of $f(x) \equiv 0 \bmod (5)$ are $x \equiv \pm 1 \bmod (5)$. Let us take $x_1 = -1$ as our starting point, so $f(x_1) = 5q_1$ with $q_1 = -1$. To find a corresponding solution of $f(x) \equiv 0 \bmod (5^2)$, we put $x_2 \equiv x_1 + 5k_1 \equiv -1 + 5k_1 \bmod (5^2)$. Then

$$
\begin{aligned}
f(x_2) &\equiv (x_1 + 5k_1)^3 - (x_1 + 5k_1)^2 + 4(x_1 + 5k_1) + 1 \\
&\equiv (x_1^3 - x_1^2 + 4x_1 + 1) + (3x_1^2 - 2x_1 + 4)5k_1 \\
&\equiv 5q_1 + 9.5k_1 \quad \bmod (5^2),
\end{aligned}
$$

where we have used the Binomial Theorem to expand each power of $x_1 + 5k_1$; we have included only the first two terms in each binomial expansion, since any subsequent terms are multiples of 5^2 and hence congruent to 0. Thus $f(x_2) \equiv 0 \bmod (5^2)$ if and only if $q_1 + 9k_1 \equiv 0 \bmod (5)$; since $q_1 = -1$, this is equivalent to $k_1 \equiv -1 \bmod (5)$, so $x \equiv x_2 \equiv x_1 + 5k_1 \equiv -6 \bmod (5^2)$ is the unique solution of $f(x) \equiv 0 \bmod (5^2)$ satisfying $x \equiv -1 \bmod (5)$.

Repeating this process, we have $f(x_2) = -275 = 5^2 q_2$ where $q_2 = -11$. If we put $x_3 \equiv x_2 + 5^2 k_2 \equiv -6 + 5^2 k_2 \bmod (5^3)$ then

$$
\begin{aligned}
f(x_3) &\equiv (x_2 + 5^2 k_2)^3 - (x_2 + 5^2 k_2)^2 + 4(x_2 + 5^2 k_2) + 1 \\
&\equiv (x_2^3 - x_2^2 + 4x_2 + 1) + (3x_2^2 - 2x_2 + 4)5^2 k_2 \\
&\equiv 5^2 q_2 + 124.5^2 k_2 \quad \bmod (5^3),
\end{aligned}
$$

so we require $q_2 + 124k_2 \equiv 0 \bmod (5)$, that is, $k_2 \equiv -1 \bmod (5)$. This gives $x \equiv x_3 \equiv x_2 + 5^2 k_2 \equiv -31 \bmod (5^3)$ as the unique solution of $f(x) \equiv 0 \bmod (5^3)$ satisfying $x \equiv -1 \bmod (5)$.

Notice that in both steps of this iteration, the expression in the second line of the displayed congruences has the form $f(x_i) + f'(x_i).5^i k_i$, where $i = 1, 2$ and $f'(x) = 3x^2 - 2x + 4$ is the derivative of $f(x)$. The same thing happens in

Example 4.9, where $f(x) = 2x - 3$ has derivative $f'(x) = 2$, and we can write the i-th step of the iteration as

$$f(x_i) + f'(x_i).5^i k_i \equiv (2x_i - 3) + 2.5^i k_i \equiv 0 \mod (5^{i+1}).$$

In each example, we divide through by 5^i (which divides $f(x_i)$) to obtain a linear congruence for $k_i \mod (5)$, in which the coefficient of k_i is $f'(x_i)$. We can solve this (uniquely) provided $f'(x_i) \not\equiv 0 \mod (5)$. To see what can happen when this condition fails, let us return to Example 4.10, but now taking the solution $x_1 = 1$ of $f(x) \equiv 0 \mod (5)$ as our starting point. We now have $f(x_1) = 5q_1$ with $q_1 = 1$. Putting $x_2 \equiv x_1 + 5k_1 \equiv 1 + 5k_1 \mod (5^2)$ we find that

$$f(x_2) \equiv f(x_1) + f'(x_1).5k_1 \equiv 5q_1 + 5^2 k_1 \mod (5^2),$$

so we need to solve $5k_1 \equiv -q_1 \equiv -1 \mod (5)$, which is impossible. Thus the solution $x \equiv 1 \mod (5)$ does not give rise to any solution of $f(x) \equiv 0 \mod (5^2)$, and consequently for each $e \geq 2$ there is no solution of $f(x) \equiv 0 \mod (5^e)$ such that $x \equiv 1 \mod (5)$. This difficulty arises because $f'(x) = 3x^2 - 2x + 4$ has a root in $\mathbb{Z}_5$ at $x \equiv 1 \mod (5)$. To summarise, we have shown that the roots of $f(x) \equiv 0 \mod (5^e)$ are $x \equiv \pm 1$ for $e = 1$, whereas for $e = 2$ and 3 the only roots are $x \equiv -6$ and $x \equiv -31$ respectively.

We now consider the general situation. Let $f(x) = \sum_j a_j x^j$ be a polynomial with integer coefficients, and let the congruence $f(x) \equiv 0 \mod (p^i)$ have a solution $x \equiv x_i \mod (p^i)$. If $x_{i+1} = x_i + p^i k_i$ then the Binomial Theorem gives

$$\begin{aligned} f(x_{i+1}) &= \sum_j a_j (x_i + p^i k_i)^j \\ &\equiv \sum_j a_j x_i^j + \sum_j j a_j x_i^{j-1}.p^i k_i \\ &= f(x_i) + f'(x_i).p^i k_i \mod (p^{i+1}), \end{aligned}$$

where we ignore multiples of p^{i+1}. Putting $f(x_i) = p^i q_i$ and dividing through by p^i, we see that $f(x_{i+1}) \equiv 0 \mod (p^{i+1})$ if and only if

$$q_i + f'(x_i)k_i \equiv 0 \mod (p). \tag{4.3}$$

There are now three possibilities:

(a) if $f'(x_i) \not\equiv 0 \mod (p)$, then (4.3) has a unique solution $k_i \mod (p)$, so x_i gives rise to a unique solution $x_{i+1} \in \mathbb{Z}_{p^{i+1}}$ of $f(x) \equiv 0 \mod (p^{i+1})$;

(b) if $f'(x_i) \equiv 0 \not\equiv q_i \mod (p)$, then (4.3) has no solution k_i, and x_i gives no solution $x_{i+1} \in \mathbb{Z}_{p^{i+1}}$;

(c) if $f'(x_i) \equiv 0 \equiv q_i \mod (p)$, then every $k_i \in \mathbb{Z}_p$ satisfies (4.3), so x_i gives rise to p solutions $x_{i+1} \in \mathbb{Z}_{p^{i+1}}$.

This principle is part of a much more general result known as *Hensel's Lemma*. Cases (a) and (b) are illustrated by Example 4.10 (with $x_1 = -1$ and 1 respectively), and the following exercise illustrates case (c):

Exercise 4.15

Find the solutions of $f(x) = x^3 + 4x^2 + 19x + 1 \equiv 0 \bmod (5^2)$.

Exercise 4.16

Solve $f(x) = x^3 - x - 1 \equiv 0 \bmod (5^e)$ for $e = 1, 2$ and 3.

There is a close analogy with Newton's method in Calculus, where a solution $x \in \mathbb{R}$ of an equation $f(x) = 0$ is found as the limit of a convergent sequence of approximations x_i given by the recurrence relation

$$x_{i+1} = x_i - \frac{f(x_i)}{f'(x_i)}.$$

In our case, we have $x_{i+1} = x_i + p^i k_i$, where $q_i + f'(x_i)k_i \equiv 0 \bmod (p)$ and $f(x_i) = p^i q_i$, so writing $k_i = -q_i/f'(x_i)$ and substituting for k_i we get the same recurrence relation (though the arithmetic used is modular, rather than real). In Newton's method, convergence means that terms x_i and x_j become close together, in the sense that $|x_i - x_j| \to 0$ as $i, j \to \infty$; in our case, however, we regard x_i and x_j as close (in modular arithmetic) if $x_i \equiv x_j \bmod (p^e)$ where e is large. Just as the real numbers can be constructed as the limits of convergent sequences of rational numbers, this new concept of convergence gives rise to a new number system, namely the field $\mathbb{Q}_p$ of p-adic numbers (one field for each prime p). The importance of this number system is that it allows algebraic, analytic and topological methods to be applied to the study of congruences mod (p^e). For the details, which are beyond the scope of this book, see Ebbinghaus *et al.* (1991) and Serre (1973).

We will use this method again in Chapter 7, when we consider congruences of the form $x^2 - a \equiv 0$.

4.4 Supplementary exercises

Exercise 4.17

A function f from $\mathbb{Z}_p$ to $\mathbb{Z}_p$ is a *polynomial function* if there is a polynomial $g(x)$, with integer coefficients, such that $f(x) = g(x)$ in $\mathbb{Z}_p$ for all $x \in \mathbb{Z}_p$. Two distinct polynomials can define the same function on $\mathbb{Z}_p$: for instance, the polynomials x and x^p, by Corollary 4.4. Show that there are exactly p^p polynomial functions $\mathbb{Z}_p \to \mathbb{Z}_p$, and deduce that every function $\mathbb{Z}_p \to \mathbb{Z}_p$ is a polynomial function.

Exercise 4.18

Show that if p and q are primes, then the cyclotomic polynomial $\Phi_q(x) = 1 + x + \cdots + x^{q-1}$ has $q - 1$ roots in $\mathbb{Z}_p$ if $p \equiv 1 \bmod (q)$, has one if $p = q$, and has none otherwise.

Exercise 4.19

Find all the roots of $x^{18} + 4x^{14} + 3x + 10 \equiv 0 \bmod (21)$.

Exercise 4.20

Prove that if p is prime then $(p - 1)! \equiv -1 \bmod (p)$ (as in Wilson's Theorem, Corollary 4.5) by pairing off non-zero classes $a, b \in \mathbb{Z}_p$ such that $ab = 1$ in $\mathbb{Z}_p$.

Exercise 4.21

Show that if $p \geq 5$, then p is prime if and only if $6(p - 4)! \equiv 1 \bmod (p)$.

Exercise 4.22

Show that 10585 is a Carmichael number.

Exercise 4.23

Find two Carmichael numbers of the form $13.61.p$, where p is prime.

5

Euler's Function

One of the most important functions in number theory is Euler's function $\phi(n)$, which gives the number of congruence classes $[a] \in \mathbb{Z}_n$ which have an inverse under multiplication. We shall see how to evaluate this function, study its basic properties, and see how it can be applied to various problems such as the calculation of large powers and the encoding of secret messages.

5.1 Units

Many of the results in Chapter 4 depended on the simple but important fact that if p is prime, and $ab \equiv 0 \bmod (p)$, then $a \equiv 0$ or $b \equiv 0 \bmod (p)$. This makes the arithmetic of $\mathbb{Z}_p$ similar to that of $\mathbb{Z}$, in which the equation $ab = 0$ implies that $a = 0$ or $b = 0$. Unfortunately, this property fails when the modulus is composite: if $n = ab$ with $1 < a < n$ and $1 < b < n$, then $ab \equiv 0 \bmod (n)$ but $a, b \not\equiv 0 \bmod (n)$. Because of technical problems like this, we have to work a little harder to extend results from prime to composite moduli.

As an example, an important result in Chapter 4 was Fermat's Little Theorem, that if p is prime then $a^{p-1} \equiv 1 \bmod (p)$ for all integers $a \not\equiv 0 \bmod (p)$. We would like a similar result for composite moduli, but if we simply replace p with a composite integer n, then the resulting congruence $a^{n-1} \equiv 1 \bmod (n)$ is not generally true: if $\gcd(a, n) = d > 1$ then any positive power of a is divisible by d, so it cannot be congruent to 1 mod (n). This suggests that we should restrict attention to those integers a coprime to n, but even then the congruence can fail: if $n = 4$ and $a = 3$ then $a^{n-1} = 27 \not\equiv 1 \bmod (4)$, for example. We need a different exponent $e(n)$ such that $a^{e(n)} \equiv 1 \bmod (n)$ for all a coprime to n. The simplest function with this property turns out to be Euler's function $\phi(n)$, the main subject of this chapter, and one of the most important functions in number theory. In order to define this function, we first need to consider division in $\mathbb{Z}_n$.

We saw in Chapter 3 how to do arithmetic with congruence classes: $\mathbb{Z}_n$ has addition, subtraction and multiplication, but if n is composite then division by non-zero classes is not always possible. (Algebraists would say here that $\mathbb{Z}_n$ is a ring, but not a field.) In $\mathbb{Z}_4$, for instance, the class $[1]/[2]$ cannot be defined, since no class $[b]$ satisfies $[2][b] = [1]$. The following definition picks out those classes $[a] \in \mathbb{Z}_n$ for which there is a class $[1]/[a]$.

Definition

A *multiplicative inverse* for a class $[a] \in \mathbb{Z}_n$ is a class $[b] \in \mathbb{Z}_n$ such that $[a][b] = [1]$. A class $[a] \in \mathbb{Z}_n$ is a *unit* if it has a multiplicative inverse in $\mathbb{Z}_n$. (In this case, we sometimes say that the integer a is a *unit* mod (n), meaning that $ab \equiv 1 \bmod (n)$ for some integer b.)

Lemma 5.1

$[a]$ is a unit in $\mathbb{Z}_n$ if and only if $\gcd(a, n) = 1$.

Proof

If $[a]$ is a unit then $ab = 1 + qn$ for some integers b and q; any common factor of a and n would therefore divide 1, so $\gcd(a, n) = 1$. Conversely, if $\gcd(a, n) = 1$ then $1 = au + nv$ for some u and v by Theorem 1.7, so $[u]$ is a multiplicative inverse of $[a]$.   □

Example 5.1

The units in $\mathbb{Z}_8$ are $[1], [3], [5]$ and $[7]$: in fact $[1][1] = [3][3] = [5][5] = [7][7] = [1]$, so each of these units is its own multiplicative inverse. In $\mathbb{Z}_9$, the units are $[1], [2], [4], [5], [7]$ and $[8]$: for instance $[2][5] = [1]$, so $[2]$ and $[5]$ are inverses of each other.

Exercise 5.1

List the units in $\mathbb{Z}_{12}$ and in $\mathbb{Z}_{15}$; in each case, find the inverse of each unit.

We let U_n denote the set of units in $\mathbb{Z}_n$. Thus $U_8 = \{[1], [3], [5], [7]\}$ and $U_9 = \{[1], [2], [4], [5], [7], [8]\}$. The next result allows us to study units algebraically.

Theorem 5.2

For each integer $n \geq 1$, the set U_n forms a group under multiplcation mod (n), with identity element $[1]$.

Proof

We have to show that U_n satisfies the group axioms (listed in Appendix B), namely closure, associativity, existence of an identity and of inverses. To prove closure, we have to show that the product $[a][b] = [ab]$ of two units $[a]$ and $[b]$ is also a unit. If $[a]$ and $[b]$ are units, they have inverses $[u]$ and $[v]$ such that $[a][u] = [au] = [1]$ and $[b][v] = [bv] = [1]$; then $[ab][uv] = [abuv] = [aubv] = [au][bv] = [1]^2 = [1]$, so $[ab]$ has inverse $[uv]$, and is therefore a unit. This proves closure. Associativity asserts that $[a]([b][c]) = ([a][b])[c]$ for all units $[a], [b]$ and $[c]$; the left- and right-hand sides are the classes $[a(bc)]$ and $[(ab)c]$, so this follows from the associativity property $a(bc) = (ab)c$ in $\mathbb{Z}$. The identity element of U_n is $[1]$, since $[a][1] = [a] = [1][a]$ for all $[a] \in U_n$. Finally, if $[a] \in U_n$ then by definition there exists $[u] \in \mathbb{Z}_n$ such that $[a][u] = [1]$; now $[u] \in U_n$ (because the class $[a]$ satisfies $[u][a] = [1]$), so $[u]$ is the inverse of $[a]$ in U_n. $\square$

Exercise 5.2

Show that the group U_n is abelian.

5.2 Euler's function

Definition

We define $\phi(n) = |U_n|$, the number of units in $\mathbb{Z}_n$; by Lemma 5.1 this is the number of integers $a = 1, 2, \ldots, n$ such that $\gcd(a, n) = 1$. This function ϕ is called *Euler's function*. For small n, its values are as follows:

$$n \;\; = \;\; 1, 2, 3, 4, 5, 6, 7, 8, 9, 10, 11, 12, \ldots$$
$$\phi(n) \;\; = \;\; 1, 1, 2, 2, 4, 2, 6, 4, 6, \;\; 4, \;\; 10, \;\; 4, \ldots$$

We define a subset R of $\mathbb{Z}$ to be a *reduced set of residues* mod (n) if it contains one element from each of the $\phi(n)$ congruence classes in U_n. For instance, $\{1, 3, 5, 7\}$ and $\{\pm 1, \pm 3\}$ are both reduced sets of residues mod (8).

Exercise 5.3

Show that if R is a reduced set of residues mod (n), and if an integer a is a unit mod (n), then the set $aR = \{ar \mid r \in R\}$ is also a reduced set of residues mod (n).

In 1760, Euler proved the following generalisation of Fermat's Little Theorem, often called *Euler's Theorem*:

Theorem 5.3

If $\gcd(a, n) = 1$ then $a^{\phi(n)} \equiv 1$ mod (n).

Proof

Both Proof A and Proof B of Theorem 4.3 can easily be adapted to this situation; we will merely outline the arguments, and leave the details as an exercise. In Proof A we use the fact that U_n is a group under multiplication (Theorem 5.2). Since this group has order $\phi(n)$, Lagrange's Theorem (see Appendix B) implies that $[a]^{\phi(n)} = [1]$ for all $[a] \in U_n$. In Proof B, we replace the integers $1, 2, \ldots, p - 1$ of Theorem 4.3 with a reduced set $R = \{r_1, r_2, \ldots, r_{\phi(n)}\}$ of residues mod (n); if $\gcd(a, n) = 1$ then aR is also a reduced set of residues mod (n) (see Exercise 5.3), so the product of all the elements of aR must be congruent to the product of all the elements of R. This gives $a^{\phi(n)} r_1 r_2 \ldots r_{\phi(n)} \equiv r_1 r_2 \ldots r_{\phi(n)}$, and since the factors r_i are all units they can be cancelled to give $a^{\phi(n)} \equiv 1$. $\square$

Example 5.2

Fermat's Little Theorem is a special case of this result: if n is a prime p, then by Lemma 5.1 the units in $\mathbb{Z}_p$ are the classes $[1], [2], \ldots, [p-1]$, so $\phi(p) = p - 1$ and hence $a^{p-1} \equiv 1$ mod (p).

Example 5.3

If we take $n = 12$ then $U_{12} = \{\pm[1], \pm[5]\}$, and $\phi(12) = 4$; we have $(\pm 1)^4 = 1$ and $(\pm 5)^4 = 625 \equiv 1$ mod (12), so $a^4 \equiv 1$ mod (12) for each a coprime to 12.

Exercise 5.4

Find $\phi(14)$, and verify that $a^{\phi(14)} \equiv 1 \bmod (14)$ for each a coprime to 14.

We aim now to find a general formula for $\phi(n)$. We have just seen that $\phi(p) = p - 1$ for all primes p, and a simple extension of this deals with the case where n is a prime-power:

Lemma 5.4

If $n = p^e$ where p is prime, then

$$\phi(n) = p^e - p^{e-1} = p^{e-1}(p-1) = n\left(1 - \frac{1}{p}\right).$$

Proof

$\phi(p^e)$ is the number of integers in $\{1, \ldots, p^e\}$ which are coprime to p^e, that is, not divisible by p; this set has p^e members, of which $p^e/p = p^{e-1}$ are multiples of p, so $\phi(p^e) = p^e - p^{e-1} = p^{e-1}(p-1)$. $\square$

One can interpret this result in terms of probabilities. An integer a is a unit mod (p^e) if and only if it is not divisible by p. If we choose a randomly, then it will be divisible by p with probability $1/p$, and hence it will be coprime to p^e with probability $1 - 1/p$. Thus the proportion $\phi(n)/n$ of classes in $\mathbb{Z}_n$ which are units must be $1 - 1/p$, so $\phi(n) = n(1 - 1/p)$ for $n = p^e$.

We need a result which combines the information given in Lemma 5.4 for different prime-powers, to give a statement about $\phi(n)$ valid for all natural numbers n. Theorem 5.6 will do this, but to prove it we first need the following technical result about complete sets of residues (introduced in Chapter 3):

Lemma 5.5

If A is a complete set of residues mod (n), and if m and c are integers with m coprime to n, then the set $Am + c = \{am + c \mid a \in A\}$ is also a complete set of residues mod (n).

Proof

If $am + c \equiv a'm + c \bmod (n)$, where $a, a' \in A$, then by subtracting c and then cancelling the unit m, we see that $a \equiv a' \bmod (n)$, and hence $a = a'$. Thus the

n elements $am + c$ ($a \in A$) all lie in different congruence classes, so they form a complete set of residues mod (n).   □

Theorem 5.6

If m and n are coprime, then $\phi(mn) = \phi(m)\phi(n)$.

Proof

We may assume that $m, n > 1$, for otherwise the result is trivial since $\phi(1) = 1$. Let us arrange the mn integers $1, 2, \ldots, mn$ into an array with n rows and m columns, as follows:

$$
\begin{array}{cccc}
1 & 2 & 3 & \cdots \quad m \\
m+1 & m+2 & m+3 & \cdots \quad 2m \\
\vdots & \vdots & \vdots & \vdots \\
(n-1)m+1 & (n-1)m+2 & (n-1)m+3 & \cdots \quad nm
\end{array}
$$

These integers i form a complete set of residues mod (mn), so $\phi(mn)$ is the number of them coprime to mn, or equivalently satisfying $\gcd(i, m) = \gcd(i, n) = 1$. The integers in a given column are all congruent mod (m), and the m columns correspond to the m congruence classes mod (m); thus exactly $\phi(m)$ of the columns consist of integers i coprime to m, and the other columns consist of integers with $\gcd(i, m) > 1$. Now each column of integers coprime to m has the form $c, m + c, 2m + c, \ldots, (n-1)m + c$ for some c; by Lemma 5.5 this is a complete set of residues mod (n), since $A = \{0, 1, 2, \ldots, n-1\}$ is and since $\gcd(m, n) = 1$. Such a column therefore contains $\phi(n)$ integers coprime to n, so these $\phi(m)$ columns yield $\phi(m)\phi(n)$ integers i coprime to both m and n. Thus $\phi(mn) = \phi(m)\phi(n)$, as required.   □

Example 5.4

The integers $m = 3$ and $n = 4$ are coprime, with $\phi(3) = \phi(4) = 2$; here $mn = 12$ and $\phi(12) = 2.2 = 4$.

Exercise 5.5

Form the array in the above proof with $m = 5$ and $n = 4$; by finding the entries coprime to 20, verify that $\phi(20) = \phi(5)\phi(4)$.

The result in Theorem 5.6 fails if $\gcd(m, n) > 1$: for instance $2^2 = 4$, but $\phi(2)^2 \neq \phi(4)$.

Corollary 5.7

If n has prime-power factorisation $n = p_1^{e_1} \ldots p_k^{e_k}$ then

$$\phi(n) = \prod_{i=1}^{k}(p_i^{e_i} - p_i^{e_i-1}) = \prod_{i=1}^{k} p_i^{e_i-1}(p_i - 1) = n \prod_{i=1}^{k}\left(1 - \frac{1}{p_i}\right).$$

Proof

We prove the first expression by induction on k (the other expressions follow easily). Lemma 5.4 deals with the case $k = 1$, so assume that $k > 1$ and that the result is true for all integers divisible by fewer than k primes. We have $n = p_1^{e_1} \ldots p_{k-1}^{e_{k-1}} \cdot p_k^{e_k}$, where $p_1^{e_1} \ldots p_{k-1}^{e_{k-1}}$ and $p_k^{e_k}$ are coprime, so Theorem 5.6 gives

$$\phi(n) = \phi(p_1^{e_1} \ldots p_{k-1}^{e_{k-1}})\phi(p_k^{e_k}).$$

The induction hypothesis gives

$$\phi(p_1^{e_1} \ldots p_{k-1}^{e_{k-1}}) = \prod_{i=1}^{k-1}(p_i^{e_i} - p_i^{e_i-1}),$$

and Lemma 5.4 gives

$$\phi(p_k^{e_k}) = (p_k^{e_k} - p_k^{e_k-1}),$$

so by combining these two results we get

$$\phi(n) = \prod_{i=1}^{k}(p_i^{e_i} - p_i^{e_i-1}).$$

$\square$

We can write this result more concisely as $\phi(n) = n \prod_{p|n}(1 - \frac{1}{p})$, where $\prod_{p|n}$ denotes the product over all primes p dividing n.

Example 5.5

The primes dividing 60 are $2, 3$ and 5, so

$$\phi(60) = 60\left(1 - \frac{1}{2}\right)\left(1 - \frac{1}{3}\right)\left(1 - \frac{1}{5}\right) = 60.\frac{1}{2}.\frac{2}{3}.\frac{4}{5} = 16.$$

We can confirm this by writing down the integers $i = 1, 2, \ldots, 60$, and then deleting those with $\gcd(i, 60) > 1$. Initially there are 60 terms; deleting the multiples of 2 removes half of them, then deleting the multiples of 3 removes a third of the remaining terms, and finally deleting the multiples of 5 removes a fifth of those left. The remaining 16 terms, namely 1, 7, 11, 13, 17, 19, 23, 29, 31, 37, 41, 43, 47, 49, 53, 59, form a reduced set of residues mod (60).

Exercise 5.6

Calculate $\phi(42)$, and confirm it by finding a reduced set of residues mod (42).

Exercise 5.7

For which values of n is $\phi(n)$ odd? Show that there are integers n with $\phi(n) = 2, 4, 6, 8, 10$ and 12, but not 14.

Exercise 5.8

Show that for each integer m, there are only finitely many integers n such that $\phi(n) = m$.

Exercise 5.9

Find the smallest integer n such that $\phi(n)/n < 1/4$.

Exercise 5.10

The *Inclusion–Exclusion Principle* states that if $A_1, \ldots, A_m$ are finite sets, then

$$|A_1 \cup \cdots \cup A_m| = \sum_i |A_i| - \sum_{i<j} |A_i \cap A_j| + \sum_{i<j<k} |A_i \cap A_j \cap A_k|$$
$$- \cdots + (-1)^{m+1} |A_1 \cap \cdots \cap A_m|,$$

where $\sum_{i<j}$ denotes summation over all pairs i, j with $i < j$, and similarly for $\sum_{i<j<k}$ etc. Use this to find an alternative proof that $\phi(n) = n \prod_{p|n}(1 - 1/p)$, by considering the multiples of p in $\mathbb{Z}_n$ for each prime $p|n$.

The final expression for $\phi(n)$ in Corollary 5.7 has a probabilistic interpretation similar to that for Lemma 5.4. An integer a is a unit mod (n) if and only if it is coprime to each of the primes p_i dividing n. If we choose a randomly, then a is coprime to p_i with probability $1 - 1/p_i$. For distinct primes p_i these events are independent, so we multiply their probabilities, giving $\prod(1 - 1/p_i)$ for the probability that a is coprime to n. This must equal the proportion $\phi(n)/n$ of congruence classes $[a]$ which are units in $\mathbb{Z}_n$, so $\phi(n)/n = \prod(1 - 1/p_i)$. If $n > 1$ then $0 < \phi(n)/n < 1$; the next exercise shows that one can choose n so that this probability is arbitrarily close to 1.

Exercise 5.11

Show that if $\varepsilon > 0$, then there exists an integer $n > 1$ such that $\phi(n)/n > 1 - \varepsilon$.

Exercise 9.3 will show that, with a different choice of n, the probability $\phi(n)/n$ can also be made arbitrarily close to 0.

The following result will prove very useful in later chapters.

Theorem 5.8

If $n \geq 1$ then

$$\sum_{d|n} \phi(d) = n.$$

(Here, as always, $\sum_{d|n}$ denotes the sum over all *positive* divisors d of n.)

Proof

Let $S = \{1, 2, \ldots, n\}$, and for each d dividing n let $S_d = \{a \in S \mid \gcd(a, n) = n/d\}$. These sets S_d partition S into disjoint subsets, since if $a \in S$ then $\gcd(a, n) = n/d$ for some unique divisor d of n. Thus $\sum_{d|n} |S_d| = |S| = n$, so it is sufficient to prove that $|S_d| = \phi(d)$ for each d. Now

$$a \in S_d \iff a \in \mathbb{Z} \quad \text{with} \quad 1 \leq a \leq n \quad \text{and} \quad \gcd(a, n) = n/d.$$

If we define $a' = ad/n$ for each integer a, then a' is an integer since $n/d = \gcd(a, n)$ divides a. Dividing on the right-hand side by n/d, we can therefore rewrite the above condition as

$$a \in S_d \iff a = \frac{n}{d}.a' \quad \text{where} \quad a' \in \mathbb{Z} \text{ with } 1 \leq a' \leq d \text{ and } \gcd(a', d) = 1.$$

Thus $|S(d)|$ is the number of integers a', between 1 and d inclusive, which are coprime to d; this is the definition of $\phi(d)$, so $|S(d)| = \phi(d)$ as required. □

Example 5.6

If $n = 10$, then the divisors are $d = 1, 2, 5$ and 10. We find that $S_1 = \{10\}, S_2 = \{5\}, S_5 = \{2, 4, 6, 8\}$ and $S_{10} = \{1, 3, 7, 9\}$, containing $\phi(d) = 1, 1, 4$ and 4 elements respectively. These four sets form a partition of $S = \{1, 2, \ldots, 10\}$, so $\phi(1) + \phi(2) + \phi(5) + \phi(10) = 10$.

Exercise 5.12

Verify the equation $\sum_{d|n} \phi(d) = n$ in the case $n = 12$, and find the corresponding sets S_d.

Exercise 5.13

What form does the equation $\sum_{d|n} \phi(d) = n$ take if n is a prime-power p^e?

5.3 Applications of Euler's function

Having seen how to calculate Euler's function $\phi(n)$, we now look for some applications of it. We saw in Chapter 4 how to use Fermat's Little Theorem $a^{p-1} \equiv 1$ to simplify congruences mod (p), where p is prime, and we can now make similar use of Euler's Theorem $a^{\phi(n)} \equiv 1$ to simplify congruences mod (n) when n is composite.

Example 5.7

Let us find the last two decimal digits of 3^{1492}. This is equivalent to finding the least non-negative residue of 3^{1492} mod (100). Now 3 is coprime to 100, so Theorem 5.3 (with $a = 3$ and $n = 100$) gives $3^{\phi(100)} \equiv 1$ mod (100). The primes dividing 100 are 2 and 5, so Corollary 5.7 gives $\phi(100) = 100.(1/2).(4/5) = 40$, and hence we have $3^{40} \equiv 1$ mod (100). Since $1492 \equiv 12$ mod (40), it follows that $3^{1492} \equiv 3^{12}$ mod (100). Now $3^4 = 81 \equiv -19$ mod (100), so $3^8 \equiv (-19)^2 = 361 \equiv -39$ and hence $3^{12} \equiv -19. - 39 = 741 \equiv 41$. The last two digits are therefore 41.

Exercise 5.14

Show that if a positive integer a is coprime to 10, then the last three decimal digits of a^{2001} are the same as those of a.

We close this chapter with some applications of number theory to cryptography. Secret codes have been used since ancient times to send messages securely, for instance in times of war or diplomatic tension. Nowadays sensitive information of a medical or financial nature is often stored in computers, and it is important to keep it secret.

Many codes are based on number theory. A simple one is to replace each letter of the alphabet with its successor. Mathematically, we can do this by representing the letters as integers, say A $= 0$, B $= 1, \ldots$, Z $= 25$, and then adding 1 to each. In order to encode Z as A, we must add mod (26), so that $25 + 1 \equiv 0$. Similar codes are obtained by adding some fixed integer k (known as the key), rather than 1: Julius Caesar used the key $k = 3$. To decode, we simply apply the reverse transformation, subtracting k mod (26).

These codes are easy to break. We could either try all possible values of k in turn until we get a comprehensible message, or we could compare the most frequent letter in the message with the known most frequent letters in the original language (E, and then T, in English), to find k.

Exercise 5.15

Which mathematician is encoded in the above way as LBSLY, and what is the value of k?

A slightly more secure class of codes uses affine transformations of the form $x \mapsto ax + b$ mod (26), for various integers a and b. To decode successfully, we need to be able to recover the value of x uniquely from $ax + b$; this is possible if and only if a is a unit mod (26), so by counting the pairs a, b we see that there are $\phi(26).26 = 12.26 = 312$ such codes. Breaking such a code by trying all the possibilities for a and b would be tedious by hand (though simple with a computer), but again frequency searches can make the task much easier.

Exercise 5.16

If the encoding transformation is $x \mapsto 7x + 3$ mod (26), encode GAUSS and decode MFSJDG.

We can do rather better with codes based on Fermat's Little Theorem. The idea is as follows. We choose a large prime p, and an integer e coprime to $p - 1$. For encoding, we use the transformation $\mathbb{Z}_p \to \mathbb{Z}_p$ given by $x \mapsto x^e$ mod (p). (We saw in Chapter 4 how to calculate large powers efficiently in $\mathbb{Z}_p$.) If $0 < x < p$ then x will be coprime to p, so $x^{p-1} \equiv 1$ mod (p). To decode, we first find the multiplicative inverse f of e mod $(p - 1)$, that is, we solve the congruence $ef \equiv 1$ mod $(p - 1)$, using the method described in Chapter 3; this is possible since e is a unit mod $(p-1)$. Then $ef = (p-1)k + 1$ for some integer k, so $(x^e)^f = x^{(p-1)k+1} = (x^{p-1})^k.x \equiv x$ mod (p). Thus we can determine x from x^e, simply by raising it to the f-th power, so the message can be decoded efficiently.

Example 5.8

Suppose that $p = 29$ (unrealistically small, but useful for a simple illustration). We must choose e coprime to $p - 1 = 28$, and then find f such that $ef \equiv 1$ mod (28). If we take $e = 5$, for example, so that encoding is given by $x \mapsto x^5$ mod (29), then $f = 17$ and decoding is given by $x \mapsto x^{17}$ mod (29). Note that $(x^5)^{17} = x^{85} = (x^{28})^3 . x \equiv x$ mod (29) since $x^{28} \equiv 1$ mod (29) for all x coprime to 29, so decoding is the inverse of encoding.

Exercise 5.17

In Example 5.8, encode 9 and decode 11.

Representing individual letters as numbers tends to be insecure, since an eavesdropper could use known frequencies of letters. A better method is to group the letters into blocks of length k, and to represent each block as an integer x. (If the length of the message is not divisible by k, one can always add extra meaningless letters at the end.) We choose p sufficiently large that the distinct blocks of length k can be represented by different congruence classes $x \not\equiv 0$ mod (p), and then the encoding and decoding are given as before by $x \mapsto x^e$ and $x \mapsto x^f$ mod (p).

Breaking this code seems to be very difficult. Suppose, for instance, that an eavesdropper has discovered the value of p being used, and also knows one pair x and $y \equiv x^e$ mod (p). To break the code, he needs to know the value of f (or equivalently e), but if p is sufficiently large (say a hundred or more decimal digits) then there is no known efficient algorithm for calculating e from the congruence $y \equiv x^e$ mod (p), where x, y and p are known. This is sometimes called the *discrete logarithm problem*, since we can regard this congruence as a modular version of the equation $e = \log_x(y)$. The whole point of this code is that, while exponentials are easy to calculate in modular arithmetic, logarithms are apparently difficult.

Exercise 5.18

Find a value of e coprime to 28 such that $27 \equiv 10^e$ mod (29).

The one weakness of this type of code is that the sender and receiver must first agree on the values of p and e (called the *key* of the code) before they can use it. How can they do this secretly, bearing in mind that they will probably need to change the key from time to time for security? They could, of course exchange this information in encoded form, but then they would have to agree about the details of the code used for discussing the key, so they are no nearer solving the problem.

One can avoid this difficulty by using a *public-key cryptographic system*. Each person using the system publishes numerical information which enables any other user to encode messages, without giving away sufficient information to allow anyone but himself to decode them. Specifically, each person chooses a pair of large primes p and q, calculates $n = pq$, and publishes its value. If p and q are sufficiently large, then n cannot be factorised in a reasonable amount of time, so the values of p and q are effectively secret. Now $\phi(n) = (p-1)(q-1)$ by Corollary 5.7, so he (alone) can easily calculate $\phi(n)$; keeping $\phi(n)$ secret, he then finds and publishes an integer e coprime to $\phi(n)$. Anyone wishing to communicate with him looks up his published values for n and e (this pair is the *public key*), and encodes the message by the method of exponentiation described earlier; the only difference is that the calculations are now done in $\mathbb{Z}_n$, rather than $\mathbb{Z}_p$, so that the encoding transformation is $x \mapsto x^e \bmod (n)$. Since e is coprime to $\phi(n)$, the receiver (alone) can easily find f such that $ef \equiv 1 \bmod (\phi(n))$; if x is coprime to n (and this is easily arranged), then $(x^e)^f \equiv x \bmod (n)$ by Euler's Theorem, so he can use exponentiation to decode the message.

Example 5.9

Suppose that $p = 89$ and $q = 97$ are chosen, so $n = 89.97 = 8633$ is published, while $\phi(n) = 88.96 = 8448 = 2^8.3.11$ is kept secret. The receiver chooses and publishes an integer e coprime to $\phi(n)$, say $e = 71$. He then finds (and keeps secret) the multiplicative inverse $f = 119$ of $71 \bmod (8448)$; to check this, note that $71.119 = 8449 \equiv 1 \bmod (8448)$. To send a message, anyone can look up the pair $n = 8633, e = 71$, and use the encoding $x \mapsto x^{71} \bmod (8633)$. The receiver uses the decoding transformation $x \mapsto x^{119} \bmod (8633)$, which is not available to anyone who does not know that $f = 119$. An eavesdropper would need to factorise $n = 8633$ in order to find $\phi(n)$ and then f. Of course, factorising 8633 is not so difficult, but this is just a simple illustration of the method, and significantly larger primes p and q would pose a much harder problem.

Exercise 5.19

If my public key is the pair $n = 10147, e = 119$, then what is my decoding transformation?

This system also gives a way of 'signing' a message, to prove to a receiver that it comes from you and from nobody else. First decode your name, using your n and f (the latter being secret to you). Then encode the result, using the receiver's n and e (which are public knowledge), and send it to him. He will decode this message with his own n and f, and then encode the result with your n and e (which are also public knowledge). At the end of this, the receiver should have your name, since he has inverted the two transformations which you applied to it. Only you could have correctly applied the first transformation, so he knows that the message must have come from you.

5.4 Supplementary exercises

Exercise 5.20

Show that $\phi(mn) \geq \phi(m)\phi(n)$ for all m and n, with equality if and only if m and n are coprime.

Exercise 5.21

Show that if d divides n then $\phi(d)$ divides $\phi(n)$.

Exercise 5.22

For which n is $\phi(n) \equiv 2 \bmod (4)$?

Exercise 5.23

Find all n such that $\phi(n) = 16$.

Exercise 5.24

(a) Find all n such that $\phi(n) = n/2$.

(b) Find all n such that $\phi(n) = n/3$.

6
The Group of Units

We saw in Chapter 5 that for each n, the set U_n of units in $\mathbb{Z}_n$ forms a group under multiplication. Our aim in this chapter is to understand more about multiplication and division in $\mathbb{Z}_n$ by studying the structure of this group. An important result is that if $n = p^e$, where p is an odd prime, then U_n is cyclic; following a commonly-used strategy, we shall prove this first for $n = p$, and then deduce it for $n = p^e$. As often happens in number theory, the prime 2 is exceptional: although U_2 and U_4 are cyclic, we shall see that the group U_{2^e} is not cyclic for $e \geq 3$, although in a certain sense it is nearly cyclic. Using the Chinese Remainder Theorem, we can use our knowledge of the prime-power case to deduce the structure of U_n for arbitrary n. As an application, we will continue the study of Carmichael numbers, begun in Chapter 4.

From now on, for notational simplicity we will often omit the square brackets when using congruence classes. Thus we will sometimes regard an integer a as an element of $\mathbb{Z}_n$ or of U_n, when we should really write $[a]$. The context should make our meaning clear.

6.1 The group U_n

We say that a group G is *abelian* if its elements commute, that is, $gh = hg$ for all $g, h \in G$.

Lemma 6.1

U_n is an abelian group under multiplication mod (n).

Proof

Theorem 5.2 shows that U_n is a group, and Exercise 5.2 shows that it is abelian.

□

If G is a finite group with an identity element e, the *order* of an element $g \in G$ is the least integer $k > 0$ such that $g^k = e$; then the integers l such that $g^l = e$ are the multiples of k.

Example 6.1

In U_5 the element 2 has order 4: its powers are $2^1 \equiv 2$, $2^2 \equiv 4$, $2^3 \equiv 3$ and $2^4 \equiv 1$ mod (5), so $k = 4$ is the least positive exponent such that $2^k = 1$ (the identity element) in U_5. Similarly, the element 1 has order 1, while the elements 3 and 4 have orders 4 and 2 respectively.

Example 6.2

In U_8, the elements $1, 3, 5, 7$ have orders $1, 2, 2, 2$ respectively.

Exercise 6.1

Find the orders of the elements of U_9 and of U_{10}.

In Lemma 2.12 we showed that distinct Fermat numbers are coprime; as an application of the group structure of U_n we can now prove the corresponding result for the Mersenne numbers. First we need:

Lemma 6.2

If l and m are coprime positive integers, then $2^l - 1$ and $2^m - 1$ are coprime.

Proof

Let n be the highest common factor of $2^l - 1$ and $2^m - 1$. Clearly n is odd, so 2 is a unit mod (n). Let k be the order of the element 2 in the group U_n. Since n divides $2^l - 1$ we have $2^l = 1$ in U_n, so k divides l. Similarly k divides m, so

k divides $\gcd(l, m) = 1$. Thus $k = 1$, so the element 2 has order 1 in U_n. This means that $2^1 \equiv 1 \bmod (n)$, so $n = 1$, as required. $\qquad\qquad\square$

Exercise 6.2

Show that if l and m are positive integers with highest common factor h, then $\gcd(2^l - 1, 2^m - 1)$ divides $2^h - 1$.

Corollary 6.3

Distinct Mersenne numbers are coprime.

Proof

In Lemma 6.2, if we take l and m to be distinct primes we see that $M_l = 2^l - 1$ and $M_m = 2^m - 1$ are coprime. $\qquad\qquad\square$

6.2 Primitive roots

Our aim is to describe the structure of the group U_n for all n. To do this, it is not sufficient simply to know its order $\phi(n)$. For example, since $\phi(5) = 4 = \phi(8)$, the groups U_5 and U_8 both have order 4. However, these two groups are not isomorphic, since U_5 has elements of order 4, namely 2 and 3, whereas U_8 has none (see Examples 6.1 and 6.2). In group-theoretic terminology and notation, U_5 is a cyclic group of order 4 ($U_5 \cong C_4$), generated by 2 or by 3, whereas U_8 is a Klein four-group ($U_8 \cong V_4 = C_2 \times C_2$).

Exercise 6.3

The groups U_{10} and U_{12} both have order 4; show that exactly one of them is cyclic.

Definition

If U_n is cyclic then any generator g for U_n is called a *primitive root* mod (n). This means that g has order equal to the order $\phi(n)$ of U_n, so that the powers of g yield all the elements of U_n. For instance, 2 and 3 are primitive roots mod (5), but there are no primitive roots mod (8) since U_8 is not cyclic.

Finding primitive roots in U_n (if they exist) is a non-trivial problem, and there is no simple solution. One obvious but tedious method is to try each of the $\phi(n)$ units $a \in U_n$ in turn, each time computing powers a^i mod (n) to find the order of a in U_n; if we find an element a of order $\phi(n)$ then we know that this must be a primitive root. The following result is a rather more efficient test for primitive roots:

Lemma 6.4

An element $a \in U_n$ is a primitive root if and only if $a^{\phi(n)/q} \neq 1$ in U_n for each prime q dividing $\phi(n)$.

Proof

($\Rightarrow$) If a is a primitive root, then it has order $|U_n| = \phi(n)$, so $a^i \neq 1$ for all i such that $1 \leq i < \phi(n)$; in particular, this applies to $i = \phi(n)/q$ for each prime q dividing $\phi(n)$.

($\Leftarrow$) If a is not a primitive root, then its order k must be a proper factor of $\phi(n)$, so $\phi(n)/k > 1$. If q is any prime factor of $\phi(n)/k$, then k divides $\phi(n)/q$, so that $a^{\phi(n)/q} = 1$ in U_n, against our hypothesis. Thus a must be a primitive root. $\square$

Example 6.3

Let $n = 11$, and let us see whether $a = 2$ is a primitive root mod (11). Lemma 5.4 gives $\phi(11) = 11 - 1 = 10$, which is divisible by the primes $q = 2$ and $q = 5$, so we take $\phi(n)/q$ to be 5 and 2 respectively. Now $2^5, 2^2 \not\equiv 1$ mod (11), so Lemma 6.4 implies that 2 is a primitive root mod (11). To verify this, note that in U_{11} we have

$$2^1 = 2, \ 2^2 = 4, \ 2^3 = 8, \ 2^4 = 5, \ 2^5 = 10,$$
$$2^6 = 9, \ 2^7 = 7, \ 2^8 = 3, \ 2^9 = 6, \ 2^{10} = 1;$$

thus 2 has order 10, and its powers give all the elements of U_{11}. If we apply Lemma 6.4 with $a = 3$, however, we find that $3^5 = 243 \equiv 1$ mod (11), so 3 is not a primitive root mod (11): its powers are $3, 9, 5, 4$ and 1.

Example 6.4

Let us find a primitive root mod (17). We have $\phi(17) = 16$, which has only $q = 2$ as a prime factor. Lemma 6.4 therefore implies that an element $a \in U_{17}$ is a primitive root if and only if $a^8 \neq 1$ in U_{17}. Trying $a = 2$ first, we have

$2^8 = 256 \equiv 1 \bmod (17)$, so 2 is not a primitive root. However, $3^8 = (3^4)^2 \equiv (-4)^2 = 16 \not\equiv 1 \bmod (17)$, so 3 is a primitive root.

Example 6.5

To demonstrate that Lemma 6.4 also applies when n is composite, let us take $n = 9$. We have $\phi(9) = 6$, which is divisible by the primes $q = 2$ and $q = 3$, so that $\phi(n)/q$ is 3 and 2 respectively. Thus an element $a \in U_9$ is a primitive root if and only if $a^2, a^3 \neq 1$ in U_9. Since $2^2, 2^3 \not\equiv 1 \bmod (9)$, we see that 2 is a primitive root.

Exercise 6.4

Find primitive roots in U_n for $n = 18, 23, 27$ and 31.

Exercise 6.5

Show that if U_n has a primitive root then it has $\phi(\phi(n))$ of them.

We will show that U_n contains primitive roots if n is prime. This follows from the next theorem.

Theorem 6.5

If p is prime, then the group U_p has $\phi(d)$ elements of order d for each d dividing $p - 1$.

Before proving this, we deduce:

Corollary 6.6

If p is prime then the group U_p is cyclic.

Proof

Putting $d = p - 1$ in Theorem 6.5, we see that there are $\phi(p-1)$ elements of order $p - 1$ in U_p. Since $\phi(p-1) \geq 1$, the group contains at least one element of this order. Now U_p has order $\phi(p) = p - 1$, so such an element is a generator for U_p, and hence this group is cyclic. $\square$

Example 6.6

Let $p = 7$, so $U_p = U_7 = \{1, 2, 3, 4, 5, 6\}$. The divisors of $p - 1 = 6$ are $d = 1, 2, 3$ and 6, and the sets of elements of order d in U_7 are respectively $\{1\}$, $\{6\}$, $\{2, 4\}$ and $\{3, 5\}$; thus the numbers of elements of order d are $1, 1, 2$ and 2 respectively, agreeing with the values of $\phi(d)$. To verify that 3 is a generator, note that

$$3^1 = 3, \quad 3^2 = 2, \quad 3^3 = 6, \quad 3^4 = 4, \quad 3^5 = 5, \quad 3^6 = 1$$

in U_7, so every element of U_7 is a power of 3.

Exercise 6.6

Verify that the element 5 is a generator of U_7.

Exercise 6.7

Find the elements of order d in U_{11}, for each d dividing 10; which elements are generators?

Proof (Proof of Theorem 6.5.)

(In reading this proof, it may help to check each of its steps in a specific example, for instance by taking $p = 7$ or $p = 11$ throughout.) For each d dividing $p - 1$ let us define

$$\Omega_d = \{a \in U_p \mid a \text{ has order } d\} \quad \text{and} \quad \omega(d) = |\Omega_d|,$$

the number of elements of order d in U_p. Our aim is to prove that $\omega(d) = \phi(d)$ for all such d. Theorem 4.3 implies that the order of each element of U_p divides $p - 1$, so the sets Ω_d form a partition of U_p and hence

$$\sum_{d \mid p-1} \omega(d) = p - 1.$$

If we put $n = p - 1$ in Theorem 5.8 we get

$$\sum_{d \mid p-1} \phi(d) = p - 1,$$

so

$$\sum_{d \mid p-1} (\phi(d) - \omega(d)) = 0.$$

If we can show that $\omega(d) \leq \phi(d)$ for all d dividing $p - 1$, then each summand in this expression is non-negative; since their sum is 0, the summands must all be 0, so $\omega(d) = \phi(d)$, as required.

The inequality $\omega(d) \leq \phi(d)$ is obvious if Ω_d is empty, so assume that Ω_d contains an element a. By the definition of Ω_d, the powers $a^i = a, a^2, \ldots, a^d$ $(= 1)$ are all distinct, and they satisfy $(a^i)^d = 1$, so they are d distinct roots of the polynomial $f(x) = x^d - 1$ in $\mathbb{Z}_p$; by Theorem 4.1, $f(x)$ has at most $\deg(f) = d$ roots in $\mathbb{Z}_p$, so these are a complete set of roots of $f(x)$. We shall show that Ω_d consists of those roots a^i with $\gcd(i, d) = 1$. If $b \in \Omega_d$ then b is a root of $f(x)$, so $b = a^i$ for some $i = 1, 2, \ldots, d$. If we let j denote $\gcd(i, d)$, then

$$b^{d/j} = a^{id/j} = (a^d)^{i/j} = 1^{i/j} = 1$$

in U_p; but d is the order of b, so no lower positive power of b than b^d can be equal to 1, and hence $j = 1$. Thus every element b of order d has the form a^i where $1 \leq i \leq d$ and i is coprime to d. The number of such integers i is $\phi(d)$, so the number $\omega(d)$ of such elements b is at most $\phi(d)$, and the proof is complete. $\square$

Comments

1   The method of proof of Theorem 6.5 and Corollary 6.6 can be adapted slightly to prove a much stronger result, that if F is any field, so that $F^* = F \setminus \{0\}$ is a group under multiplication, then every finite subgroup G of F^* is cyclic. The idea is to let $|G| = n$, and to replace $p - 1$ with n in the above proof. Thus we use Theorem 5.8, that $\sum_{d|n} \phi(d) = n$, to show that for each d dividing n, G has $\phi(d)$ elements of order d; taking $d = n$ we see that G is cyclic. In number theory, the main interest is in the present case, where $F = \mathbb{Z}_p$ for some prime p and $G = \mathbb{Z}_p^* = U_p$, but the general result is also particularly useful in algebra, for instance when F is the field $\mathbb{C}$ of complex numbers.

2   The converse of Corollary 6.6 is false: for example the group U_4 is cyclic (generated by 3). We aim eventually to determine all the values of n for which U_n is cyclic, since cyclic groups are the easiest to work with. Having dealt with prime values of n, we next consider prime-powers, treating the odd case first.

6.3 The group U_{p^e}, where p is an odd prime

Theorem 6.7

If p is an odd prime, then U_{p^e} is cyclic for all $e \geq 1$.

Proof

Corollary 6.6 deals with the case $e = 1$, so we may assume that $e \geq 2$. We use the following strategy to find a primitive root mod p^e:

(a) first we pick a primitive root g mod (p) (possible by Corollary 6.6);

(b) next we show that either g or $g + p$ is a primitive root mod (p^2);

(c) finally we show that if h is any primitive root mod p^2, then h is a primitive root mod p^e for all $e \geq 2$.

Corollary 6.6 covers step (a), giving us a primitive root g mod (p). Thus $g^{p-1} \equiv 1$ mod (p), but $g^i \not\equiv 1$ mod (p) for $1 \leq i < p - 1$. We now proceed to step (b).

Since $\gcd(g, p) = 1$ we have $\gcd(g, p^2) = 1$, so we can consider g as an element of U_{p^2}. If d denotes the order of g mod (p^2), then Euler's Theorem implies that d divides $\phi(p^2) = p(p - 1)$. By definition of d, we have $g^d \equiv 1$ mod (p^2), so $g^d \equiv 1$ mod (p); but g has order $p - 1$ mod (p), so $p - 1$ divides d. Since p is prime, these two facts imply that either $d = p(p - 1)$ or $d = p - 1$. If $d = p(p - 1)$ then g is a primitive root mod (p^2), as required, so assume that $d = p - 1$. Let $h = g + p$. Since $h \equiv g$ mod (p), h is a primitive root mod (p), so arguing as before we see that h has order $p(p - 1)$ or $p - 1$ in U_{p^2}. Since $g^{p-1} \equiv 1$ mod (p^2), the Binomial Theorem gives

$$h^{p-1} = (g + p)^{p-1} = g^{p-1} + (p - 1)g^{p-2}p + \cdots \equiv 1 - pg^{p-2} \text{ mod } (p^2),$$

where the dots represent terms divisible by p^2. Since g is coprime to p, we have $pg^{p-2} \not\equiv 0$ mod (p^2) and hence $h^{p-1} \not\equiv 1$ mod (p^2). Thus h does not have order $p - 1$ in U_{p^2}, so it must have order $p(p - 1)$ and is therefore a primitive root. This completes step (b), but before proceeding to step (c), we look at an example of step (b).

Example 6.7

Let $p = 5$. We have seen that $g = 2$ is a primitive root mod (5), since it has order $\phi(5) = 4$ as an element of U_5. If we regard $g = 2$ as an element of $U_{p^2} = U_{25}$, then by the above argument its order d in U_{25} must be either $p(p - 1) = 20$ or $p - 1 = 4$. Now $2^4 = 16 \not\equiv 1$ mod (25), so $d \neq 4$ and hence $d = 20$. Thus $g = 2$ is a primitive root mod (25). (One can check this directly by computing the powers $2, 2^2, \ldots, 2^{20}$ mod (25), using $2^{10} = 1024 \equiv -1$ mod(25) to simplify the calculations.) Suppose instead that we had chosen $g = 7$; this is also a primitive root mod (5), since $7 \equiv 2$ mod (5), but it is *not* a primitive root mod (25): we have $7^2 = 49 \equiv -1$ mod (25), so $7^4 \equiv 1$ and hence 7 has order 4 in U_{25}. Step (b) guarantees that in this case, $g + p = 12$ must be a primitive root.

Exercise 6.8

Verify that 2 is a primitive root mod (25) by calculating its powers.

Proof (Continued.)

Now we consider step (c). Let h be any primitive root mod (p^2). We will show, by induction on e, that h is a primitive root mod (p^e) for all $e \geq 2$. Suppose, then, that h is a primitive root mod (p^e) for some $e \geq 2$, and let d be the order of h mod (p^{e+1}). An argument similar to that at the beginning of step (b) shows that d divides $\phi(p^{e+1}) = p^e(p-1)$ and is divisible by $\phi(p^e) = p^{e-1}(p-1)$, so $d = p^e(p-1)$ or $d = p^{e-1}(p-1)$. In the first case, h is a primitive root mod (p^{e+1}), as required, so it is sufficient to eliminate the second case by showing that $h^{p^{e-1}(p-1)} \not\equiv 1 \bmod (p^{e+1})$.

Since h is a primitive root mod (p^e), it has order $\phi(p^e) = p^{e-1}(p-1)$ in U_{p^e}, so $h^{p^{e-2}(p-1)} \not\equiv 1 \bmod (p^e)$. However $p^{e-2}(p-1) = \phi(p^{e-1})$, so $h^{p^{e-2}(p-1)} \equiv 1 \bmod (p^{e-1})$ by Euler's Theorem. Combining these two results, we see that $h^{p^{e-2}(p-1)} = 1 + kp^{e-1}$ where k is coprime to p, so the Binomial Theorem gives

$$
\begin{aligned}
h^{p^{e-1}(p-1)} &= (1 + kp^{e-1})^p \\
&= 1 + \binom{p}{1}kp^{e-1} + \binom{p}{2}(kp^{e-1})^2 + \cdots \\
&= 1 + kp^e + \frac{1}{2}k^2 p^{2e-1}(p-1) + \cdots .
\end{aligned}
$$

The dots here represent terms divisible by $(p^{e-1})^3$ and hence by p^{e+1}, since $3(e-1) \geq e+1$ for $e \geq 2$, so

$$
h^{p^{e-1}(p-1)} \equiv 1 + kp^e + \frac{1}{2}k^2 p^{2e-1}(p-1) \bmod (p^{e+1}).
$$

Now p is odd, so the third term $k^2 p^{2e-1}(p-1)/2$ is also divisible by p^{e+1}, since $2e - 1 \geq e+1$ for $e \geq 2$. Thus

$$
h^{p^{e-1}(p-1)} \equiv 1 + kp^e \bmod (p^{e+1}).
$$

Since p does not divide k, we therefore have $h^{p^{e-1}(p-1)} \not\equiv 1 \bmod (p^{e+1})$, so step (c) is complete. (Notice where we need p to be odd: if $p = 2$ then the third term $k^2 p^{2e-1}(p-1)/2 = k^2 2^{2e-2}$ is not divisible by 2^{e+1} when $e = 2$, so the first step of the induction argument fails.) $\square$

Comment

If g is a primitive root mod (p), where p is an odd prime, then g is *usually* a primitive root mod (p^2), in which case g is *always* a primitive root mod (p^e) for all e. For instance, $g = 2$ is a primitive root mod (5^e) for $e = 2$, and hence for all e.

Exercise 6.9

Show that 2 is a primitive root mod (3^e) for all $e \geq 1$.

Exercise 6.10

Find an integer which is a primitive root mod (7^e) for all $e \geq 1$.

6.4 The group U_{2^e}

We now deal with the powers of 2: in contrast with Theorem 6.7, we find that U_{2^e} is cyclic only for $e \leq 2$; in this sense, at least, the prime 2 is very odd!

Theorem 6.8

The group U_{2^e} is cyclic if and only if $e = 1$ or $e = 2$.

Proof

The groups $U_2 = \{1\}$ and $U_4 = \{1, 3\}$ are cyclic, generated by 1 and by 3, so it is sufficient to show that U_{2^e} is not cyclic for $e \geq 3$. We show that U_{2^e} has no elements of order $\phi(2^e) = 2^{e-1}$ by showing that

$$a^{2^{e-2}} \equiv 1 \bmod (2^e) \tag{6.1}$$

for all odd a. We prove this by induction on e. For the lowest value $e = 3$, (6.1) says that $a^2 \equiv 1 \bmod (8)$ for all odd a, and this is true since if $a = 2b + 1$ then $a^2 = 4b(b + 1) + 1 \equiv 1 \bmod (8)$. If we assume (6.1) for some exponent $e \geq 3$, then for each odd a we have

$$a^{2^{e-2}} = 1 + 2^e k$$

for some integer k. Squaring, we get

$$a^{2^{(e+1)-2}} = (1 + 2^e k)^2 = 1 + 2^{e+1} k + 2^{2e} k^2 = 1 + 2^{e+1}(k + 2^{e-1} k^2) \equiv 1 \bmod (2^{e+1}),$$

which is the required form of (6.1) for exponent $e + 1$. Thus (6.1) is true for all integers $e \geq 3$, and the proof is complete.   □

Exercise 6.11

Find the order of each element of U_{16}.

Exercise 6.12

Show that in U_{2^e} ($e \geq 3$), the elements of order 2 are $2^{e-1} \pm 1$ and -1.

Despite Theorem 6.8, we will show that U_{2^e} is nearly cyclic for $e \geq 3$ in the sense that the element 5 is almost a primitive root. First we need some notation and a lemma.

Notation. Recall that if p is prime, then $p^e \parallel n$ means that n is divisible by p^e but not by p^{e+1}. Thus $2^2 \parallel 20$, $5 \parallel 20$, and so on.

Lemma 6.9

$2^{n+2} \parallel 5^{2^n} - 1$ for all $n \geq 0$.

Proof

We use induction on n. The result is trivial for $n = 0$. Suppose it is true for some $n \geq 0$. Now

$$5^{2^{n+1}} - 1 = \left(5^{2^n}\right)^2 - 1 = \left(5^{2^n} - 1\right)\left(5^{2^n} + 1\right),$$

with $2^{n+2} \parallel 5^{2^n} - 1$ by the induction hypothesis, and with $2 \parallel 5^{2^n} + 1$ since $5^{2^n} \equiv 1 \bmod (4)$. Combining the powers of 2 we get $2^{n+3} \parallel 5^{2^{n+1}} - 1$ as required. $\square$

Theorem 6.10

If $e \geq 3$ then $U_{2^e} = \{\pm 5^i \mid 0 \leq i < 2^{e-2}\}$.

Proof

Let m be the order of the element 5 in U_{2^e}. By Euler's Theorem, m divides $\phi(2^e) = 2^{e-1}$, so $m = 2^k$ for some $k \leq e - 1$. Theorem 6.8 implies that U_{2^e} has no elements of order 2^{e-1}, so $k \leq e - 2$. By putting $n = e - 3$ in Lemma 6.9 we see that $2^{e-1} \parallel 5^{2^{e-3}} - 1$, so $5^{2^{e-3}} \not\equiv 1 \bmod (2^e)$ and hence $k > e - 3$. Thus $k = e - 2$, so $m = 2^{e-2}$. This means that 5 has 2^{e-2} distinct powers 5^i ($0 \leq i < 2^{e-2}$) in U_{2^e}. Since $5 \equiv 1 \bmod (4)$, these are all represented by integers congruent to 1 mod (4). This accounts for exactly half of the 2^{e-1} elements $1, 3, 5, \ldots, 2^e - 1$ of U_{2^e}, and the other half, represented by integers congruent to -1 mod (4), must be the elements of the form -5^i. This shows that every element has the form $\pm 5^i$ for some $i = 0, 1, \ldots, 2^{e-2} - 1$, as required.

□

Comment

The proof shows that the group U_{2^e} is generated by its elements -1 and 5, which individually generate cyclic subgroups of orders 2 and $m = 2^{e-2}$. These subgroups commute, and intersect in the identity subgroup, so they generate their direct product. Thus $U_{2^e} \cong C_2 \times C_{2^{e-2}}$ for $e \geq 3$, with the factors C_2 and $C_{2^{e-2}}$ generated by -1 and by 5 respectively. In terms of elements, this means that each $a \in U_{2^e}$ can be written uniquely in the form $a = (-1)^j 5^i$, where $j = 0, 1$ and $i = 0, 1, \ldots, 2^{e-2} - 1$.

Example 6.8

U_{16} consists of $1 = 5^4$, $3 = -5^3$, $5 = 5^1$, $7 = -5^2$, $9 = 5^2$, $11 = -5^1$, $13 = 5^3$, $15 = -5^4$.

Exercise 6.13

Show that if $e \geq 3$ then $U_{2^e} = \{\pm 3^i \mid 0 \leq i < 2^{e-2}\}$.

6.5 The existence of primitive roots

Having dealt with prime powers, we can now determine all the integers n for which there exist primitive roots mod (n).

Theorem 6.11

The group U_n is cyclic if and only if

$$n = 1, 2, 4, p^e \text{ or } 2p^e,$$

where p is an odd prime.

Proof

($\Leftarrow$) The cases $n = 1, 2$ and 4 are trivial, and Theorem 6.7 deals with the odd prime-powers, so we may assume that $n = 2p^e$ where p is an odd prime. Then Corollary 5.7 gives $\phi(n) = \phi(2)\phi(p^e) = \phi(p^e)$. By Theorem 6.7 there is a primitive root g mod (p^e). Then $g + p^e$ is also a primitive root mod (p^e), and one of g and $g + p^e$ is odd, so there is an odd primitive root h mod (p^e). We will show that h is a primitive root mod $(2p^e)$. By its construction, h is coprime to both 2 and p^e, so h is a unit mod $(2p^e)$. If $h^i \equiv 1$ mod $(2p^e)$, then certainly $h^i \equiv 1$ mod (p^e); since h is a primitive root mod (p^e), this implies that $\phi(p^e)$ divides i. Since $\phi(p^e) = \phi(2p^e)$, this shows that $\phi(2p^e)$ divides i, so h has order $\phi(2p^e)$ in U_{2p^e} and is therefore a primitive root. Before proving the converse part of the theorem, let us consider an example.

Example 6.9

We know that $g = 2$ is a primitive root mod (5^e) for all $e \geq 1$ (this follows from Example 6.7 and step (c) of the proof of Theorem 6.7). Now g is even, so $h = 2 + 5^e$ is an odd primitive root mod (5^e). The above argument then shows that h is also a primitive root mod (2.5^e). For instance, 7 is a primitive root mod (10), and 27 is a primitive root mod (50).

Proof (Continued.)

($\Rightarrow$) If $n \neq 1, 2, 4, p^e$ or $2p^e$, then either

(a) $n = 2^e$ where $e \geq 3$, or

(b) $n = 2^e p^f$ where $e \geq 2$, $f \geq 1$ and p is an odd prime, or

(c) n is divisible by at least two odd primes.

Theorem 6.8 shows that in case (a), U_n is not cyclic. Cases (b) and (c) are covered by the following result:

Lemma 6.12

If $n = rs$ where r and s are coprime and are both greater than 2, then U_n is not cyclic.

Proof

Since $\gcd(r,s) = 1$ we have $\phi(n) = \phi(r)\phi(s)$ by Theorem 5.6. Since $r, s > 2$, both $\phi(r)$ and $\phi(s)$ are even (see Exercise 5.7), so $\phi(n)$ is divisible by 4. It follows that the integer $e = \phi(n)/2$ is divisible by both $\phi(r)$ and $\phi(s)$. If a is a unit mod (n), then a is a unit mod (r) and also a unit mod (s), so $a^{\phi(r)} \equiv 1$ mod (r) and $a^{\phi(s)} \equiv 1$ mod (s) by Euler's Theorem. Since $\phi(r)$ and $\phi(s)$ divide e, we therefore have $a^e \equiv 1$ mod (r) and $a^e \equiv 1$ mod (s). Since r and s are coprime, this implies that $a^e \equiv 1$ mod (rs), that is, $a^e \equiv 1$ mod (n). Thus every element of U_n has order dividing e, and since $e < \phi(n)$, this means that there is no primitive root mod (n). $\square$

Proof (Proof of Theorem 6.11, concluded.)

In case (b) we can take $r = 2^e$ and $s = p^f$, while in case (c) we can take $r = p^e || n$ for some odd prime p dividing n, and $s = n/r$. In either case, $n = rs$ where r and s are coprime and greater than 2, so Lemma 6.12 shows that U_n is not cyclic. $\square$

Exercise 6.14

Find an integer which is a primitive root mod (2.3^e) for all $e \geq 1$. Find an integer which is a primitive root mod (2.7^e) for all $e \geq 1$.

Theorem 6.11 tells us when U_n has a primitive root, and its proof, together with the proof of Theorem 6.7, shows us how to find one, provided we can first find a primitive root in U_p where p is an odd prime. Unfortunately, although Corollary 6.6 proves that U_p has a primitive root, it does not give us a specific example of one; the best we can do is to keep applying Lemma 6.4 to elements $a \in U_p$ until we find a primitive root.

6.6 Applications of primitive roots

In this chapter, we have determined when there is a primitive root mod (n), and in those cases where one does exist, we have shown how to find one. We will now consider some applications of primitive roots, specifically to solving congruences of the form $x^m \equiv c$ mod (n), where m, c and n are given, and x has to be found. We will do this by considering some typical examples, and then explaining how our methods extend to more general situations.

Example 6.10

Consider the congruence $x^4 \equiv 13 \bmod (17)$. First note that any solution x must be a unit mod (17), so x, like 13, is an element of U_{17}. By Corollary 6.6, this group is cyclic, so both x and 13 can be expressed as powers of a primitive root $g \bmod (17)$. We saw earlier that 3 is a primitive root mod (17), so we will take $g = 3$. In general, there is no efficient way of expressing an arbitrary element, like 13, as a power of a primitive element g: we simply have to compute powers of g until the required element appears. In this case, $3^2 = 9, 3^3 = 27 \equiv 10$ and $3^4 = 81 \equiv 13 \bmod (17)$, so we have $13 = 3^4$ in U_{17}. We now write $x = 3^i$, where the exponent i is unknown. Then $x^4 = 3^{4i}$, so our congruence becomes $3^{4i} = 3^4$ in U_{17}. Now 3, being a primitive root, has order $\phi(17) = 16$, so $3^{4i} = 3^4$ if and only if $4i \equiv 4 \bmod (16)$, or equivalently $i \equiv 1 \bmod (4)$. The relevant values of i (between 0 and 15) are therefore $1, 5, 9$ and 13, so the solutions of the original congruence are $x \equiv 3, 3^5, 3^9$ and $3^{13} \bmod (17)$. We have seen that $3^4 \equiv 13$, so $3^5 \equiv 39 \equiv 5$. Instead of computing 3^9 and 3^{13}, we can take a short cut, and notice that if x is a solution then so is $-x$, so the remaining two classes of solutions must be $x \equiv -3 \equiv 14$ and $x \equiv -5 \equiv 12$. To summarise, there are four congruence classes of solutions, namely $x \equiv 3, 5, 12$ and $14 \bmod (17)$.

This example is typical of cases where there is a primitive root $g \bmod (n)$: by writing $x = g^i$ and $c = g^b$, we convert the original non-linear congruence $x^m \equiv c \bmod (n)$ into a linear congruence $mi \equiv b \bmod \phi(n)$ of the type considered in Chapter 3. The techniques described there allow us to find all the relevant values of i, and hence to find all the solutions x of the original congruence. The only difficulties tend to be the rather tedious problems of finding a primitive root g, and then expressing c as a power of g.

Exercise 6.15

Solve the congruence $x^6 \equiv 4 \bmod (23)$.

The next example illustrates the methods available when there is not a primitive root mod (n).

Example 6.11

Consider the congruence $x^3 \equiv 1 \bmod (63)$. Since 63 factorises as $3^2 \times 7$, Theorem 6.11 shows that there is no primitive root mod (63), so the method of the previous example does not work here. Instead, we note that this congruence is equivalent to the pair of simultaneous congruences $x^3 \equiv 1 \bmod (9)$ and $x^3 \equiv 1 \bmod (7)$; we find the solutions of each of these congruences, using primitive roots mod (9) and mod (7), and then we use Theorem 3.10 (the Chinese Remainder Theorem) to combine these solutions to get solutions of the original congruence.

We have seen that 2 is a primitive root mod (9). Writing $x = 2^i$ we see that $x^3 \equiv 1$ mod (9) is equivalent to $2^{3i} \equiv 1$ mod (9), and thus to $3i \equiv 0$ mod (6), since 2 has order $\phi(9) = 6$ in U_9. The general solution of this is $i \equiv 0$ mod (2), so $x^3 \equiv 1$ mod (9) has general solution $x \equiv 2^0, 2^2, 2^4 \equiv 1, 4, 7$ mod (9).

We have also seen that 3 is a primitive root mod (7), so by putting $x = 3^i$ we can rewrite $x^3 \equiv 1$ mod (7) as $3^{3i} \equiv 1$ mod (7); since 3 has order $\phi(7) = 6$ in U_7, this is equivalent to $3i \equiv 0$ mod (6), so again $i \equiv 0$ mod (2). Thus $x^3 \equiv 1$ mod (7) has general solution $x \equiv 3^0, 3^2, 3^4 \equiv 1, 2, 4$ mod (7).

We thus have three classes of solutions mod (9), and three classes mod (7). Since these moduli are coprime, the Chinese Remainder Theorem implies that each of these nine pairs of solutions gives rise to a single class of solutions mod (63): for instance, the pair of solutions $x \equiv 1$ mod (9) and $x \equiv 1$ mod (7) clearly correspond to the solution $x \equiv 1$ mod (63) of the original congruence. By using the method of Chapter 3, we can solve the other eight pairs of simultaneous congruences (try this as an exercise!), and we find that the general solution is $x \equiv 1, 4, 16, 22, 25, 37, 43, 46, 58$ mod (63). This gives another illustration of how Lagrange's Theorem (Theorem 4.1) on polynomials does not extend to composite moduli: the cubic polynomial $f(x) = x^3 - 1$ has nine roots in $\mathbb{Z}_{63}$.

Exercise 6.16

Solve the congruence $x^4 \equiv 4$ mod (99).

Example 6.11 is typical of those cases where there is no primitive root: we factorise the modulus n, giving a set of simultaneous congruences modulo various prime-powers p^e; we solve these individually, and then combine their solutions by means of the Chinese Remainder Theorem. We have seen how to use primitive roots to solve congruences of the form $x^m \equiv c$ mod (p^e) when p is an odd prime; however, Theorem 6.8 shows that if $p = 2$ and $e \geq 3$ then there is no primitive root, so in this case we need another method. Again, we will illustrate this with a typical example.

Example 6.12

Consider the congruence $x^3 \equiv 3$ mod (16). Since $16 = 2^4$, Theorem 6.8 implies that there is no primitive root mod (16); however, we know from Theorem 6.10 that every element of U_{16} has a unique expression of the form $\pm 5^i$ where $0 \leq i \leq 3$. By trial and error, we find that $5^3 = 125 \equiv -3$ mod (16), so $3 = -5^3$ in U_{16}. If we write $x = \pm 5^i$ then the congruence becomes $(\pm 5^i)^3 \equiv -5^3$, that is,

$\pm 5^{3i} \equiv -5^3$. If we take the plus sign (so that $x = 5^i$), then we have $5^{3i} = -5^3$ in U_{16}; this is impossible, since the powers of 5 are all congruent to 1 mod (4). If we take the minus sign (so that $x = -5^i$), then $5^{3i} = 5^3$ in U_{16}; since 5 has order $\phi(16)/2 = 4$ in U_{16}, this is equivalent to $3i \equiv 3$ mod (4), that is, $i \equiv 1$ mod (4), so $x \equiv -5^1 \equiv 11$ mod (16). Thus there is a unique class of solutions, namely $x \equiv 11$ mod (16).

Exercise 6.17

Solve the congruence $x^{11} \equiv 7$ mod (32).

When solving congruences $x^m \equiv c$ mod (2^e), it is sometimes more convenient to write each element in the form $\pm 3^i$ (see Exercise 6.13), rather than $\pm 5^i$: for instance, in Example 6.12 it is a little easier to express $c = 3$ in the form $\pm 3^i$ than in the form $\pm 5^i$!

6.7 The algebraic structure of U_n

Theorem 6.11 tells us the integers n for which U_n is cyclic. For most values of n, this group is not cyclic, and it is also useful to determine its structure in these cases; indeed, we have already done this for $n = 2^e$ in Theorem 6.10 and the subsequent comment. We will show that the ring $\mathbb{Z}_n$ and the group U_n each have a factorisation as a direct product, which imitates the prime-power factorisation of the integer n (see Appendix B for rings and direct products). This reduces the study of U_n to the prime-power case, which we have already considered.

First we need to understand the relationship between the rings $\mathbb{Z}_l$ and $\mathbb{Z}_n$ when l divides n. If $l|n$ then $a \equiv a'$ mod (n) implies $a \equiv a'$ mod (l), so $[a]_n \subseteq [a]_l$. In fact, it is easy to verify that if $n = lm$ then

$$[a]_l = [a]_n \cup [a + l]_n \cup [a + 2l]_n \cup \cdots \cup [a + (m-1)l]_n \,,$$

so each class of $\mathbb{Z}_l$ is the disjoint union of m classes of $\mathbb{Z}_n$. For instance, if $l = 2$ and $n = 6$ (so $m = 3$) then

$$[0]_2 = [0]_6 \cup [2]_6 \cup [4]_6 \qquad \text{and} \qquad [1]_2 = [1]_6 \cup [3]_6 \cup [5]_6 \,.$$

We can therefore define an m-to-1 function $\phi = \phi_{n,l} : \mathbb{Z}_n \to \mathbb{Z}_l$ by sending each class of $\mathbb{Z}_n$ to the unique class of $\mathbb{Z}_l$ which contains it, that is, $\phi([a]_n) = [a]_l$. Now

$$\phi([a]_n \pm [b]_n) = \phi([a]_n) \pm \phi([b]_n) \quad \text{and} \quad \phi([a]_n.[b]_n) = \phi([a]_n).\phi([b]_n)$$

for all $[a]_n, [b]_n \in \mathbb{Z}_n$; for instance, $[a]_n + [b]_n = [a+b]_n$, so $\phi([a]_n + [b]_n) = \phi([a+b]_n) = [a+b]_l$, while $\phi([a]_n) + \phi([b]_n) = [a]_l + [b]_l = [a+b]_l$ also, with similar proofs for subtraction and multiplication. Thus ϕ takes sums, differences and products in $\mathbb{Z}_n$ to the corresponding operations in $\mathbb{Z}_l$; in algebra, one says that ϕ is a *homomorphism* between these two rings. If a is coprime to n then it is also coprime to l, so $\phi(U_n) \subseteq U_l$; the restriction of ϕ to U_n takes products in U_n to products in U_l, so it is a homomorphism $U_n \to U_l$ of groups. This situation is symmetric with respect to l and m, so we also obtain a ring-homomorphism $\phi' = \phi_{n,m} : \mathbb{Z}_n \to \mathbb{Z}_m$, $[a]_n \mapsto [a]_m$, which restricts to a group-homomorphism $U_n \to U_m$.

The direct product $\mathbb{Z}_l \times \mathbb{Z}_m$ is the set of all ordered pairs $([a]_l, [b]_m)$ where $[a]_l \in \mathbb{Z}_l$ and $[b]_m \in \mathbb{Z}_m$. We define addition, subtraction and multiplication of such ordered pairs by performing these operations on their components:

$$([a]_l, [b]_m) + ([a']_l, [b']_m) = ([a + a']_l, [b + b']_m),$$

and so on. This makes $\mathbb{Z}_l \times \mathbb{Z}_m$ into a ring, and its subset $U_l \times U_m$ into a group. There is a ring-homomorphism $\theta : \mathbb{Z}_n \to \mathbb{Z}_l \times \mathbb{Z}_m$ given by $\theta([a]_n) = ([a]_l, [a]_m)$, which restricts to a group-homomorphism $U_n \to U_l \times U_m$.

We now show if that l and m are coprime (with $n = lm$ as before), then θ is an isomorphism, that is, in addition to being a homomorphism, it is a bijection. We have $n = \mathrm{lcm}(l, m)$, so the Chinese Remainder Theorem (Theorem 3.10) implies that, for each pair $([a]_l, [b]_m) \in \mathbb{Z}_l \times \mathbb{Z}_m$, there is a single congruence class x mod (n) of solutions of the simultaneous congruences $x \equiv a$ mod (l) and $x \equiv b$ mod (m). This means that there is exactly one class $[x]_n \in \mathbb{Z}_n$ such that $\theta([x]_n) = ([a]_l, [b]_m)$, so θ is a bijection. Thus θ is an ring-isomorphism

$$\mathbb{Z}_n \cong \mathbb{Z}_l \times \mathbb{Z}_m,$$

and it restricts to a group-isomorphism

$$U_n \cong U_l \times U_m.$$

An obvious extension of this argument, either by induction on k or using the full strength of the Chinese Remainder Theorem, proves the following theorem:

Theorem 6.13

If $n = n_1 \ldots n_k$ where $n_1, \ldots, n_k$ are mutually coprime, then there is a ring-isomorphism $\theta : \mathbb{Z}_n \to \mathbb{Z}_{n_1} \times \cdots \times \mathbb{Z}_{n_k}$ given by $\theta([a]_n) = ([a]_{n_1}, \ldots, [a]_{n_k})$, which restricts to a group-isomorphism $U_n \to U_{n_1} \times \cdots \times U_{n_k}$. In particular, if $n = p_1^{e_1} \ldots p_k^{e_k}$ where $p_1, \ldots, p_k$ are distinct primes, then

$$\mathbb{Z}_n \cong \mathbb{Z}_{p_1^{e_1}} \times \cdots \times \mathbb{Z}_{p_k^{e_k}} \quad \text{and} \quad U_n \cong U_{p_1^{e_1}} \times \cdots \times U_{p_k^{e_k}}.$$

For instance, in solving the congruence $x^3 \equiv 1 \bmod (63)$ in Example 6.11, we used a pair of simultaneous congruences mod (9) and mod (7). In effect, we were using the isomorphism $U_{63} \cong U_9 \times U_7$, and working simultaneously in the two direct factors U_9 and U_7.

Theorem 6.13 describes the structure of U_n in terms of that of U_{p^e} for various prime-powers p^e. We know from Lemma 5.4 that U_{p^e} has order $\phi(p^e) = p^{e-1}(p-1)$ for all $e \geq 1$; if p is odd then U_{p^e} is cyclic, by Theorem 6.7, while Theorem 6.10 implies that $U_{2^f} \cong C_2 \times C_{2^{f-2}}$ for all $f \geq 2$, and U_2 is the identity group. Putting all this information together, we get the following description of U_n as a direct product of cyclic groups:

Corollary 6.14

If $2^f \| n$ then

$$
U_n \cong
\begin{cases}
C_2 \times C_{2^{f-2}} \times \prod_{p^e \| n} C_{p^{e-1}(p-1)} & \text{if } f \geq 2, \text{ and} \\[2ex]
\prod_{p^e \| n} C_{p^{e-1}(p-1)} & \text{if } f \leq 1,
\end{cases}
$$

where $\prod_{p^e \| n}$ denotes the direct product as p^e ranges over the odd prime-powers appearing in the prime-power factorisation of n.

Example 6.13

If $n = 784 = 2^4.7^2$, then $f = 4$ and there is a unique odd prime-power $p^e = 7^2$ in the factorisation of n, so $U_{784} \cong C_2 \times C_4 \times C_{42}$.

In general, one can further factorise the cyclic groups $C_{p^{e-1}(p-1)}$ appearing in Corollary 6.14 into direct products of cyclic groups of prime-power order, using the factorisation of $p^{e-1}(p-1)$: this depends on the group-theoretic result that $C_m \cong \prod_q C_q$, where q ranges over all the prime-powers in the factorisation of m (this follows by applying the isomorphism $\mathbb{Z}_m \cong \prod_q \mathbb{Z}_q$, given by Theorem 6.13, to the additive groups of these rings). For instance, $C_{42} \cong C_2 \times C_3 \times C_7$, so in Example 6.13 we have $U_{784} \cong C_2 \times C_4 \times C_2 \times C_3 \times C_7$. Note, however, that a cyclic group C_q of prime-power order cannot be factorised further as a direct product: for instance $C_4 \not\cong C_2 \times C_2$, since C_4 has elements of order 4 whereas $C_2 \times C_2$ has none.

6.8 The universal exponent

The factorisation of U_n can be used to simplify large powers, in the same way as we used Euler's Theorem for this in Chapter 5. The *exponent* $e(G)$ of a finite group G is the least integer $e > 0$ satisfying $a^e = 1$ for all $a \in G$; the other integers with this property are the multiples of $e(G)$. Lagrange's Theorem implies that $a^{|G|} = 1$ for all $a \in G$, so $e(G)$ divides $|G|$. If we put $G = U_n$, of order $|U_n| = \phi(n)$, we get Euler's Theorem, that $a^{\phi(n)} = 1$ for all $a \in U_n$. The exponent $e(U_n)$ of U_n is called the *universal exponent* $e(n)$ of n; it divides $\phi(n)$, and it is the least positive integer e such that $a^e = 1$ for all $a \in U_n$. (Some authors use the notation $\lambda(n)$ for the universal exponent, but we will need the symbol λ for a different function in Chapter 9, Section 7.) If $e(n) < \phi(n)$ then the identity $a^{e(n)} = 1$ for all $a \in U_n$ is stronger, and often more useful than Euler's Theorem. Fortunately, it is easy to compute the exponent of U_n, or indeed of any finite abelian group G: simply express G as a direct product of cyclic groups, and take the least common multiple of their orders. In the case of U_n, Corollary 6.14 shows that $e(n)$ is the least common multiple of the numbers $e(2^f)$ and $e(p^e)$, where $2^f \| n$ and p^e ranges over the odd prime-powers in the factorisation of n; here $e(p^e) = \phi(p^e) = p^{e-1}(p-1)$ by Theorem 6.7, and $e(2^f) = 1, 2$ or 2^{f-2} as $f \leq 1, f = 2$ or $f \geq 3$ by Theorem 6.10.

Example 6.14

In Example 6.13 we saw that $U_{784} \cong C_2 \times C_4 \times C_{42}$; this group has order $\phi(784) = 2 \times 4 \times 42 = 336$, and exponent $e(784) = \mathrm{lcm}(2, 4, 42) = 84$. In place of Euler's Theorem $a^{336} = 1$ we therefore have the stronger result $a^{84} = 1$ for all $a \in U_{784}$. For instance, if we want to calculate 3^{256} mod (784), then Euler's Theorem cannot be used directly, since $1 \leq 256 < 336$; however, $256 \equiv 4$ mod (84), so putting $a = 3$ we get $3^{256} \equiv 3^4 \equiv 81$ mod (784).

Exercise 6.18

Express U_{520} as a direct product of cyclic groups of prime-power order. Find $e(520)$, and hence calculate 11^{123} mod (520).

Exercise 6.19

Show that a finite abelian group G satisfies $e(G) = |G|$ if and only if G is cyclic. For which integers n is $e(n) = \phi(n)$?

Recall that a Carmichael number is a composite integer n such that $a^{n-1} \equiv 1 \bmod (n)$ for every $a \in U_n$. Lemma 4.8 states that if n is square-free, and if $p - 1$ divides $n - 1$ for each prime p dividing n, then n is either a prime or a Carmichael number. We can now use $e(n)$ to prove the converse of this, as promised in Chapter 4. Clearly any prime number has the stated properties, so we need to prove that Carmichael numbers also have them.

Theorem 6.15

If n is a Carmichael number then n is square-free, and $p - 1$ divides $n - 1$ for each prime p dividing n.

Proof

By the definition of a Carmichael number, $a^{n-1} \equiv 1 \bmod (n)$ for all $a \in U_n$, so $n - 1$ is a multiple of $e(n)$. If $p^f \| n$ for some prime p and $f \geq 1$, then $e(n)$ is divisible by $e(p^f) = \phi(p^f) = p^{f-1}(p - 1)$, so $p - 1$ divides $n - 1$. If $f > 1$, then this argument also shows that p divides $n - 1$, which is impossible since p divides n; thus n must be square-free. $\square$

Comment

This proof also shows that a Carmichael number n must be odd: since n is composite, we have $n > 2$, so $e(n)$ is even; since $e(n)$ divides $n - 1$, this shows that n is odd.

Exercise 6.20

Show that a Carmichael number must be a product of at least three distinct primes.

6.9 Supplementary exercises

Exercise 6.21

Show that if there exists $a \in \mathbb{Z}$ such that $a^{p-1} \equiv 1 \bmod (p)$, whereas $a^{(p-1)/q} \not\equiv 1 \bmod (p)$ for each prime q dividing $p - 1$, then p is prime and a is a primitive root mod (p). Hence show that the Fermat number $F_4 = 2^{2^4} + 1 = 65537$ is prime.

Exercise 6.22

Show that if p is prime, then $(p - 2)! \equiv 1 \bmod (p)$. Show that if p is an odd prime, then $(p - 3)! \equiv (p - 1)/2 \bmod (p)$.

Exercise 6.23

Use Corollary 6.3 to show that there are infinitely many primes. (Take care to avoid a circular argument!)

Exercise 6.24

For which Fermat primes and Mersenne primes is 2 a primitive root?

Exercise 6.25

Find all the primitive roots for the integers $n = 18$ and 27. (Hint: see Exercises 6.4 and 6.5.)

Exercise 6.26

(a) Show that if p is an odd prime, and g is a primitive root mod (p) but not mod (p^2), then $g + rp$ is a primitive root mod (p^2) for $r = 1, 2, \ldots, p - 1$. By counting primitive roots, deduce that if g is a primitive root mod (p) then exactly one of $g, g + p, g + 2p, \ldots, g + (p - 1)p$ is not a primitive root mod (p^2).

(b) Find elements of U_{25} congruent to $2, 3 \bmod (5)$ respectively, which are not primitive roots mod (25).

7
Quadratic Residues

In this chapter, we will consider the general question of whether an integer a has a square root mod (n), and if so, how many there are and how one can find them. One of the main applications of this is to the solution of quadratic congruences, but we will also deduce a proof that there are infinitely many primes $p \equiv 1 \bmod (4)$, and we will give a useful primality test for Fermat numbers.

7.1 Quadratic congruences

To provide some motivation for what follows, we first briefly consider quadratic congruences. Just as in the case of quadratic equations, solving quadratic congruences can be reduced to the problem of finding square roots. Consider the formula

$$x = \frac{-b \pm \sqrt{b^2 - 4ac}}{2a}$$

for the roots of a quadratic equation $ax^2 + bx + c = 0$, where a, b and c are real or complex numbers. If we want this to apply to the case where $a, b, c \in \mathbb{Z}_n$, we will clearly need $2a$ to be a unit mod (n), so that we can divide by $2a$. Let us therefore assume, for the moment, that n is odd and that $a \in U_n$. Then $4a \in U_n$, so the quadratic equation is equivalent to

$$4a^2x^2 + 4abx + 4ac = 0.$$

Now we know that

$$(2ax + b)^2 = 4a^2x^2 + 4abx + b^2,$$

so we can write the equation in the form

$$(2ax + b)^2 = b^2 - 4ac.$$

If we can find all the square roots s of $b^2 - 4ac$ in $\mathbb{Z}_n$, we can then find all the solutions $x \in \mathbb{Z}_n$ of the quadratic equation in the form $2ax + b = s$, or equivalently $x = (-b + s)/2a$. In looking for square roots in $\mathbb{Z}_n$, however, we should be prepared for surprises (if that is not a contradiction): for instance, in $\mathbb{Z}_{15}$ the elements 1 and 4 each have four square roots (namely $\pm 1, \pm 4$ and $\pm 2, \pm 7$ respectively), while the other units have none.

Exercise 7.1

Find all the solutions in $\mathbb{Z}_{15}$ of the congruence $x^2 - 3x + 2 \equiv 0 \bmod (15)$.

Exercise 7.2

What square roots do the elements 5 and 16 have in $\mathbb{Z}_{21}$? Hence find all solutions of the congruences $x^2 + 3x + 1 \equiv 0 \bmod (21)$ and $x^2 + 2x - 3 \equiv 0 \bmod (21)$.

7.2 The group of quadratic residues

Definition

An element $a \in U_n$ is a *quadratic residue* mod (n) if $a = s^2$ for some $s \in U_n$; the set of such quadratic residues is denoted by Q_n. For small n one can determine Q_n simply by squaring all the elements $s \in U_n$.

Example 7.1

$Q_7 = \{1, 2, 4\} \subset U_7$, while $Q_8 = \{1\} \subset U_8$.

Exercise 7.3

Find Q_n for each $n \leq 12$.

We now determine how many square roots an element $a \in Q_n$ can have.

Lemma 7.1

Let k denote the number of distinct primes dividing n. If $a \in Q_n$, then the number N of elements $t \in U_n$ such that $t^2 = a$ is given by

$$N = \begin{cases} 2^{k+1} & \text{if } n \equiv 0 \bmod (8), \\ 2^{k-1} & \text{if } n \equiv 2 \bmod (4), \\ 2^k & \text{otherwise.} \end{cases}$$

Proof. If $a \in Q_n$ then $s^2 = a$ for some $s \in U_n$. Any element $t \in U_n$ has the form $t = sx$ for some unique $x \in U_n$, and we have $t^2 = a$ if and only if $x^2 = 1$ in U_n. Thus N is the number of solutions of $x^2 = 1$ in U_n, and Example 3.18 gives the required formula for N.   □

Comment

The number N of square roots depends only on n, and not on the element $a \in Q_n$. Moreover, if we have one square root s of a, then we can find all its other square roots $t = sx$ by finding all solutions of $x^2 = 1$, using the method of Example 3.18.

Exercise 7.4

Show that $|Q_n| = \phi(n)/N$, where N is given by Lemma 7.1.

Exercise 7.5

Find the elements of Q_{60}, together with their square roots.

Lemma 7.2

Q_n is a subgroup of U_n.

Proof

We need to show that Q_n contains the identity element of U_n, and is closed under taking products and inverses. Firstly, $1 \in Q_n$ since $1 = 1^2$ with $1 \in U_n$. If $a, b \in Q_n$ then $a = s^2$ and $b = t^2$ for some $s, t \in U_n$, so $ab = (st)^2$ with $st \in U_n$, giving $ab \in Q_n$. Finally, if $a \in Q_n$ then $a = s^2$ for some $s \in U_n$; since a and s

are units mod (n) they have inverses a^{-1} and s^{-1} in U_n, and $a^{-1} = (s^{-1})^2$ so that $a^{-1} \in Q_n$.   □

Algebraic comment

The function $\theta : U_n \to Q_n$, given by $\theta(s) = s^2$, is a homomorphism of groups, since $\theta(st) = (st)^2 = s^2 t^2 = \theta(s)\theta(t)$ for all $s, t \in U_n$. It is onto, by definition of Q_n, and its kernel K (the set of elements $x \in U_n$ such that $\theta(x) = 1$) is the subgroup of U_n consisting of the N solutions of $x^2 = 1$. For each $a \in Q_n$, the N square roots of a form a coset $\theta^{-1}(a)$ of K in U_n. For instance, the square roots of the elements $1, 2, 4$ of Q_7 form the cosets $\{\pm 1\}, \{\pm 3\}, \{\pm 2\}$ of $K = \{\pm 1\}$ in U_7.

In the special cases where U_n is cyclic (see Theorem 6.11), we have a simple description of Q_n:

Lemma 7.3

Let $n > 2$, and suppose that there is a primitive root g mod (n); then Q_n is a cyclic group of order $\phi(n)/2$, generated by g^2, consisting of the even powers of g.

Proof

Since $n > 2$, Exercise 5.7 implies that $\phi(n)$ is even. The elements $a \in U_n$ are the powers g^i for $i = 1, \ldots, \phi(n)$, with $g^{\phi(n)} = 1$. If i is even, then $a = g^i = (g^{i/2})^2 \in Q_n$. Conversely, if $a \in Q_n$ then $a = (g^j)^2$ for some j, so $i \equiv 2j$ mod $(\phi(n))$ for some j; since $\phi(n)$ is even, this implies that i is even. Thus Q_n consists of the even powers of g, so it is the cyclic group of order $\phi(n)/2$ generated by g^2.   □

Warning

We need the condition $n > 2$ to ensure that the cyclic group U_n has even order. In any cyclic group of *odd* order m, every element is a square and can be written as an even power of a generator g: for each i we have $g^i = g^{i+m}$, with one of i and $i + m$ even, so g^i is a square.

Example 7.2

If $n = 7$ then we can take $g = 3$ as a primitive root. The powers of g in U_7 are $g = 3$, $g^2 = 2$, $g^3 = 6$, $g^4 = 4$, $g^5 = 5$ and $g^6 = 1$; of these, the quadratic residues $a = 1, 2$ and 4 correspond to the even powers of g. Thus Q_7 is the cyclic group of order 3 generated by $g^2 = 2$.

Exercise 7.6

Use a primitive root to find the elements of Q_{25}.

7.3 The Legendre symbol

We now consider the problem of determining whether or not a given element $a \in U_n$ is a quadratic residue. Unfortunately, Lemma 7.3 is not very effective here: U_n is not always cyclic, and even when it is, it can be difficult to find a primitive root g and then express a as a power of g (see Chapter 6). We therefore need more powerful techniques. Quadratic residues are easiest to determine in the case of prime moduli; the case $n = 2$ is trivial, so we assume for the time being that n is an odd prime p. The following piece of notation greatly simplifies the problem of determining the elements of Q_p:

Definition

For an odd prime p, the *Legendre symbol* of any integer a is

$$\left(\frac{a}{p}\right) = \begin{cases} 0 & \text{if } p|a, \\ 1 & \text{if } a \in Q_p, \\ -1 & \text{if } a \in U_p \setminus Q_p. \end{cases}$$

Clearly this depends only on the congruence class of $a \bmod (p)$, so we can regard it as being defined either on $\mathbb{Z}$ or on $\mathbb{Z}_p$.

Example 7.3

Let $p = 7$. Then as in Example 7.2 we have

$$\left(\frac{a}{7}\right) = \begin{cases} 0 & \text{if } a \equiv 0 \bmod (7), \\ 1 & \text{if } a \equiv 1, 2 \text{ or } 4 \bmod (7), \\ -1 & \text{if } a \equiv 3, 5 \text{ or } 6 \bmod (7). \end{cases}$$

Recall that by Corollary 6.6 there is a primitive root g mod (p), so that each $a \in U_p$ has the form g^i for some i. This justifies the next result:

Corollary 7.4

If p is an odd prime, and g is a primitive root mod (p), then

$$\left(\frac{g^i}{p}\right) = (-1)^i.$$

Proof

Both $\left(\frac{g^i}{p}\right)$ and $(-1)^i$ are equal to ± 1, and Lemma 7.3 shows that $\left(\frac{g^i}{p}\right) = 1$ if and only if i is even, which is also the condition for $(-1)^i$ to be 1. $\qquad\square$

The next result is very useful for calculations with the Legendre symbol:

Theorem 7.5

If p is an odd prime, then

$$\left(\frac{ab}{p}\right) = \left(\frac{a}{p}\right)\left(\frac{b}{p}\right)$$

for all integers a and b.

Proof

If p divides a or b then each side is equal to 0, so we may assume that $a, b \in U_p$. If we put $a = g^i$ and $b = g^j$ for some primitive root $g \in U_p$, so that $ab = g^{i+j}$, then Corollary 7.4 gives

$$\left(\frac{ab}{p}\right) = (-1)^{i+j} = (-1)^i(-1)^j = \left(\frac{a}{p}\right)\left(\frac{b}{p}\right).$$

$\qquad\square$

Example 7.4

Let $p = 17$. Then $-1 \equiv 4^2$, so $\left(\frac{-1}{17}\right) = 1$ and hence Theorem 7.5 gives $\left(\frac{a}{17}\right) = \left(\frac{-a}{17}\right)$ for all $a \in U_{17}$, that is, $a \in Q_{17}$ if and only if $-a \in Q_{17}$. For instance, $13 \in Q_{17}$ since $-13 \equiv 2^2 \in Q_{17}$.

Warning

The values of $\left(\frac{-a}{p}\right)$ and $-\left(\frac{a}{p}\right)$ may well be different: for instance, $\left(\frac{-1}{17}\right) = 1$ but $-\left(\frac{1}{17}\right) = -1$.

There is an obvious extension of Theorem 7.5: for all integers $a_1, \ldots, a_k$ we have

$$\left(\frac{a_1 \ldots a_k}{p}\right) = \left(\frac{a_1}{p}\right) \ldots \left(\frac{a_k}{p}\right).$$

Exercise 7.7

By factorising 28, show that $-1 \in Q_{29}$.

In algebraic terms, Theorem 7.5 states that the function $U_p \to \{\pm 1\}$, which sends each unit a to $\left(\frac{a}{p}\right)$, is a group-homomorphism; its kernel is Q_p. The next result is known as *Euler's criterion*:

Theorem 7.6

If p is an odd prime, then for all integers a we have

$$\left(\frac{a}{p}\right) \equiv a^{(p-1)/2} \bmod (p).$$

(This is slightly more effective than Corollary 7.4 for determining quadratic residues, since one is not required to find a primitive root, but it can nevertheless be a little tedious to compute $a^{(p-1)/2} \bmod (p)$.)

Proof

The result is trivial if p divides a, so we may assume that $a \in U_p$. Thus $a = g^i$ for some primitive root $g \in U_p$. Define $h = g^{(p-1)/2}$. Then $h^2 = g^{p-1} = 1$ in U_p, so $h = \pm 1$ (either apply Lagrange's Theorem (Theorem 4.1) to the polynomial $x^2 - 1$, or note that p divides $h^2 - 1 = (h-1)(h+1)$). Since g has order $p - 1 > (p-1)/2$ we cannot have $h = 1$, so $h = -1$. Then using Corollary 7.4 we have

$$a^{(p-1)/2} = \left(g^i\right)^{(p-1)/2} = \left(g^{(p-1)/2}\right)^i = h^i = (-1)^i = \left(\frac{g^i}{p}\right) = \left(\frac{a}{p}\right)$$

in $\mathbb{Z}_p$, which proves the result. $\qquad \square$

Example 7.5

Let $p = 23$ and $a = 5$. To determine whether $5 \in Q_{23}$ we need to compute $5^{11} \bmod (23)$. Now $5^2 \equiv 2$, so $5^{11} \equiv 2^5.5 \equiv 9.5 \equiv -1$. Thus $\left(\frac{5}{23}\right) = -1$ and so $5 \notin Q_{23}$.

Exercise 7.8

Determine whether 3 and 5 are quadratic residues mod (29).

Corollary 7.7

Let p be an odd prime. Then $-1 \in Q_p$ if and only if $p \equiv 1 \bmod (4)$.

Proof

If we take $a = -1$ in Theorem 7.6, we see that

$$\left(\frac{-1}{p}\right) \equiv (-1)^{(p-1)/2} \bmod (p),$$

so $-1 \in Q_p$ if and only if $(p-1)/2$ is even, that is, $p \equiv 1 \bmod (4)$.   □

Example 7.6

We have $-1 = 2^2$ in $\mathbb{Z}_5$ and $-1 = 5^2$ in $\mathbb{Z}_{13}$, but -1 is not a square in $\mathbb{Z}_3$ or $\mathbb{Z}_7$.

We showed in Theorem 2.9 that there are infinitely many primes $p \equiv 3$ mod (4). We can now fulfil the promise made then to prove the same result for primes $p \equiv 1 \bmod (4)$.

Corollary 7.8

There are infinitely many primes $p \equiv 1 \bmod (4)$.

Proof

If there are only finitely many primes $p \equiv 1 \bmod (4)$, say $p_1, \ldots, p_k$, then define $m = (2p_1 \ldots p_k)^2 + 1$. Being odd, m must be divisible by some odd prime p. Then $(2p_1 \ldots p_k)^2 \equiv -1 \bmod (p)$, so $-1 \in Q_p$, and hence $p \equiv 1 \bmod (4)$ by Corollary 7.7. By our hypothesis, this implies that $p = p_i$ for some $i = 1, \ldots, k$, so p divides $m - (2p_1 \ldots p_k)^2 = 1$, which is impossible. Hence there must be infinitely many primes $p \equiv 1 \bmod (4)$.   □

In order to state the next result we need some more notation. If we use the set $\{\pm 1, \pm 2, \ldots, \pm(p-1)/2\}$ as a reduced set of residues mod (p), then we can partition U_p into two subsets

$$P = \{1, 2, \ldots, (p-1)/2\} \subset U_p \quad \text{and} \quad N = \{-1, -2, \ldots, -(p-1)/2\} \subset U_p,$$

represented as shown by positive and negative integers. For each $a \in U_p$ we define

$$aP = \{ax \mid x \in P\} = \{a, 2a, \ldots, (p-1)a/2\} \subset U_p.$$

Thus $N = (-1)P$, for example.

A more effective test for quadratic residues is given by *Gauss's Lemma*:

Theorem 7.9

If p is an odd prime and $a \in U_p$, then $\left(\frac{a}{p}\right) = (-1)^\mu$ where $\mu = |aP \cap N|$.

Before proving this, let us consider an example:

Example 7.7

Let $p = 19$, so $P = \{1, 2, \ldots, 9\}$, and let $a = 11$. If we multiply each element of P by 11 mod (19), and then represent it by an element of $P \cup N$, we get

$$aP = 11P = \{-8, 3, -5, 6, -2, 9, 1, -7, 4\}.$$

This contains four elements of N (the terms with minus signs), so $\mu = 4$, which is even; thus $\left(\frac{11}{19}\right) = 1$, so $11 \in Q_{19}$. In fact, $11 \equiv 7^2 \bmod (19)$.

Proof (Proof of Theorem 7.9.)

If x and y are distinct elements of P then $ax \neq \pm ay$ in U_p: for if $ax \equiv \pm ay$ in $\mathbb{Z}$ then $p \mid a(x \mp y)$, so $p \mid (x \mp y)$, which is impossible since x and y are distinct elements of $\{1, 2, \ldots, (p-1)/2\}$. This means that the elements of aP lie in distinct sets

$$\{\pm 1\}, \ \{\pm 2\}, \ \ldots, \{\pm(p-1)/2\}.$$

There are $(p-1)/2$ such sets, and there are $(p-1)/2$ elements of aP, so each set contains exactly one element of aP; thus

$$aP = \{\varepsilon_i i \mid i = 1, 2, \ldots, (p-1)/2\}$$

where each $\varepsilon_i = \pm 1$. Note that $\varepsilon_i = 1$ if $\varepsilon_i i \in P$, and $\varepsilon_i = -1$ if $\varepsilon_i i \in N$. Since aP is contained in the abelian group U_p, we can multiply all its elements together in any order, and get the same result, so

$$a^{(p-1)/2}((p-1)/2)! = \left(\prod_i \varepsilon_i\right).((p-1)/2)!$$
$$= (-1)^\mu.((p-1)/2)!$$

in U_p, where $\mu = |aP \cap N|$ is the number of i such that $\varepsilon_i = -1$. Cancelling the unit $((p-1)/2)!$, we see that $a^{(p-1)/2} = (-1)^\mu$ in U_p, so that

$$a^{(p-1)/2} \equiv (-1)^\mu \bmod (p)$$

in $\mathbb{Z}$. Now Euler's criterion (Theorem 7.6) gives

$$a^{(p-1)/2} \equiv \left(\frac{a}{p}\right) \bmod (p),$$

so

$$\left(\frac{a}{p}\right) \equiv (-1)^\mu \bmod (p).$$

Both sides of this congruence are equal to ± 1, so they must be equal to each other since $p > 2$.   □

Exercise 7.9

Apply Gauss's Lemma to Exercise 7.8 ($p = 29$ and $a = 3, 5$). Does 10 belong to Q_{29}?

Corollary 7.10

If p is an odd prime then

$$\left(\frac{2}{p}\right) = (-1)^{(p^2-1)/8};$$

thus $2 \in Q_p$ if and only if $p \equiv \pm 1 \bmod (8)$.

Proof

Putting $a = 2$ in Gauss's Lemma, we get

$$aP = 2P = \{2, 4, 6, \ldots, p-1\}.$$

First suppose that $p \equiv 1 \bmod (4)$. Then

$$2P = \{2, 4, \ldots, (p-1)/2, (p+3)/2, \ldots, p-1\}$$

with the first $(p-1)/4$ elements $2, 4, \ldots, (p-1)/2$ in P, and the remaining $(p-1)/4$ elements $(p+3)/2, \ldots, p-1$ in N. Thus $\mu = |2P \cap N| = (p-1)/4$, so Gauss's Lemma gives

$$\left(\frac{2}{p}\right) = (-1)^{(p-1)/4} = \left((-1)^{(p-1)/4}\right)^{(p+1)/2} = (-1)^{(p^2-1)/8},$$

where we have used the fact that $(p+1)/2$ is odd. Now suppose that $p \equiv -1$ mod (4). Then

$$2P = \{2, 4, \ldots, (p-3)/2, (p+1)/2, \ldots, p-1\}$$

with the first $(p-3)/4$ elements $2, 4, \ldots, (p-3)/2$ in P, and the remaining $(p+1)/4$ elements $(p+1)/2, \ldots, p-1$ in N. Thus $\mu = (p+1)/4$ and hence

$$\left(\frac{2}{p}\right) = (-1)^{(p+1)/4} = \left((-1)^{(p+1)/4}\right)^{(p-1)/2} = (-1)^{(p^2-1)/8},$$

where we have now used the fact that $(p-1)/2$ is odd.

This proves the first part of the theorem, and for the second part we have

$$
\begin{aligned}
2 \in Q_p \quad &\Longleftrightarrow \quad \left(\frac{2}{p}\right) = +1 \\
&\Longleftrightarrow \quad (p^2-1)/8 \text{ is even} \\
&\Longleftrightarrow \quad 16 | p^2 - 1 \\
&\Longleftrightarrow \quad 16 | (p-1)(p+1) \\
&\Longleftrightarrow \quad 8 | p-1 \text{ or } 8 | p+1 \\
&\Longleftrightarrow \quad p \equiv \pm 1 \text{ mod (8)},
\end{aligned}
$$

completing the proof.   □

Example 7.8

2 is a quadratic residue mod (p) for $p = 7, 17, 23, 31, \ldots$, with square roots $\pm 3, \pm 6, \pm 5, \pm 8, \ldots$; however, 2 is not a quadratic residue mod (p) for $p = 3, 5, 11, 13, \ldots$.

Exercise 7.10

For which primes p is $-2 \in Q_p$?

7.4 Quadratic reciprocity

To determine whether or not an integer a is a quadratic residue mod (p), we need to evaluate $(\frac{a}{p})$; by Theorem 7.5, $(\frac{a}{p})$ is the product of the Legendre symbols $(\frac{q}{p})$ where q ranges over the primes dividing a (with repetitions, as necessary). It is therefore sufficient to evaluate $(\frac{q}{p})$ for each prime q. We have just dealt with the case $q = 2$, so we can assume that q is an odd prime. If we calculate $(\frac{q}{p})$ for small primes p and q (for instance by Gauss's Lemma) we get the following table, with rows and columns indexed by the values of p and q respectively:

	$q = 3$	5	7	11	13	17	19
$p = 3$	0	−1	1	−1	1	−1	1
5	−1	0	−1	1	−1	−1	1
7	−1	−1	0	1	−1	−1	−1
11	1	1	−1	0	−1	−1	−1
13	1	−1	−1	−1	0	1	−1
17	−1	−1	−1	−1	1	0	1
19	−1	1	1	1	−1	1	0

Values of the Legendre symbol $(\frac{q}{p})$ for odd primes $p, q \leq 19$

We notice that the table is nearly symmetric, that is, $(\frac{q}{p}) = (\frac{p}{q})$ for most p and q, the exceptions occuring when p and q are distinct primes congruent to 3 mod (4). This is a general result, called the *Law of Quadratic Reciprocity*, conjectured by Euler in 1783. Legendre gave several incomplete proofs, but in 1795 Gauss (aged 18) discovered the law for himself and provided the first correct proof. This is one of the central theorems of number theory, and many different proofs have subsequently been published.

Theorem 7.11

If p and q are distinct odd primes, then

$$\left(\frac{q}{p}\right) = \left(\frac{p}{q}\right)$$

except when $p \equiv q \equiv 3 \mod (4)$, in which case

$$\left(\frac{q}{p}\right) = -\left(\frac{p}{q}\right).$$

Comment

An equivalent form of this is the elegant result, due to Legendre, that

$$\left(\frac{q}{p}\right)\cdot\left(\frac{p}{q}\right) = (-1)^{(p-1)(q-1)/4}$$

for all distinct odd primes p and q.

Before proving this theorem, let us look at some applications.

Example 7.9

Is $83 \in Q_{103}$? Since 83 and 103 are distinct odd primes, we have

$$
\begin{aligned}
\left(\frac{83}{103}\right) &= -\left(\frac{103}{83}\right) && \text{(by Theorem 7.11, since } 83 \equiv 103 \equiv 3 \bmod (4)) \\
&= -\left(\frac{20}{83}\right) && \text{(since } 103 \equiv 20 \bmod (83)) \\
&= -\left(\frac{2}{83}\right)^2\left(\frac{5}{83}\right) && \text{(by Theorem 7.5, since } 20 = 2^2.5) \\
&= -\left(\frac{5}{83}\right) && \text{(since } \left(\frac{2}{83}\right) = \pm 1) \\
&= -\left(\frac{83}{5}\right) && \text{(by Theorem 7.11)} \\
&= -\left(\frac{3}{5}\right) && \text{(since } 83 \equiv 3 \bmod (5)) \\
&= -\left(\frac{5}{3}\right) && \text{(by Theorem 7.11)} \\
&= -\left(\frac{2}{3}\right) && \text{(since } 5 \equiv 2 \bmod (3)) \\
&= 1, && \text{(since } 2 \notin Q_3)
\end{aligned}
$$

so that $83 \in Q_{103}$. (In fact $83 \equiv 17^2 \bmod (103)$.)

Exercise 7.11

Is 219 a quadratic residue mod (383)?

Example 7.10

For which primes p is $3 \in Q_p$? Since $3 \in Q_2$ and $3 \notin Q_3$, we may assume that $p > 3$. If $p \equiv 1 \bmod (4)$ then the Law of Quadratic Reciprocity gives

$$\left(\frac{3}{p}\right) = \left(\frac{p}{3}\right) = \begin{cases} +1 & \text{if } p \equiv 1 \bmod (3), \text{ that is, if } p \equiv 1 \bmod (12), \\ -1 & \text{if } p \equiv 2 \bmod (3), \text{ that is, if } p \equiv 5 \bmod (12). \end{cases}$$

If $p \equiv 3 \bmod (4)$ then it gives

$$\left(\frac{3}{p}\right) = -\left(\frac{p}{3}\right) = \begin{cases} -1 & \text{if } p \equiv 1 \bmod (3), \text{ that is, if } p \equiv 7 \bmod (12), \\ +1 & \text{if } p \equiv 2 \bmod (3), \text{ that is, if } p \equiv 11 \bmod (12). \end{cases}$$

Putting these results together, we see that $3 \in Q_p$ if and only if $p = 2$ or $p \equiv \pm 1$ mod (12).

Exercise 7.12

For each of the following integers a, determine the primes p for which $a \in Q_p$: $a = -3, 5, 6, 7, 10, 169$.

Example 7.10 leads to *Pepin's test* for primality of Fermat numbers (see Chapter 2). Pepin proved this in 1877, and in recent years it has been implemented on computers to show that several Fermat numbers are composite.

Corollary 7.12

If $n \geq 1$, then the Fermat number $F_n = 2^{2^n} + 1$ is prime if and only if

$$3^{(F_n-1)/2} \equiv -1 \bmod (F_n).$$

Proof

It is easily seen that $F_n \equiv 5 \bmod (12)$, so if F_n is a prime p then $3 \notin Q_p$ by Example 7.10; then Euler's criterion gives $3^{(p-1)/2} \equiv -1 \bmod (p)$, as required. For the converse, suppose that $3^{(F_n-1)/2} \equiv -1 \bmod (F_n)$; then squaring, we get $3^{F_n-1} \equiv 1 \bmod (F_n)$ and hence $3^{F_n-1} \equiv 1 \bmod (p)$ for any prime p dividing F_n. As an element of the group U_p, 3 therefore has order m dividing $F_n - 1 = 2^{2^n}$, so $m = 2^i$ for some $i \leq 2^n$. Now

$$3^{2^{2^{n-1}}} = 3^{(F_n-1)/2} \equiv -1 \not\equiv 1 \bmod (p)$$

(since p is odd, because F_n is), so $i = 2^n$ and $m = 2^{2^n} = F_n - 1$. However, $m \leq |U_p| = p - 1$ by definition of m, so $F_n \leq p$ and hence $F_n = p$, showing that F_n is prime.   □

Comment

This proof shows that 3 is a primitive root for any Fermat prime p, since 3 has order $p - 1$ as an element of U_p.

Example 7.11

Let $n = 2$, so that $F_n = 17$. Then $3^{(F_n - 1)/2} = 3^8 = (3^4)^2 \equiv (-4)^2 \equiv -1$ mod (17), confirming that 17 is prime.

Exercise 7.13

Use Pepin's test to show that $F_3 = 257$ is prime.

Proof (Proof of Theorem 7.11.)

There are many proofs, none of them entirely straightforward. Perhaps the neatest is that due to Eisenstein, using trigonometric functions, given in Serre (1973), but it is so slick that it doesn't really explain why the result should be true. The following proof is a little longer, but it is fairly elementary and somewhat more illuminating.

Let $P = \{1, 2, \ldots, (p-1)/2\} \subset U_p$ and $N = (-1)P$ as before, and similarly let $Q = \{1, 2, \ldots, (q-1)/2\} \subset U_q$. If we put $a = q$ in Gauss's Lemma, then

$$\left(\frac{q}{p}\right) = (-1)^\mu$$

where $\mu = |qP \cap N|$ is the number of elements $x \in P$ such that $qx \equiv n \bmod (p)$ for some $n \in N$; this congruence is equivalent to $qx - py \in N$ for some integer y, that is,

$$-\frac{p}{2} < qx - py < 0$$

for some integer y. We now look for the possible values of y satisfying this condition.

Given any $x \in P$, the values of $qx - py$ for $y \in \mathbb{Z}$ differ by multiples of p, so $-p/2 < qx - py < 0$ for at most one integer y. If such an integer y exists, then

$$0 < \frac{qx}{p} < y < \frac{qx}{p} + \frac{1}{2}.$$

Now $x \leq (p-1)/2$, so

$$y < \frac{qx}{p} + \frac{1}{2} \leq \frac{q(p-1)}{2p} + \frac{1}{2} < \frac{q+1}{2}.$$

Thus y is an integer strictly between 0 and $(q+1)/2$, so $y \in \{1, 2, \ldots, (q-1)/2\} = Q$. We have therefore shown that μ is the number of pairs $(x, y) \in P \times Q$ such that

$$-\frac{p}{2} < qx - py < 0.$$

Interchanging the roles of p and q, we also have

$$\left(\frac{p}{q}\right) = (-1)^{\nu}$$

where ν is the number of pairs $(y,x) \in Q \times P$ such that $-q/2 < py - qx < 0$, or equivalently the number of pairs $(x,y) \in P \times Q$ such that

$$0 < qx - py < \frac{q}{2}.$$

It follows that

$$\left(\frac{q}{p}\right) \cdot \left(\frac{p}{q}\right) = (-1)^{\mu+\nu},$$

where $\mu + \nu$ is the number of pairs $(x,y) \in P \times Q$ such that

$$-\frac{p}{2} < qx - py < 0 \quad \text{or} \quad 0 < qx - py < \frac{q}{2}.$$

There are no pairs $(x,y) \in P \times Q$ satisfying $qx - py = 0$, since p and q are coprime, so this condition can be simplified to

$$-\frac{p}{2} < qx - py < \frac{q}{2}.$$

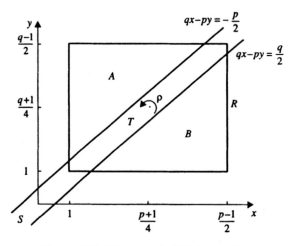

Figure 7.1. The proof of Theorem 7.11.

Figure 7.1 shows $P \times Q$ as the set of integer points (x,y) (points with integer coordinates) in the rectangle R in the xy-plane given by

$$1 \le x \le \frac{p-1}{2}, \quad 1 \le y \le \frac{q-1}{2}.$$

The inequalities $-p/2 < qx - py < q/2$ define the strip S between the two parallel straight lines $qx - py = -p/2$ and $qx - py = q/2$, so $\mu + \nu$ is the number of integer points in the region $T = R \cap S$. Now the number of integer points $(x, y) \in R$ is $|P \times Q| = |P|.|Q| = (p-1)(q-1)/4$, so

$$\mu + \nu = \frac{(p-1)(q-1)}{4} - (\alpha + \beta)$$

where α and β are the numbers of integer points in the subsets A and B of R above and below S. If we can show that $\alpha = \beta$, then $\mu + \nu \equiv (p-1)(q-1)/4$ mod (2), and hence

$$\left(\frac{q}{p}\right).\left(\frac{p}{q}\right) = (-1)^{(p-1)(q-1)/4}$$

as required.

We prove that $\alpha = \beta$ by using the half-turn of R about its midpoint $((p+1)/4, (q+1)/4)$ to pair off the integer points in A and B. This half-turn is the rotation ρ given by

$$\rho(x,y) = (x', y') = \left(\frac{p+1}{2} - x, \frac{q+1}{2} - y\right),$$

a formula which shows that ρ sends integer points to integer points. Moreover, it is straightforward to check that $qx - py < -p/2$ if and only if $qx' - py' > q/2$, so $\rho(A) = B$ and $\rho(B) = A$. Thus ρ induces the required bijection between the integer points in A and B, so $\alpha = \beta$ and the proof is complete. $\qquad\square$

7.5 Quadratic residues for prime-power moduli

Having dealt with quadratic residues for prime moduli, we now consider prime-power moduli, dealing with the odd case first.

Theorem 7.13

Let p be an odd prime, let $e \geq 1$, and let $a \in \mathbb{Z}$. Then $a \in Q_{p^e}$ if and only if $a \in Q_p$.

Proof

We know from Theorem 6.7 that there is a primitive root g mod (p^e), so by applying Lemma 7.3 with $n = p^e$ we see that Q_{p^e} consists of the even powers of g. Now g, regarded as an element of U_p, is also a primitive root mod (p), and by applying Lemma 7.3 with $n = p$ we know that Q_p also consists of the even powers of g. Thus $a \in Q_{p^e}$ if and only if $a \in Q_p$. $\qquad\square$

For odd primes p, we can find square roots in U_{p^e} for $e \geq 2$ by applying the iterative method in Chapter 4, Section 3 to the polynomial $f(x) = x^2 - a$: we use a square root of a mod (p^i) to find the square roots mod (p^{i+1}). Suppose that $a \in Q_p$, and r is a square root of a mod (p^i) for some $i \geq 1$; thus $r^2 \equiv a$ mod (p^i), say $r^2 = a + p^i q$. If we put $s = r + p^i k$, where k is as yet unknown, then $s^2 = r^2 + 2rp^i k + p^{2i}k^2 \equiv a + (q + 2rk)p^i$ mod (p^{i+1}), since $2i \geq i+1$. Now $\gcd(2r, p) = 1$, so we can choose k to satisfy the linear congruence $q + 2rk \equiv 0$ mod (p), giving $s^2 \equiv a$ mod (p^{i+1}) as required. By Lemma 7.1, an element $a \in Q_{p^{i+1}}$ has just two square roots in $U_{p^{i+1}}$ for odd p, so these must be $\pm s$. It follows that if we have a square root for a in U_p, then we can iterate this process to find its square roots in U_{p^e} for all e.

Example 7.12

Let us take $a = 6$ and $p^e = 5^2$. In U_5 we have $a = 1 = 1^2$, so we can take $r = 1$ as a square root mod (5). Then $r^2 = 1 = 6 + 5.(-1)$, so $q = -1$ and we need to solve the linear congruence $-1 + 2k \equiv 0$ mod (5). This has solution $k \equiv 3$ mod (5), so we take $s = r + p^i k = 1 + 5.3 = 16$, and the square roots of 6 in $\mathbb{Z}_{5^2}$ are given by ± 16, or equivalently ± 9 mod (5^2). If we want the square roots of 6 in $\mathbb{Z}_{5^3}$ we repeat the process: we can take $r = 9$ as a square root mod (5^2), with $r^2 = 81 = 6 + 5^2.3$, so $q = 3$; solving $3 + 18k \equiv 0$ mod (5) we have $k \equiv -1$, so $s = 9 + 5^2.(-1) = -16$, giving square roots ± 16 mod (5^3).

Exercise 7.14

Find the square roots of 6 mod (5^4).

Exercise 7.15

Find the square roots of -3 mod (7^2) and mod (7^3).

It should not be a surprise to learn that the situation for $p = 2$ is similar but slightly more complicated:

Theorem 7.14

Let a be an odd integer. Then

(a) $a \in Q_2$;

(b) $a \in Q_4$ if and only if $a \equiv 1$ mod (4);

(c) if $e \geq 3$, then $a \in Q_{2^e}$ if and only if $a \equiv 1$ mod (8).

Proof

Parts (a) and (b) are obvious: squaring the elements of $U_2 = \{1\} \subset \mathbb{Z}_2$ and of $U_4 = \{1, 3\} \subset \mathbb{Z}_4$, we see that $Q_2 = \{1\}$ and $Q_4 = \{1\}$. For part (c) we use Theorem 6.10, which states that the elements of U_{2^e} all have the form $\pm 5^i$ for some i; squaring, we see that the quadratic residues are the even powers of 5. Since $5^2 \equiv 1 \mod (8)$, these are all represented by integers $a \equiv 1 \mod (8)$. Now both the even powers of 5 and the elements $a \equiv 1 \mod (8)$ account for exactly one quarter of the classes in Q_{2^e}; since the first set is contained in the second, these two sets are equal. $\square$

Example 7.13

$Q_8 = \{1\}$, $Q_{16} = \{1, 9\}$, $Q_{32} = \{1, 9, 17, 25\}$, and so on.

One can find square roots in Q_{2^e} by adapting the iterative algorithm given earlier for odd prime-powers. Suppose that $a \in Q_{2^i}$ for some $i \geq 3$, say $r^2 = a + 2^i q$. If we put $s = r + 2^{i-1}k$, then $s^2 = r^2 + 2^i rk + 2^{2(i-1)}k^2 \equiv a + (q + rk)2^i \mod (2^{i+1})$, since $2(i - 1) \geq i + 1$. Now r is odd, so we can choose $k = 0$ or 1 to make $q + rk$ even, giving $s^2 \equiv a \mod(2^{i+1})$. Thus s is a square root of a in $U_{2^{i+1}}$. By Lemma 7.1 there are four square roots of a in $U_{2^{i+1}}$, and these have the form $t = sx$, where $x = \pm 1$ or $2^i \pm 1$ is a square root of 1. Since $a \equiv 1 \mod (8)$, we can start with a square root $r = 1$ for a in U_{2^3}, and then by iterating this process we can find the square roots of a in U_{2^e} for any e.

Example 7.14

Let us find the square roots of $a = 17 \mod (2^5)$; these exist since $17 \equiv 1 \mod (8)$. First we find a square root mod (2^4). Taking $r = 1$ we have $r^2 = 1^2 = 17 + 2^3.(-2)$, so $q = -2$; taking $k = 0$ makes $q + rk = -2$ even, so $s = r + 2^2 k = 1$ is a square root of 17 mod (2^4). (This is obvious, but it is worth illustrating the process first in a simple case.) Now we repeat this process, using $r = 1$ as a square root mod (2^4) to find a square root s mod (2^5). We have $r^2 = 1 = 17 + 2^4.(-1)$, so now $q = -1$; taking $k = 1$ makes $q + rk = 0$ even, so $s = r + 2^3 k = 9$ is a square root of 17 mod (2^5). The remaining square roots t are found by multiplying $s = 9$ by -1 and by $2^4 \pm 1 = \pm 15$, so we have $\pm 7, \pm 9$ as the complete set of square roots of 17 mod (2^5).

Exercise 7.16

Find the square roots of 41 mod (2^6).

7.6 Quadratic residues for arbitrary moduli

The following result allows us to combine our characterisations of Q_{p^e} for different prime-powers:

Theorem 7.15

Let $n = n_1 n_2 \ldots n_k$, where the integers n_i are mutually coprime. Then $a \in Q_n$ if and only if $a \in Q_{n_i}$ for each i.

Proof

If $a \in Q_n$ then $a \equiv s^2 \bmod (n)$ for some $s \in U_n$. Clearly $a \equiv s^2 \bmod (n_i)$ for each i, with s coprime to n_i, so $a \in Q_{n_i}$. Conversely, if $a \in Q_{n_i}$ for each i then there exist elements $s_i \in U_{n_i}$ such that $a \equiv s_i^2 \bmod (n_i)$. By the Chinese Remainder Theorem (Theorem 3.10) there is an element $s \in \mathbb{Z}_n$ such that $s \equiv s_i \bmod (n_i)$ for all i. Then $s^2 \equiv s_i^2 \equiv a \bmod (n_i)$ for all i, and hence $s^2 \equiv a \bmod (n)$ since the moduli n_i are coprime, so $a \in Q_n$. $\qquad\square$

This result can be expressed in algebraic terms as giving a direct product decomposition

$$Q_n \cong Q_{n_1} \times \cdots \times Q_{n_k}.$$

This is analogous to the decomposition $U_n \cong U_{n_1} \times \cdots \times U_{n_k}$ given in Theorem 6.13, and indeed it can be deduced directly from it by noting that an element of U_n is a square if and only if its component in each factor U_{n_i} is a square.

We can now answer the question of whether $a \in Q_n$ for arbitrary moduli n:

Theorem 7.16

Let $a \in U_n$. Then $a \in Q_n$ if and only if

(1) $a \in Q_p$ for each odd prime p dividing n, and

(2) $a \equiv 1 \bmod (4)$ if $2^2 \,\|\, n$, and $a \equiv 1 \bmod (8)$ if $2^3 \,|\, n$.

(Note that condition (2) is relevant only when n is divisible by 4; in all other cases we can ignore it.)

Proof

By Theorem 7.15, $a \in Q_n$ if and only if $a \in Q_{p^e}$ for each prime-power p^e in the factorisation of n. For odd primes p this is equivalent to $a \in Q_p$, by

Theorem 7.13, giving condition (1); for $p = 2$ it is equivalent to condition (2), by Theorem 7.14. $\square$

Example 7.15

Let $n = 144 = 2^4.3^2$. An element $a \in U_{144}$ is a quadratic residue if and only if $a \in Q_3$ and $a \equiv 1 \bmod (8)$; since $Q_3 = \{1\} \subset \mathbb{Z}_3$, this is equivalent to $a \equiv 1 \bmod (24)$, so $Q_{144} = \{1, 25, 49, 73, 97, 121\} \subset U_{144}$. Any $a \in Q_{144}$ must have $N = 8$ square roots, by Lemma 7.1. To find these, we first find its four square roots mod (2^4) and its two square roots mod (3^2) by the methods described in Section 7.5, and then we use the Chinese Remainder Theorem to convert each of these eight pairs of roots into a square root mod (144). For instance, let $a = 73$; then $a \equiv 9 \bmod (2^4)$, with square roots $s \equiv \pm 3, \pm 5 \bmod (2^4)$, and similarly $a \equiv 1 \bmod (3^2)$, with square roots $s \equiv \pm 1 \bmod (3^2)$; solving these eight pairs of simultaneous congruences for s, we get the square roots $s \equiv \pm 19, \pm 35, \pm 37, \pm 53 \bmod (144)$.

Exercise 7.17

Find the square roots of 49 mod (144).

Exercise 7.18

Find the square roots of 25 mod (168).

Example 7.16

As an application of the results in this chapter, let us return to Example 3.8. We claimed there (without proof) that if

$$f(x) = (x^2 - 13)(x^2 - 17)(x^2 - 221),$$

then for each integer $n \geq 1$ there is a solution $x \in \mathbb{Z}$ of the congruence $f(x) \equiv 0 \bmod (n)$. (This is despite the fact that the equation $f(x) = 0$ clearly has no integer solutions.) To prove this, it is sufficient by the Chinese Remainder Theorem to show that for each prime-power p^e there is a solution of $f(x) \equiv 0 \bmod (p^e)$, and for this, it is sufficient to show that at least one of $13, 17$ and 221 is a quadratic residue mod (p^e). If $p = 2$, then since $17 \equiv 1 \bmod (8)$ we have $17 \in Q_{2^e}$ for all e by Theorem 7.14. If $p = 13$, then since $17 \equiv 2^2 \bmod (13)$ we have $17 \in Q_{13}$, and hence $17 \in Q_{13^e}$ for all e by Theorem 7.13. If $p = 17$, then a similar argument based on $13 \equiv 8^2 \bmod (17)$ gives $13 \in Q_{17^e}$ for all e.

Finally, if $p \neq 2, 13$ or 17, then since $221 = 13 \times 17$ we have

$$\left(\frac{221}{p}\right) = \left(\frac{13}{p}\right)\left(\frac{17}{p}\right)$$

by Theorem 7.5, with each of these three terms equal to ± 1; at least one of them must therefore be equal to 1, so at least one of $13, 17$ or 221 must be in Q_p and hence in Q_{p^e} for all e by Theorem 7.13.

Exercise 7.19

Show that the polynomial $g(x) = (x^2 - 5)(x^2 - 41)(x^2 - 205)$ has no integer roots, but the congruence $g(x) \equiv 0$ has a solution mod (n) for every integer $n \geq 1$.

7.7 Supplementary exercises

Exercise 7.20

Show that, for each $r \geq 1$, there are infinitely many primes $p \equiv 1$ mod (2^r).

Exercise 7.21

For which values of n is -1 a quadratic residue mod (n)?

Exercise 7.22

Show that if q and r are distinct primes, with $q \equiv r \equiv 1$ mod (4) and $\left(\frac{q}{r}\right) = 1$, then the polynomial $h(x) = (x^2 - q)(x^2 - r)(x^2 - qr)$ has no integer roots, but the congruence $h(x) \equiv 0$ has a solution mod (n) for every integer $n \geq 1$.

Exercise 7.23

Show that if $n > 2$ then a quadratic residue mod (n) cannot also be a primitive root mod (n).

Exercise 7.24

Show that if p is a Fermat prime F_n, then each element of U_p is either a primitive root or a quadratic residue, but not both. Show that the Fermat primes are the only primes with this property.

Exercise 7.25

Is 43 a quadratic residue mod (923)?

Exercise 7.26

Find the square roots of 7 mod (513).

Exercise 7.27

Show that $\sum_{a=1}^{p-1}(\frac{a}{p}) = 0$ for each odd prime p. Show that $\sum_{a \in Q_p} a \equiv 0$ mod (p) for each prime $p > 3$.

8

Arithmetic Functions

In Chapter 5 we studied Euler's function ϕ. Two of its most important properties are Theorem 5.6, that if m and n are coprime then $\phi(mn) = \phi(m)\phi(n)$, and Theorem 5.8, that $\sum_{d|n} \phi(d) = n$ for all n. In this chapter we will meet other examples of functions with similar properties. Some of these, such as the divisor functions and the Möbius function, have important applications, including the study of perfect numbers and various enumeration problems.

8.1 Definition and examples

Definition

An *arithmetic** *function* is a function $f(n)$ defined for all $n \in \mathbb{N}$; it is usually taken to be complex-valued, so that it is a function $f : \mathbb{N} \to \mathbb{C}$, or equivalently a sequence (a_n) of complex numbers $a_n = f(n)$.

In many of the more important cases, $f(n)$ is an integer, describing some number-theoretic property of n; examples include

$$\phi(n) \quad = \quad |U_n|, \quad \text{the number of units mod } (n),$$

---

* The stress is on the third syllable, to indicate that the word is an adjective, like *algebraic*.

143

$$\tau(n) \;=\; \sum_{d|n} 1, \quad \text{the number of divisors of } n,$$

$$\sigma(n) \;=\; \sum_{d|n} d, \quad \text{the sum of the divisors of } n.$$

For instance, the divisors of 12 are $1, 2, 3, 4, 6$ and 12, so $\tau(12) = 6$ and $\sigma(12) = 28$. The functions τ and σ are called *divisor functions*; they are the special cases $k = 0$ and 1 of the function

$$\sigma_k(n) = \sum_{d|n} d^k.$$

In some books, the function $\tau(n)$ is written $d(n)$, but we shall avoid this notation since we often use d to denote a divisor of n.

Definition

An arithmetic function f is *multiplicative* if

$$f(mn) = f(m)f(n)$$

whenever $\gcd(m, n) = 1$. A simple induction argument shows that if f is multiplicative and $n_1, \ldots, n_k$ are mutually coprime, then

$$f(n_1 \ldots n_k) = f(n_1) \ldots f(n_k);$$

in particular, if n has prime-power factorisation $p_1^{e_1} \ldots p_k^{e_k}$, then

$$f(n) = f(p_1^{e_1}) \ldots f(p_k^{e_k}).$$

In many cases, it is straightforward to evaluate $f(p^e)$ for prime-powers p^e, so one can deduce the value of $f(n)$ for all n.

For instance, Theorem 5.6 shows that ϕ is multiplicative, and we used this property to evaluate $\phi(n)$ in Corollary 5.7. We will prove later that τ and σ are also multiplicative. Theorem 3.11 shows that the number of solutions in $\mathbb{Z}_n$ of a given polynomial congruence is a multiplicative function of n; we used this property in Example 3.18 to count solutions of $x^2 \equiv 1 \bmod (n)$ for composite n.

Exercise 8.1

Prove that $|Q_n|$, the number of quadratic residues mod (n), is a multiplicative function of n.

The following result is very useful for proving that functions are multiplicative:

Lemma 8.1

If g is a multiplicative function and $f(n) = \sum_{d|n} g(d)$ for all n, then f is multiplicative.

Proof

To show that f is multiplicative, suppose that m and n are coprime. Then the divisors d of mn are the products $d = ab$ where $a \mid m$ and $b \mid n$; each such pair a and b determines a unique divisor $d = ab$, and conversely, since m and n are coprime, each divisor d of mn determines a unique pair $a = \gcd(m, d)$ and $b = \gcd(n, d)$ of divisors of m and n. Thus there is a bijection between divisors d of mn and pairs a, b of divisors of m and n, so

$$
\begin{aligned}
f(mn) &= \sum_{d|mn} g(d) \\
&= \sum_{a|m} \sum_{b|n} g(ab).
\end{aligned}
$$

Now g is multiplicative, and a and b are coprime, so $g(ab) = g(a)g(b)$, giving

$$
\begin{aligned}
f(mn) &= \sum_{a|m} \sum_{b|n} g(a)g(b) \\
&= \left(\sum_{a|m} g(a) \right) \cdot \left(\sum_{b|n} g(b) \right) \\
&= f(m)f(n),
\end{aligned}
$$

as required. $\square$

To apply this, we first introduce two more arithmetic functions u and N, defined by

$$
u(n) = 1 \quad \text{and} \quad N(n) = n
$$

for all n. The function u is sometimes called the unit function. Clearly, u and N are both multiplicative. These functions may look rather trivial, but they can be very useful, as the next result shows.

Theorem 8.2

The divisor functions τ and σ are multiplicative.

Proof

We have $\tau(n) = \sum_{d|n} 1 = \sum_{d|n} u(d)$ and $\sigma(n) = \sum_{d|n} d = \sum_{d|n} N(d)$. Since u and N are multiplicative, so are τ and σ by Lemma 8.1. $\qquad\square$

Exercise 8.2

Give direct proofs that τ and σ are multiplicative, using the definitions of these functions.

Exercise 8.3

Show that for each k, the function $\sigma_k(n) = \sum_{d|n} d^k$ is multiplicative.

We can use Theorem 8.2 to evaluate the divisor functions, by first evaluating them at the prime-powers p^e. Since the divisors of p^e are $d = 1, p, p^2, \ldots, p^e$, we have

$$\tau(p^e) = e + 1 \quad \text{and} \quad \sigma(p^e) = 1 + p + p^2 + \cdots + p^e = \frac{p^{e+1} - 1}{p - 1};$$

now τ and σ are multiplicative, so we immediately deduce

Theorem 8.3

If n has prime-power factorisation $n = p_1^{e_1} \ldots p_k^{e_k}$, then

$$\tau(n) = \prod_{i=1}^{k}(e_i + 1) \quad \text{and} \quad \sigma(n) = \prod_{i=1}^{k}\left(\frac{p_i^{e_i+1} - 1}{p_i - 1}\right).$$

Exercise 8.4

For which integers n is $\tau(n)$ odd?

8.2 Perfect numbers

Definition

A positive integer n is *perfect* if n is the sum of its proper divisors (the positive divisors $d \neq n$). Since $\sigma(n)$ is the sum of all the positive divisors of n, this condition can be written as $n = \sigma(n) - n$, or equivalently

$$\sigma(n) = 2n.$$

The perfect numbers were believed by the Ancient Greeks to have particular aesthetic and religious significance. The first two examples are

$$6 = 1 + 2 + 3 \quad \text{and} \quad 28 = 1 + 2 + 4 + 7 + 14,$$

and the next is 496.

Exercise 8.5

Verify that 496 is perfect.

Most of what is known about perfect numbers is embodied in the following theorem; the first part is in Euclid's *Elements*, and the second is due to Euler.

Theorem 8.4

(a) If $n = 2^{p-1}(2^p - 1)$ where p and $2^p - 1$ are both prime (so that $2^p - 1$ is a Mersenne prime M_p), then n is perfect.

(b) If n is even and perfect, then n has the form given in (a).

This theorem shows that there is a one-to-one correspondence between even perfect numbers and the Mersenne primes M_p which we met in Chapter 2; for instance the perfect numbers 6, 28 and 496 correspond to the Mersenne primes $M_2 = 3$, $M_3 = 7$ and $M_5 = 31$. No example of an odd perfect number is known, and it is conjectured that they do not exist; if there is one, it must be very large.

Proof

(a) If $n = 2^{p-1}(2^p - 1)$ as described, then Theorem 8.2 gives $\sigma(n) = \sigma(2^{p-1})\sigma(2^p - 1)$. Now Theorem 8.3 gives $\sigma(2^{p-1}) = (2^p - 1)/(2 - 1) = 2^p - 1$, and since $2^p - 1$ is prime we have $\sigma(2^p - 1) = (2^p - 1) + 1 = 2^p$. Thus $\sigma(n) = (2^p - 1)2^p = 2n$, so n is perfect.

(b) Since n is even, we can write $n = 2^{p-1}q$ for some integer $p \geq 2$, where q is odd. Now σ is multiplicative, so $\sigma(n) = \sigma(2^{p-1})\sigma(q) = (2^p - 1)\sigma(q)$. Since n is perfect we also have $\sigma(n) = 2n = 2^p q$, so

$$(2^p - 1)\sigma(q) = 2^p q.$$

Thus $2^p - 1$ divides $2^p q$, and hence divides q, say $q = (2^p - 1)r$, so substituting for q and then cancelling $2^p - 1$ we get

$$\sigma(q) = 2^p r.$$

Now q and r are distinct divisors of q, with $q + r = (2^p - 1)r + r = 2^p r = \sigma(q)$, which is the sum of all the divisors of q; thus q and r must be the only divisors of q, so q is prime and $r = 1$. Thus $q = 2^p - 1$ and so $n = 2^{p-1}(2^p - 1)$; since $2^p - 1$ is prime, Theorem 2.13 implies that p must be prime. $\square$

Exercise 8.6

Is $2^{10}(2^{11} - 1)$ perfect?

Exercise 8.7

Find two more perfect numbers, other than the examples given above.

Exercise 8.8

Show that n is perfect if and only if $\sigma_{-1}(n) = 2$.

8.3 The Möbius Inversion Formula

The multiplicative property can be useful in proving identities between arithmetic functions.

Lemma 8.5

Let f and g be multiplicative functions, with $f(p^e) = g(p^e)$ for all primes p and integers $e \geq 0$. Then $f = g$.

Proof

If n has prime-power factorisation $\prod_i p_i^{e_i}$, then

$$f(n) = f\left(\prod_i p_i^{e_i}\right) = \prod_i f(p_i^{e_i}) = \prod_i g(p_i^{e_i}) = g\left(\prod_i p_i^{e_i}\right) = g(n).$$

$\square$

This gives us another proof of Theorem 5.8, that $\sum_{d|n} \phi(d) = n$. Since ϕ is multiplicative (by Theorem 5.6), so is the function $f(n) = \sum_{d|n} \phi(d)$, by Lemma 8.1. We have seen that the function $N(n) = n$ is multiplicative, so to prove Theorem 5.8 it is sufficient (by Lemma 8.5) to show that f and N agree on all prime-powers p^e. Now the divisors d of p^e are $d = p^i$ ($i = 0, 1, \ldots, e$), with $\phi(1) = 1$ and $\phi(p^i) = p^i - p^{i-1}$ for $i > 0$, so

$$f(p^e) = 1 + \sum_{i=1}^{e} (p^i - p^{i-1}) = p^e = N(p^e),$$

as required.

We have seen several instances where pairs of arithmetic functions f and g are related by an identity $f(n) = \sum_{d|n} g(d)$: for instance, we can take $f = N$ and $g = \phi$, or $f = \sigma$ and $g = N$. In this situation, it is often useful to be able to invert the roles of f and g, that is, to find a similar formula expressing g in terms of f. The result which allows us to do this is the Möbius Inversion Formula, but before proving this, we need to study one of its main ingredients, the Möbius function μ. First we define the *identity function* I, given by

$$I(n) = \begin{cases} 1 & \text{if } n = 1, \\ 0 & \text{if } n > 1. \end{cases}$$

Clearly I is multiplicative. The name of this function can be a little confusing, since I is *not* the identity function in the set-theoretic sense of sending every n to itself (N does that). We will later introduce an algebraic operation $*$ on arithmetic functions, and show that $f * I = f = I * f$ for all f; thus I is the identity with respect to $*$, whereas N is the identity with respect to the operation $\circ$ of composition, since $f \circ N = f = N \circ f$ for all f. A useful alternative formula for I is

$$I(n) = \left\lfloor \frac{1}{n} \right\rfloor,$$

the integer part of $1/n$.

We define the *Möbius function* μ by the formula

$$\sum_{d|n} \mu(d) = I(n) = \begin{cases} 1 & \text{if } n = 1, \\ 0 & \text{if } n > 1. \end{cases}$$

This is an example of an inductive (or recursive) definition: if $n = 1$ (with a unique divisor $d = 1$) then $\mu(1) = 1$, and if $n > 1$ then $\sum_{d|n} \mu(d) = 0$, so

$$\mu(n) = - \sum_{d|n, d<n} \mu(d),$$

which defines $\mu(n)$ in terms of the values of μ at smaller integers d. For instance, if n is prime then its only divisors are $d = 1$ and $d = n$, so $\mu(n) = -\mu(1) = -1$. A little calculation gives the values

$$\mu(n) = 1, -1, -1, 0, -1, 1, -1, 0, 0, 1, -1, 0$$

for $n = 1, 2, \ldots, 12$. In Theorem 8.8 we will give a simple formula for $\mu(n)$ in terms of the prime-power factorisation of n, which is more convenient for calculation.

Exercise 8.9

Show that $\mu(n)$ is an integer for each $n \geq 1$.

Exercise 8.10

Show that if p and q are distinct primes, then $\mu(pq) = 1$ and $\mu(p^2) = 0$.

Exercise 8.11

Calculate $\mu(n)$ for all $n \leq 30$, and make a conjecture about the values of $\mu(n)$.

The function μ derives its importance from the following major result, the *Möbius Inversion Formula*:

Theorem 8.6

Let f and g be arithmetic functions. If

$$f(n) = \sum_{d|n} g(d)$$

for all n, then

$$g(n) = \sum_{d|n} f(d) \mu\left(\frac{n}{d}\right) = \sum_{d|n} \mu(d) f\left(\frac{n}{d}\right)$$

for all n.

This shows that if f is expressed in terms of g as a sum over divisors, then one can invert their roles and define g in terms of f by a similar expression. The relationship between f and g is nearly symmetric, except that the function μ appears in the expression for g.

Proof

The two expressions for $g(n)$ are easily seen to be equal: if we put $e = n/d$ then the first summation can be written as

$$\sum_{de=n} f(d)\mu(e),$$

where the sum is over all pairs d, e with product n; transposing the names of these two dummy variables, we see that this is also equal to

$$\sum_{ed=n} f(e)\mu(d) = \sum_{de=n} \mu(d)f(e),$$

which gives the second summation. It is therefore sufficient to show that the second summation is equal to $g(n)$. Our hypothesis about f implies that

$$f\left(\frac{n}{d}\right) = \sum_{e \mid \frac{n}{d}} g(e),$$

so

$$\sum_{d \mid n} \mu(d) f\left(\frac{n}{d}\right) = \sum_{d \mid n}\left(\mu(d) \sum_{e \mid \frac{n}{d}} g(e)\right).$$

Now if e divides n/d then it divides n; conversely, for each divisor e of n, we see that e divides n/d if and only if d divides n/e, in which case d also divides n. Hence the coefficient of $g(e)$ in this expression is

$$\sum_{d \mid \frac{n}{e}} \mu(d) = \begin{cases} 1 & \text{if } n/e = 1, \\ 0 & \text{if } n/e > 1. \end{cases}$$

This means that the only term $g(e)$ with a non-zero coefficient is $g(n)$, which has coefficient 1, so the expression is equal to $g(n)$, as required.   □

Corollary 8.7

If $n \geq 1$ then

$$\phi(n) = \sum_{d \mid n} d\mu\left(\frac{n}{d}\right) = \sum_{d \mid n} \mu(d)\frac{n}{d}.$$

Proof

By Theorem 5.8 we have $\sum_{d \mid n} \phi(d) = n = N(n)$, so by applying the Möbius Inversion Formula with $f = N$ and $g = \phi$ we get the required result.   □

Example 8.1

Let $n = 12$, so $\phi(12) = 4$. The divisors of 12 are $d = 1, 2, 3, 4, 6$ and 12, and the values of μ at the first twelve integers were given on p. 150, so we find that

$$\sum_{d|12} d\mu\left(\frac{12}{d}\right) = 1.0 + 2.1 + 3.0 + 4.(-1) + 6.(-1) + 12.1 = 4$$

and also

$$\sum_{d|12} \mu(d)\frac{12}{d} = 1.\frac{12}{1} - 1.\frac{12}{2} - 1.\frac{12}{3} + 0.\frac{12}{4} + 1.\frac{12}{6} + 0.\frac{12}{12} = 4,$$

in agreement with Corollary 8.7.

Exercise 8.12

Prove that

$$\sum_{d|n} \tau(d)\mu\left(\frac{n}{d}\right) = \sum_{d|n} \mu(d)\tau\left(\frac{n}{d}\right) = 1$$

and

$$\sum_{d|n} \sigma(d)\mu\left(\frac{n}{d}\right) = \sum_{d|n} \mu(d)\sigma\left(\frac{n}{d}\right) = n$$

for all $n \geq 1$. Verify these equations for $n = 12$.

8.4 An application of the Möbius Inversion Formula

In this section we give a totally different application of the Möbius Inversion Formula. A set of n chairs are arranged regularly around a circular table. Each chair may be occupied by a woman (W) or by a man (M), giving 2^n possible patterns of sexes W or M at the table. If the people all rotate one place around the table, a pattern may change, but after n successive rotations it must recur. We say that a pattern has period d if recurs for the first time after d rotations, or equivalently, if rotating the pattern produces exactly d different patterns. Thus a single-sex pattern $WW \ldots W$ or $MM \ldots M$ has period 1, while for even n the two alternating patterns $WMWM \ldots WM$ and $MWMW \ldots MW$ each have period 2. How many different patterns of period d are there, for each d?

First note that if a pattern has period d, then d must divide n: for if $n = qd + r$ with $0 \leq r < d$ then the pattern recurs after both n and d rotations, and hence also after $r = n - qd$ rotations, so $r = 0$ by the definition of d. For each d, the number of patterns of period d depends only on d, and not on the multiple $n = qd$ of d: this is because such a pattern consists of q repetitions of a block of d symbols W or M, which is not itself a repetition of smaller blocks, and the number of such blocks of length d depends only on d. For instance, a pattern of period $d = 2$ must consist of q repetitions of the block WM or MW (but not WW or MM), so there are two such patterns for each even n. It follows that if we let $f(d)$ denote the number of patterns of period d, then $\sum_{d|n} f(d) = 2^n$, the total number of patterns for n chairs. Putting $g(n) = 2^n$ in Theorem 8.6, we deduce that

$$f(n) = \sum_{d|n} 2^d \mu\left(\frac{n}{d}\right),$$

or equivalently (changing notation)

$$f(d) = \sum_{e|d} 2^e \mu\left(\frac{d}{e}\right).$$

For instance

$$
\begin{aligned}
f(12) &= 2^1\mu(12) + 2^2\mu(6) + 2^3\mu(4) + 2^4\mu(3) + 2^6\mu(2) + 2^{12}\mu(1) \\
&= 2^2 - 2^4 - 2^6 + 2^{12} \\
&= 4020\,.
\end{aligned}
$$

The expression $2^2 - 2^4 - 2^6 + 2^{12}$ for $f(12)$ can also be obtained from the Inclusion–Exclusion Principle (Exercise 5.10). The term 2^{12} counts all the different patterns of length 12, and we need to exclude those which are repetitions of smaller blocks. If a pattern of length 12 is a repetition of smaller blocks, then it consists of either two copies of a block of length 6, or three copies of a block of length 4; any other cases are included in these, for instance four copies of a block B of length 3 can also be regarded as two copies of the block BB of length 6. Now the number of patterns of length 12 consisting of two identical blocks of length 6 is equal to 2^6, the total number of blocks of this length, so we subtract 2^6; similarly, we subtract 2^4 for those consisting of three identical blocks of length 4. In doing this, we have excluded some patterns twice, namely those which consist of two blocks of length 6 and also of three of length 4; these are the patterns $BBBBBB = (BBB)(BBB) = (BB)(BB)(BB)$ consisting of six identical blocks B of length 2; the number of such patterns is 2^2, so by adding 2^2 to compensate for this double-counting we obtain the required formula $2^{12} - 2^6 - 2^4 + 2^2$. More generally, the Möbius Inversion Formula can be regarded as an analogue of the Inclusion-Exclusion Principle, in which divisibility of integers has replaced inclusion of sets.

Let us regard two patterns as equivalent if each is a rotation of the other. Thus a pattern of period d lies in a class of d equivalent patterns, so the number of equivalence classes of patterns of period d is

$$\frac{f(d)}{d} = \frac{1}{d} \sum_{e|d} 2^e \mu\left(\frac{d}{e}\right).$$

For instance, there are $4020/12 = 335$ equivalence classes of patterns of period 12.

Although this may not seem a particularly serious application, there are in fact many mathematical situations involving similar types of cyclic symmetry, where this enumeration technique is important. For instance, by using the theory of finite fields one can show that the above formula for $f(d)/d$ also gives the number of irreducible polynomials of degree d with coefficients in $\mathbb{Z}_2$. Indeed a whole branch of mathematics, spanning number theory, combinatorics and algebra, has built itself up around the Möbius Inversion Formula, its generalisations and its applications.

Exercise 8.13

Enumerate and determine all the patterns with periods $d = 2, 3$ and 4, and show how they are divided into equivalence classes.

Exercise 8.14

How would the solution to this problem be affected if there were more than two sexes?

8.5 Properties of the Möbius function

Having seen some applications of the Möbius function, we now need a more efficient method of evaluating it than by means of its inductive definition. The evidence for small values of n (Exercise 8.11) may lead one to conjecture the following formula:

Theorem 8.8

If $n = p_1^{e_1} \ldots p_k^{e_k}$, where $p_1, \ldots, p_k$ are distinct primes and each $e_i \geq 1$, then

$$\mu(n) = \begin{cases} 0 & \text{if some } e_i > 1, \\ (-1)^k & \text{if each } e_i = 1. \end{cases}$$

(Thus $\mu(n) \neq 0$ if and only if n is square-free. This formula includes the case $\mu(1) = (-1)^0 = 1$, where we regard 1 as the product of the empty set of primes, so that $k = 0$.)

Proof

Let μ' be the function defined by the formula in the theorem, so that $\mu'(n) = (-1)^k$ if n is a product of k distinct primes, and $\mu'(n) = 0$ otherwise. We will prove that $\mu(n) = \mu'(n)$ for all n by strong induction on n. Clearly $\mu(1) = 1 = \mu'(1)$, so suppose that $n > 1$ and $\mu(d) = \mu'(d)$ for all $d < n$.

We first show that $\sum_{d|n} \mu'(d) = 0$ (so μ' satisfies the same recurrence relation $\sum_{d|n} \mu'(d) = I(n)$ as μ). If the factorisation of n is as in the theorem (with $k \geq 1$), then by definition of μ', the non-zero terms in $\sum_{d|n} \mu'(d)$ are those of the form $\mu'(d)$ where d is a product of distinct primes $p_i \in \{p_1, \ldots, p_k\}$. If d is a product of r such primes, where $0 \leq r \leq k$, then $\mu'(d) = (-1)^r$; for each r the number of ways of choosing these r primes is equal to the binomial coefficient $\binom{k}{r}$, so there are $\binom{k}{r}$ such divisors d, each contributing $(-1)^r$ to $\sum_{d|n} \mu'(d)$. Summing over all r we therefore have

$$\sum_{d|n} \mu'(d) = \sum_{r=0}^{k} \binom{k}{r}(-1)^r = \left(1 + (-1)\right)^k = 0$$

by the Binomial Theorem. (Alternatively, note that μ' is multiplicative (see Corollary 8.9), and hence by Lemma 8.1 so is the function $f(n) = \sum_{d|n} \mu'(d)$; by Lemma 8.5 it is therefore sufficient to show that $f(p^e) = 0$ for each prime power $p^e > 1$, and this follows immediately from the definition of μ'.) We can write this as

$$\mu'(n) = -\sum_{d|n, d<n} \mu'(d) ;$$

now the induction hypothesis states that $\mu(d) = \mu'(d)$ for all $d < n$, and the definition of μ implies that

$$\mu(n) = -\sum_{d|n, d<n} \mu(d) ,$$

so $\mu(n) = \mu'(n)$ as required.   □

Example 8.2

$15 = 3.5$, so $\mu(15) = (-1)^2 = 1$; $30 = 2.3.5$, so $\mu(30) = (-1)^3 = -1$; $60 = 2^2.3.5$, so $\mu(60) = 0$.

Exercise 8.15

Find a simple formula for $\sum_{d|n} |\mu(d)|$.

We can use Theorem 8.8 to give an alternative proof of Corollary 5.7, that

$$\phi(n) = n \prod_{p|n} \left(1 - \frac{1}{p}\right)$$

where p ranges over the distinct primes dividing n. If we multiply out the factors on the right-hand side, the general term has the form $n(-1)^r/p_1 \ldots p_r$ where $p_1, \ldots, p_r$ are distinct prime factors of n; by Theorem 8.8, this is equal to $n\mu(d)/d$, where $d = p_1 \ldots p_r$ is a square-free divisor of n. The remaining non-square-free divisors d of n have $\mu(d) = 0$, so the right-hand side can be written as $\sum_{d|n} n\mu(d)/d$. By Corollary 8.7, this is equal to $\phi(n)$. (This argument is not circular: although Corollary 8.7 depends on Theorem 5.8, that $\sum_{d|n} \phi(d) = n$, the proof of this does not use Corollary 5.7.)

Example 8.3

Taking $n = 12$, we have

$$\phi(12) = 12\left(1 - \frac{1}{2}\right)\left(1 - \frac{1}{3}\right) = \frac{12}{1} - \frac{12}{2} - \frac{12}{3} + \frac{12}{6} = \sum_{d|12} \mu(d)\frac{12}{d}.$$

(The divisors $d = 4$ and 12 have $\mu(d) = 0$, since they are not square-free.)

Corollary 8.9

The function μ is multiplicative.

Proof

We need to prove that $\mu(mn) = \mu(m)\mu(n)$ whenever m and n are coprime. If m and n are not both square-free, then neither is mn, so Theorem 8.8 gives $\mu(m)\mu(n) = 0 = \mu(mn)$ as required. We may assume therefore that both $m = p_1 \ldots p_k$ and $n = q_1 \ldots q_l$ are products of distinct primes. Since they are coprime, no prime p_i dividing m can appear as a prime q_j dividing n, so $mn = p_1 \ldots p_k q_1 \ldots q_l$ is also a product of distinct primes. Thus $\mu(m) = (-1)^k$, $\mu(n) = (-1)^l$ and $\mu(mn) = (-1)^{k+l} = (-1)^k(-1)^l = \mu(m)\mu(n)$. $\quad\square$

8.6 The Dirichlet product

Theorems 5.8 and 8.6, Lemma 8.1, Corollary 8.7 and our definition of μ all involve summation over the divisors d of n; these are special cases of a general theory of such sums.

Definition

If f and g are arithmetic functions, then their *Dirichlet product*, or *convolution*, is the arithmetic function $f * g$ given by

$$(f * g)(n) = \sum_{d|n} f(d)g\left(\frac{n}{d}\right) ;$$

equivalently, putting $e = n/d$, we have

$$(f * g)(n) = \sum_{de=n} f(d)g(e)$$

where $\sum_{de=n}$ denotes summation over all pairs d, e such that $de = n$.

Example 8.4

Theorem 5.8 states that $\sum_{d|n} \phi(d) = n$ for all n; using the functions $u(n) = 1$ and $N(n) = n$, we can rewrite this as

$$\sum_{d|n} \phi(d)u\left(\frac{n}{d}\right) = N(n)$$

for all n, which becomes, in our new notation, $\phi * u = N$. Similarly, our definition of μ can be written as $\mu * u = I$, while Corollary 8.7 becomes $\phi = N * \mu = \mu * N$. Lemma 8.1 states that if g is multiplicative, then so is the function $f = g * u$. Theorem 8.6 (the Möbius Inversion Formula) states that if $f = g * u$ then $g = f * \mu = \mu * f$.

Exercise 8.16

Express the divisor functions τ and σ as Dirichlet products of simpler functions.

The basic algebraic properties of the Dirichlet product are as follows:

Lemma 8.10

For all arithmetic functions f, g and h we have

(a) $f * g = g * f$,

(b) $(f * g) * h = f * (g * h)$,

(c) $f * I = I * f = f$.

(Thus $*$ is commutative and associative, and has I as an identity.)

Proof

(a) For all arithmetic functions f and g, we have

$$(f * g)(n) = \sum_{de=n} f(d)g(e) = \sum_{ed=n} g(e)f(d) = \sum_{de=n} g(d)f(e) = (g * f)(n).$$

(b) For all arithmetic functions f, g and h, we have

$$((f * g) * h)(n) = \sum_{dc=n} (f * g)(d)h(c)$$

$$= \sum_{dc=n} \left(\sum_{ab=d} f(a)g(b) \right) h(c) = \sum_{abc=n} f(a)g(b)h(c),$$

and similarly

$$(f * (g * h))(n) = \sum_{ae=n} f(a)(g * h)(e)$$

$$= \sum_{ae=n} f(a) \left(\sum_{bc=e} g(b)h(c) \right) = \sum_{abc=n} f(n)g(b)h(c).$$

$\square$

Exercise 8.17

Prove Lemma 8.10(c).

The next result shows that the arithmetic functions f satisfying $f(1) \neq 0$ have inverses with respect to the Dirichlet product:

Lemma 8.11

If f is an arithmetic function with $f(1) \neq 0$, then there exists an arithmetic function g such that $f * g = I = g * f$; it is given by

$$g(1) = \frac{1}{f(1)} \qquad \text{and} \qquad g(n) = -\frac{1}{f(1)} \sum_{\substack{d|n \\ d<n}} g(d)f\left(\frac{n}{d}\right)$$

for all $n > 1$.

(These equations define $g(n)$ for all $n \geq 1$ by induction on n.)

Proof

By the commutativity of $*$ it is sufficient to prove that the given function g satisfies $g * f = I$, that is,

$$\sum_{d|n} g(d) f\left(\frac{n}{d}\right) = \begin{cases} 1 & \text{if } n = 1, \\ 0 & \text{if } n > 1. \end{cases}$$

This is trivial for $n = 1$, since the only divisor is $d = 1$ and we have $g(1)f(1) = 1$ by definition of g. If $n > 1$ then

$$\sum_{d|n} g(d) f\left(\frac{n}{d}\right) = g(n)f(1) + \sum_{\substack{d|n \\ d<n}} g(d) f\left(\frac{n}{d}\right)$$

$$= -\frac{f(1)}{f(1)} \sum_{\substack{d|n \\ d<n}} g(d) f\left(\frac{n}{d}\right) + \sum_{\substack{d|n \\ d<n}} g(d) f\left(\frac{n}{d}\right) = 0,$$

as required.   □

Definition

The function g in Lemma 8.11 is called the *Dirichlet inverse* of f, denoted by f^{-1} (not to be confused with the inverse function or with the reciprocal of f.)

Let G denote the set of all arithmetic functions f for which $f(1) \neq 0$.

Theorem 8.12

G is an abelian group with respect to the operation $*$, with identity element I.

Proof

To prove closure, let $f, g \in G$, so $f(1), g(1) \neq 0$; then $(f * g)(1) = \sum_{d|1} f(d)g(1/d) = f(1)g(1) \neq 0$, so $f * g \in G$. Associativity, commutativity and the existence of an identity I are proved in Lemma 8.10. Finally, if $f \in G$ then its Dirichlet inverse $g = f^{-1}$ is also in G since $g(1) = 1/f(1) \neq 0$, so every element has an inverse in G.   □

Example 8.5

The equation $\mu * u = I$ (which we used to define μ) shows that μ and u are inverses of each other in G, and this helps to explain why the function μ should be so important. To illustrate the power of the new notation, we give a one-line proof of the Möbius Inversion Formula (Theorem 8.6), which states that if $f = g*u$ then $g = f*\mu = \mu*f$: if $f = g*u$, then $f*\mu = (g*u)*\mu = g*(u*\mu) = g*I = g$, and so commutativity of $*$ gives $\mu*f = g$. We can also prove the converse of Theorem 8.6: if $g = f*\mu$ then $g*u = (f*\mu)*u = f*(\mu*u) = f*I = f$. These arguments are valid for all arithmetic functions f and g, not just those in G, so we have proved the following stronger form of the Möbius Inversion Formula:

Theorem 8.13

Let f and g be arithmetic functions. Then

$$f(n) = \sum_{d|n} g(d)$$

for all n if and only if

$$g(n) = \sum_{d|n} f(d)\mu\left(\frac{n}{d}\right) = \sum_{d|n} \mu(d)f\left(\frac{n}{d}\right)$$

for all n.

Example 8.6

If we take $f = N$ and $g = \phi$, we see that Theorem 5.8 and Corollary 8.7 are equivalent to each other, that is, $N = \phi*u$ is equivalent to $\phi = N*\mu = \mu*N$.

Exercise 8.18

Which arithmetic functions are represented by $\tau*\mu$ and by $\sigma*\mu$?

We can now prove an extension of Lemma 8.1 (which is the special case $h = u$):

Theorem 8.14

If g and h are multiplicative functions, and if $f = g*h$, then f is multiplicative.

Proof

The proof is similar to that of Lemma 8.1. Instead, the first equation now becomes

$$f(mn) = \sum_{d|mn} g(d)h\left(\frac{mn}{d}\right),$$

and we carry values of h throughout the calculation. □

Exercise 8.19

Fill in the details of the proof of Theorem 8.14.

Exercise 8.20

Show that if f is multiplicative, and f is not identically zero, then $f \in G$ and the Dirichlet inverse f^{-1} is multiplicative. (Hint: if not, consider the least mn such that $\gcd(m, n) = 1$ and $f^{-1}(mn) \neq f^{-1}(m)f^{-1}(n)$.) Deduce that the set M of non-zero multiplicative functions forms a subgroup of G.

We can also prove the converse of Lemma 8.1:

Corollary 8.15

Suppose that $f(n) = \sum_{d|n} g(d)$. Then f is multiplicative if and only if g is multiplicative.

Proof

The hypothesis is that $f = g * u$. Now u is multiplicative, so if g is multiplicative then so is f, by Theorem 8.14. The converse is similar, using $g = f * \mu$ and Corollary 8.9 (that μ is multiplicative). □

Example 8.7

Theorem 5.8 gives $\sum_{d|n} \phi(n) = n = N(n)$. It is obvious that N is multiplicative, so Corollary 8.15 gives an alternative proof that ϕ is multiplicative. (This is not a circular argument, since the proof of Theorem 5.8 did not require the multiplicative property of ϕ.)

8.7 Supplementary exercises

Exercise 8.21

(a) Show that χ is multiplicative, where $\chi(n) = 0, 1$ or -1 as n is even or $n \equiv 1$ or 3 mod (4) respectively.

(b) Let $\tau_1(n)$ and $\tau_3(n)$ denote the number of divisors d of n such that $d \equiv 1$ or 3 mod (4) respectively; show that the function $g(n) = \tau_1(n) - \tau_3(n)$ is multiplicative, and hence find an expression for $g(n)$ in terms of the prime-power factorisation of n. (See Exercise 10.8 for an application of this.)

Exercise 8.22

Show that $\mu(n)$ is the sum of the primitive complex n-th roots of 1. (These are the elements $z \in \mathbb{C}$ such that $z^n = 1$ but $z^m \neq 1$ for $1 \leq m < n$.)

Exercise 8.23

Show that if g is multiplicative, then the functions $f(n) = \sum_{d^2|n} g(d^2)$ and $h(n) = \sum_{d^2|n} g(n/d^2)$ are both multiplicative.

Exercise 8.24

The *Mangoldt function* is given by $\Lambda(n) = \ln p$ if $n = p^e$ for some prime p and integer $e > 0$, and $\Lambda(n) = 0$ otherwise. Show that $\sum_{d|n} \Lambda(d) = \ln(n)$ and deduce that $\Lambda(n) = \sum_{d|n} \ln(d)\mu(n/d) = -\sum_{d|n} \ln(d)\mu(d)$.

Exercise 8.25

Show that $(f_1 * \cdots * f_k)(n) = \sum f_1(d_1) \ldots f_k(d_k)$ for all arithmetic functions $f_1, \ldots, f_k$, where the summation is over all k-tuples $(d_1, \ldots, d_k)$ with $d_1 \ldots d_k = n$.

Exercise 8.26

Show that the number of subgroups of finite index n in the group $\mathbb{Z}^2$ is equal to $\sigma(n)$. (Hint: you may assume that these subgroups correspond to integer matrices $A = \begin{pmatrix} a & b \\ 0 & d \end{pmatrix}$ where $a, d > 0$, $ad = n$ and $0 \leq b < d$.) How many subgroups of index n are there in the group $\mathbb{Z}^k$?

9

The Riemann Zeta Function

In order to make progress in number theory, it is sometimes necessary to use techniques from other areas of mathematics, such as algebra, analysis or geometry. In this chapter we give some number-theoretic applications of the theory of infinite series. These are based on the properties of the Riemann zeta function $\zeta(s)$, which provides a link between number theory and real and complex analysis. Some of the results we obtain have probabilistic interpretations in terms of random integers. For the background on convergence of infinite series, see Appendix C. For a detailed treatment of $\zeta(s)$, see Titchmarsh (1951).

9.1 Historical background

One of the most familiar examples of an infinite series is the *harmonic series*

$$\sum_{n=1}^{\infty} \frac{1}{n} = 1 + \frac{1}{2} + \frac{1}{3} + \frac{1}{4} + \cdots.$$

Since number theory is mainly about the positive integers $n = 1, 2, 3 \ldots$, it is not surprising that this series is of interest to number theorists. Unfortunately, it diverges, but only just: the sum of the first n terms is about $\ln n$, and although this tends to $+\infty$ as $n \to \infty$, it does so rather slowly. To make the series converge, without losing its important number-theoretic properties, we replace its general term $1/n$ with the smaller term $1/n^s$, where $s > 1$. This gives rise to the *Riemann zeta function*, defined by

$$\zeta(s) = \sum_{n=1}^{\infty} \frac{1}{n^s} = 1 + \frac{1}{2^s} + \frac{1}{3^s} + \cdots. \tag{9.1}$$

Although this function is named after Riemann, who wrote a fundamental paper on its properties in 1859, it was in fact introduced about 120 years earlier by Euler, who showed that it can be expanded as a product

$$\zeta(s) = \prod_p \left(\frac{1}{1 - p^{-s}} \right), \tag{9.2}$$

where p ranges over all the primes. This is a very powerful result, since it allows methods of analysis to be applied to the study of prime numbers. Euler regarded $\zeta(s)$ as a function of a real variable s, whereas Riemann's great contribution depended on allowing s to be a complex number, so that the even richer theory of complex functions could be used. One of the great unsolved problems in number theory is the Riemann Hypothesis (see Section 9), a conjecture concerning the complex zeros of $\zeta(s)$; a solution of this would resolve many important problems concerning the distribution of prime numbers.

Before dealing with questions of convergence, we will first outline a proof of (9.2), and then show a simple but effective application of this product formula. Each factor on the right-hand side of (9.2) can be expanded as a geometric series

$$\frac{1}{1 - p^{-s}} = 1 + p^{-s} + p^{-2s} + \cdots = \sum_{e=0}^{\infty} p^{-es},$$

convergent since $|p^{-s}| = p^{-s} < 1$ for all $s > 0$. If we multiply these series together (and we will justify this later), then the general term in their product has the form

$$p_1^{-e_1 s} \ldots p_k^{-e_k s} = \frac{1}{n^s},$$

where $p_1, \ldots, p_k$ are distinct primes, each $e_i \geq 0$, and $n = p_1^{e_1} \ldots p_k^{e_k}$. By the Fundamental Theorem of Arithmetic (Theorem 2.3), every integer $n \geq 1$ has a unique factorisation of this form, so it contributes exactly one summand, equal to $1/n^s$, and hence (9.2) represents $\sum 1/n^s = \zeta(s)$. (We will prove this more rigorously later in the chapter, in Theorem 9.3.)

Exercise 9.1

Use a similar argument to outline a proof that

$$\prod_p (1 - p^{-s}) = \sum_{n=1}^{\infty} \frac{\mu(n)}{n^s},$$

where μ is the Möbius function, and hence show that

$$\sum_{n=1}^{\infty} \frac{\mu(n)}{n^s} = \frac{1}{\zeta(s)}.$$

Using (9.2), we can now sketch a quick proof of Theorem 2.6, that there are infinitely many primes: if there were only finitely many primes, then $\zeta(s)$ would approach a finite limit $\prod_p (1 - p^{-1})^{-1}$ as $s \to 1$, whereas in fact $\zeta(s) \to +\infty$, as we shall shortly see.

9.2 Convergence

To justify the preceding arguments, we must first consider the convergence of the series (9.1). We will show that it converges for all real $s > 1$, and diverges for all real $s \leq 1$. Suppose first that $s > 1$. We group the terms together in blocks of length $1, 2, 4, 8, \ldots$, giving

$$\zeta(s) = 1 + \left(\frac{1}{2^s} + \frac{1}{3^s}\right) + \left(\frac{1}{4^s} + \cdots + \frac{1}{7^s}\right) + \left(\frac{1}{8^s} + \cdots + \frac{1}{15^s}\right) + \cdots.$$

Now

$$\frac{1}{2^s} + \frac{1}{3^s} \leq \frac{1}{2^s} + \frac{1}{2^s} = \frac{2}{2^s} = 2^{1-s},$$

$$\frac{1}{4^s} + \cdots + \frac{1}{7^s} \leq \frac{1}{4^s} + \cdots + \frac{1}{4^s} = \frac{4}{4^s} = (2^{1-s})^2,$$

$$\frac{1}{8^s} + \cdots + \frac{1}{15^s} \leq \frac{1}{8^s} + \cdots + \frac{1}{8^s} = \frac{8}{8^s} = (2^{1-s})^3,$$

and so on, so we can compare (9.1) with the geometric series

$$1 + 2^{1-s} + (2^{1-s})^2 + (2^{1-s})^3 + \cdots.$$

This converges since $0 < 2^{1-s} < 1$, and hence so does (9.1) by the Comparison Test (Appendix C). In fact, this argument shows that $1 \leq \zeta(s) \leq f(s)$ for all $s > 1$, where

$$f(s) = \sum_{n=0}^{\infty} (2^{1-s})^n = \frac{1}{1 - 2^{1-s}}.$$

If $s \to +\infty$ then $2^{1-s} \to 0$ and so $f(s) \to 1$, giving

$$\lim_{s \to +\infty} \zeta(s) = 1.$$

We now show that (9.1) diverges for $s \leq 1$. This is obvious if $s \leq 0$, since then $1/n^s \not\to 0$ as $n \to \infty$, so let us assume that $s > 0$. By grouping the terms of (9.1) in blocks of length $1, 1, 2, 4, \ldots$, we have

$$\zeta(s) = 1 + \frac{1}{2^s} + \left(\frac{1}{3^s} + \frac{1}{4^s}\right) + \left(\frac{1}{5^s} + \cdots + \frac{1}{8^s}\right) + \cdots.$$

If $s \leq 1$ then

$$\frac{1}{2^s} \geq \frac{1}{2},$$

$$\frac{1}{3^s} + \frac{1}{4^s} \geq \frac{1}{4} + \frac{1}{4} = \frac{1}{2},$$

$$\frac{1}{5^s} + \cdots + \frac{1}{8^s} \geq \frac{1}{8} + \cdots + \frac{1}{8} = \frac{1}{2},$$

and so on, so (9.1) diverges by comparison with the divergent series $1 + \frac{1}{2} + \frac{1}{2} + \frac{1}{2} + \cdots$. In particular, by taking $s = 1$ we see that the harmonic series $\sum 1/n$ diverges.

To summarise, we have proved:

Theorem 9.1

The series (9.1) converges for all real $s > 1$, and diverges for all real $s \leq 1$.

There is an alternative proof based on the Integral Test (Appendix C), using the fact that $\int_1^{+\infty} x^{-s}\, dx$ converges if and only if $s > 1$.

Exercise 9.2

Show that if $s > 1$ then $\zeta(s) \geq (1 + f(s))/2$, where $f(s)$ is as defined above, and deduce that $\zeta(s) \to +\infty$ as $s \to 1$.

9.3 Applications to prime numbers

We can now give a more rigorous analytic proof of Theorem 2.6, that there are infinitely many primes. Suppose that there are only finitely many primes, say $p_1, \ldots, p_k$. For each prime $p = p_i$ we have $|1/p| < 1$, so there is a convergent geometric series

$$1 + \frac{1}{p} + \frac{1}{p^2} + \frac{1}{p^3} + \cdots = \frac{1}{1 - p^{-1}}.$$

It follows that if we multiply these k different series together, their product

$$\prod_{i=1}^{k}\left(1 + \frac{1}{p_i} + \frac{1}{p_i^2} + \frac{1}{p_i^3} + \cdots\right) = \prod_{i=1}^{k}\left(\frac{1}{1 - p_i^{-1}}\right) \qquad (9.3)$$

is finite. Now these convergent series all consist of positive terms, so they are absolutely convergent. It follows (see Appendix C) that we can multiply out the series in (9.3) and rearrange the terms, without changing the product. If we take a typical term $1/p_1^{e_1}$ from the first series, $1/p_2^{e_2}$ from the second series, and so on, where each $e_i \geq 0$, then their product

$$\frac{1}{p_1^{e_1}} \cdot \frac{1}{p_2^{e_2}} \cdot \cdots \cdot \frac{1}{p_k^{e_k}} = \frac{1}{p_1^{e_1} p_2^{e_2} \cdots p_k^{e_k}}$$

will represent a typical term in the expansion of (9.3). By the Fundamental Theorem of Arithmetic, every integer $n \geq 1$ has a unique expression

$$n = p_1^{e_1} p_2^{e_2} \cdots p_k^{e_k} \qquad (e_i \geq 0)$$

as a product of powers of the primes p_i, since we are assuming that these are the *only* primes; notice that we allow $e_i = 0$, in case n is not divisible by a particular prime p_i. This uniqueness implies that each n contributes exactly one term $1/n$ to (9.3), so the expansion takes the form

$$\prod_{i=1}^{k}\left(1 + \frac{1}{p_i} + \frac{1}{p_i^2} + \frac{1}{p_i^3} + \cdots\right) = \sum_{n=1}^{\infty} \frac{1}{n}.$$

The right-hand side is the *harmonic series*, which is divergent. However, we have seen that the left-hand side is finite, so this contradiction proves that there must be infinitely many primes.

The next result, also due to Euler, develops this method a little further:

Theorem 9.2

If p_n denotes the n-th prime (in increasing order), then the series

$$\sum_{n=1}^{\infty} \frac{1}{p_n} = \frac{1}{2} + \frac{1}{3} + \frac{1}{5} + \cdots$$

diverges.

Proof

If $\sum 1/p_n$ converges to a finite sum l, then its partial sums must satisfy

$$\left|\sum_{n=1}^{N} \frac{1}{p_n} - l\right| \leq \frac{1}{2}$$

for all sufficiently large N, so that

$$\sum_{n>N} \frac{1}{p_n} \le \frac{1}{2}$$

for any such N. This implies that the series

$$\sum_{k=1}^{\infty} \left(\sum_{n>N} \frac{1}{p_n} \right)^k \tag{9.4}$$

converges by comparison with the geometric series $\sum_{k=1}^{\infty} 1/2^k$. If q denotes the product $p_1 \ldots p_N$ then no integer of the form $qr + 1$ $(r \ge 1)$ can be divisible by any of $p_1, \ldots, p_N$, so it must be a product of primes p_n for $n > N$ (possibly with repetitions), say

$$qr + 1 = p_{n_1} \ldots p_{n_k}$$

where each $n_i > N$. Then the reciprocal $1/(qr + 1)$ of each such an integer appears as a summand $1/p_{n_1} \ldots p_{n_k}$ in the expansion of

$$\left(\sum_{n>N} \frac{1}{p_n} \right)^k,$$

and hence it appears (just once) as a summand in the expansion of (9.4). Since (9.4) converges, it follows that the series

$$\sum_{r=1}^{\infty} \frac{1}{qr + 1},$$

which is contained within (9.4), also converges. However this series diverges, since its terms exceed those of the divergent series

$$\sum_{r=1}^{\infty} \frac{1}{qr + q} = \frac{1}{q} \sum_{r=1}^{\infty} \frac{1}{r + 1} = \frac{1}{q} \sum_{r=2}^{\infty} \frac{1}{r}.$$

This contradiction shows that $\sum 1/p_n$ must diverge.   □

Comments

1 It can be shown that

$$\frac{1}{p_1} + \cdots + \frac{1}{p_n} \to +\infty$$

about as fast as $\ln \ln n$, so this series diverges very slowly indeed.

2 Theorem 9.2 gives yet another proof that there are infinitely many primes, since a finite series must converge. It also shows that the primes are more densely distributed than the perfect squares: the series $\sum 1/n^2$ converges (by the Integral Test), so $1/n^2 \to 0$ faster than $1/p_n \to 0$ as $n \to \infty$, that is, primes occur more frequently than squares.

We can now use these ideas to give a rigorous proof of the product expansion (9.2):

Theorem 9.3

If $s > 1$ then

$$\zeta(s) = \prod_p \Big(\frac{1}{1 - p^{-s}}\Big),$$

where the product is over all primes p.

Proof

The method is to consider the product $P_k(s)$ of the factors corresponding to the first k primes, and to show that $P_k(s) \to \zeta(s)$ as $k \to \infty$. Let $p_1, \dots, p_k$ be the first k primes. Arguing as before with (9.3), we see that if $s > 0$ (so that the geometric series all converge) then

$$P_k(s) = \prod_{i=1}^{k} \Big(\frac{1}{1 - p_i^{-s}}\Big) = \prod_{i=1}^{k} \Big(1 + \frac{1}{p_i^s} + \frac{1}{p_i^{2s}} + \frac{1}{p_i^{3s}} + \cdots\Big).$$

If we expand this product, the general term in the resulting series is $1/n^s$ where $n = p_1^{e_1} \dots p_k^{e_k}$ and each $e_i \geq 0$. The Fundamental Theorem of Arithmetic implies that each such n contributes just one term to $P_k(s)$, so

$$P_k(s) = \sum_{n \in A_k} \frac{1}{n^s},$$

where $A_k = \{n \mid n = p_1^{e_1} \dots p_k^{e_k}, \; e_i \geq 0\}$ is the set of integers n whose prime factors are among $p_1, \dots, p_k$. Each $n \notin A_k$ is divisible by some prime $p > p_k$, and so $n > p_k$. It follows that if $s > 1$ then

$$|P_k(s) - \zeta(s)| = \sum_{n \notin A_k} \frac{1}{n^s} \leq \sum_{n > p_k} \frac{1}{n^s} = \zeta(s) - \sum_{n \leq p_k} \frac{1}{n^s}.$$

Since $s > 1$, the partial sums of the series $\sum 1/n^s$ converge to $\zeta(s)$, so in particular

$$\sum_{n \leq p_k} \frac{1}{n^s} \to \zeta(s)$$

as $k \to \infty$. Thus $|P_k(s) - \zeta(s)| \to 0$ as $k \to \infty$, so $P_k(s) \to \zeta(s)$ as required. $\square$

Exercise 9.3

Show that $P_k(1) \to +\infty$ as $k \to \infty$, and deduce that for each $\varepsilon > 0$ there exists n such that $\phi(n)/n < \varepsilon$. (See Exercise 5.11 in Chapter 5 for a similar result, and for a probabilistic interpretation of this.)

A similar method gives a rigorous prooof of the following result. We will also prove this as part of a more general result later in this chapter (see Example 9.4).

Theorem 9.4

If $s > 1$ then

$$\sum_{n=1}^{\infty} \frac{\mu(n)}{n^s} = \prod_{p} (1 - p^{-s}) = \frac{1}{\zeta(s)}.$$

Exercise 9.4

Prove Theorem 9.4.

9.4 Random integers

As an application of the Riemann zeta function, we will calculate the probability P that a pair of randomly-chosen integers are coprime. Since we do not wish to spend too much time on some of the finer details of probability theory, we will simply outline the main points. We will, in fact, use three different methods, leading to the formulae

$$\frac{1}{\zeta(2)}, \quad \prod_{p} (1 - p^{-2}), \quad \sum_{n=1}^{\infty} \frac{\mu(n)}{n^2}$$

for P, and the fact that they must all be equal gives an alternative proof of Theorem 9.4 in the case $s = 2$. (In fact, one can extend this to all integers $s \geq 2$ by calculating the probability that s randomly-chosen integers have greatest common divisor 1.) We will then show that $\zeta(2) = \pi^2/6$, so that $P = 6/\pi^2$.

There is an interesting geometric application of this result. An *integer point* in the plane $\mathbb{R}^2$ is a point with integer coordinates. Such a point A is *visible from the origin* if the line-segment AO joining A to the origin $O = (0,0)$ contains no integer points other than A and O. It is easy to show that an integer point $A \neq O$ is visible from O if and only if its coordinates are coprime, and it follows from this that P represents the probability that a randomly-chosen integer point is visible from O. (Restricting to positive coordinates does not alter the probability.) To put this another way, P is the proportion of the integer lattice $\mathbb{Z}^2$ which can be seen from any given integer point.

Exercise 9.5

Show that an integer point $(x, y) \neq (0, 0)$ is visible from O if and only if x and y are coprime.

There is an immediate problem in discussing randomly-chosen integers $x \in \mathbb{N}$ (or indeed randomly-chosen elements of any infinite set). If p_n denotes the probability $\Pr(x = n)$ that a particular integer n is chosen, then clearly

$$\sum_{n=1}^{\infty} p_n = 1.$$

However, if we want all integers to have the same status, then p_n must be a constant, independent of n, so that $\sum p_n$ is either 0 or divergent.

One way of avoiding this difficulty is to assign probabilities to certain *sets* of integers, rather than to individual integers. For any integers c and n we will assign the probability

$$\Pr(x \equiv c \bmod (n)) = \frac{1}{n},$$

so that x has the same probability $1/n$ of lying in any of the n classes $[c] \in \mathbb{Z}_n$.

Now the Chinese Remainder Theorem (Theorem 3.10) implies that if m and n are coprime, then the solutions x of any pair of simultaneous congruences

$$x \equiv b \bmod (m),$$

$$x \equiv c \bmod (n)$$

form a single congruence class mod (mn); thus

$$\Pr(x \equiv b \bmod (m) \text{ and } x \equiv c \bmod (n))$$

$$= \frac{1}{mn} = \Pr(x \equiv b \bmod (m)). \Pr(x \equiv c \bmod (n)),$$

so the pair of congruences are statistically independent. (Two events are *statistically independent* if the probability of them both happening is the product of their individual probabilities.) The same argument applies to any finite set of linear congruences with mutually coprime moduli.

Suppose now that x and y are chosen randomly from $\mathbb{N}$, as above, and that they are also chosen independently, that is, that

$$\Pr(x, y \in S) = \Pr(x \in S). \Pr(y \in S)$$

for any subset S of $\mathbb{N}$ for which these probabilites are defined. Let

$$P = \Pr\big(\gcd(x, y) = 1\big)$$

denote the probability that x and y are coprime. We will calculate P in three different ways.

Method A   For each $n \in \mathbb{N}$, we have

$$\gcd(x, y) = n \iff \begin{cases} x \equiv 0 \bmod (n) & \text{and} \\ y \equiv 0 \bmod (n) & \text{and} \\ \gcd(x/n, y/n) = 1. \end{cases}$$

Now the conditions $x \equiv 0 \bmod (n)$ and $y \equiv 0 \bmod (n)$ are each satisfied with probability $1/n$; since x and y are chosen independently, these two conditions are simultaneously satisfied with probability $1/n^2$. When they are both satisfied, we can regard x/n and y/n as randomly-chosen integers, so they will be coprime (the third condition) with conditional probability P. It follows that the three conditions are simultaneously satisfied with probability P/n^2, so

$$\Pr\big(\gcd(x, y) = n\big) = \frac{P}{n^2}.$$

Now $\gcd(x, y)$ must take a unique value $n \in \mathbb{N}$ for each pair x, y, so the sum of all these probabilities must be equal to 1. Thus

$$1 = \sum_{n=1}^{\infty} \Pr\big(\gcd(x, y) = n\big) = \sum_{n=1}^{\infty} \frac{P}{n^2} = P \sum_{n=1}^{\infty} \frac{1}{n^2} = P\zeta(2),$$

and hence

$$P = \frac{1}{\zeta(2)}.$$

Method B   We have

$$\gcd(x, y) = 1 \iff \begin{cases} x \not\equiv 0 \bmod (p) \\ \text{or} \\ y \not\equiv 0 \bmod (p) \end{cases}$$

for every prime p. For each p, the congruences $x \equiv 0$ and $y \equiv 0 \bmod (p)$ each have probability p^{-1}, so $x \equiv 0 \equiv y \bmod (p)$ with probability p^{-2}, and hence $x \not\equiv 0$ or $y \not\equiv 0 \bmod (p)$ with probability $1 - p^{-2}$. Now congruences modulo distinct primes are statistically independent, so we multiply these probabilities for all primes p to get

$$P = \prod_{p} (1 - p^{-2}).$$

(Strictly speaking, we need to justify the use of an infinite product here, since we have discussed statistical independence of finitely many congruences, but not infinitely many; for simplicity of exposition, we will omit the details of this.)

Method C. We have

$$\gcd(x,y) > 1 \iff \begin{cases} x \equiv 0 \bmod (p) \\ \text{and} \\ y \equiv 0 \bmod (p) \end{cases}$$

for some prime p. The event $\gcd(x,y) > 1$ has probability $1 - P$, so this must be the probability that $x \equiv 0 \equiv y \bmod(p)$ for at least one prime p. We will now use the Inclusion–Exclusion Principle (see Exercise 5.10) to find an alternative expression for this probability. For each p, the event $x \equiv 0 \equiv y \bmod (p)$ has probability p^{-2}, so adding these probabilities for all p we get a contribution

$$S_1 = \sum_p p^{-2}$$

to $1 - P$. From this we subtract a double sum

$$S_2 = \sum_{p<q} (pq)^{-2}$$

to compensate for the double counting in S_1 of cases in which x and y are divisible by two primes $p < q$. We then add a triple sum

$$S_3 = \sum_{p<q<r} (pqr)^{-2}$$

to allow for over-compensation in S_2 of integers divisible by three primes, and so on. Thus

$$1 - P = S_1 - S_2 + S_3 - \cdots,$$

where the general term S_k has the form

$$S_k = \sum (p_1 \ldots p_k)^{-2},$$

summing over all increasing k-tuples $p_1 < \cdots < p_k$ of distinct primes. If we define $S_0 = 1$ then we can write

$$P = \sum_{k=0}^{\infty} (-1)^k S_k.$$

In this expression for P, every square-free integer $n = p_1 \ldots p_k \in \mathbb{N}$ contributes one summand $(-1)^k n^{-2} = \mu(n) n^{-2}$, where μ is the Möbius function, while all other integers $n \geq 1$ contribute nothing and satisfy $\mu(n) = 0$; using absolute convergence to justify rearrangement, we therefore have

$$P = \sum_{n=1}^{\infty} \frac{\mu(n)}{n^2} .$$

Exercise 9.6

For each integer $s \geq 2$, let $P(s)$ denote the probability that s randomly- and independently-chosen integers have greatest common divisor 1 (so $P = P(2)$). Give three arguments to show that $P(s)$ is given by the formulae

$$\frac{1}{\zeta(s)}, \qquad \prod_p (1 - p^{-s}), \qquad \sum_{n=1}^{\infty} \frac{\mu(n)}{n^s} .$$

Exercise 9.7

Prove (in three different ways) that a single randomly-chosen integer x is square-free with probability $P = 1/\zeta(2)$. (Hint: consider $\mathrm{Sq}(x)$, the largest square factor of x.)

Exercise 9.8

For each integer $s \geq 2$, calculate (in three different ways) the probability $Q(s)$ that a randomly-chosen integer x should be s-th power-free, that is, divisible by no s-th power greater than 1.

9.5 Evaluating $\zeta(2)$

Having shown that $P = 1/\zeta(2)$, we now prove that

$$\zeta(2) = \frac{\pi^2}{6}. \tag{9.5}$$

Apostol (1983) gives an elementary proof of this, evaluating

$$\int_0^1 \int_0^1 (1 - xy)^{-1} \, dx \, dy$$

in two ways: first by writing $(1 - xy)^{-1} = \sum (xy)^n$ and integrating term by term, and second by using a change of variables to rotate the xy-plane through

$\pi/4$ and then using some straightforward trigonometric substitutions. A quicker but less elementary proof is simply to put $x = 1$ in the Fourier series expansion

$$x^2 = \frac{1}{3} + \frac{4}{\pi^2} \sum_{n \geq 1} \frac{(-1)^n}{n^2} \cos(n\pi x)$$

of the function x^2 on the interval $[-1, 1]$; we have $\cos(n\pi x) = (-1)^n$, so (9.5) follows immediately. (See Chapter IV of Churchill, 1963 for background on Fourier series.)

Instead, we will give a third proof, which has the advantage of extending to certain other values of $\zeta(s)$. We will use the infinite product expansion

$$\sin z = z \prod_{n \neq 0} \left(1 - \frac{z}{n\pi}\right) = z \prod_{n \geq 1} \left(1 - \frac{z^2}{n^2\pi^2}\right), \tag{9.6}$$

proofs of which can be found in books on complex variable theory, e.g. Jones and Singerman (1987, Chapter 3, Section 8). The first product in (9.6) is over all non-zero integers n, and the second product is obtained from the first by pairing the factors corresponding to $\pm n$. One can explain (if not rigorously prove) the first product by regarding $\sin z$ as behaving like a polynomial with infinitely many zeros at $z = n\pi$ $(n \in \mathbb{Z})$, so we have a 'factorisation'

$$\sin z = cz \prod_{n \neq 0} \left(1 - \frac{z}{n\pi}\right)$$

with

$$c = \lim_{z \to 0} \frac{\sin z}{z} = 1.$$

By expanding the second product in (9.6) and collecting powers of z, we obtain a power series for $\sin z$ which must coincide with its Taylor series expansion

$$\sin z = z - \frac{z^3}{3!} + \frac{z^5}{5!} - \cdots. \tag{9.7}$$

By comparing coefficients of z^3 in (9.6) and (9.7) we see that

$$-\sum_{n \geq 1} \frac{1}{n^2\pi^2} = -\frac{1}{3!},$$

so multiplying through by $-\pi^2$ we obtain (9.5).

With the aid of the previous section and a pocket calculator, we immediately deduce:

Theorem 9.5

The probability that two randomly- and independently-chosen integers are co-prime is given by

$$P = \frac{1}{\zeta(2)} = \frac{6}{\pi^2} = 0.607927101\ldots .$$

By Exercise 9.7, this is also the probability that a single randomly-chosen integer is square-free.

9.6 Evaluating $\zeta(2k)$

For many reasons, it would be useful to know the values of $\zeta(s)$ for *all* integers $s \geq 2$ (see Exercise 9.6, for instance). In 1978 Apéry proved a long-standing conjecture, that $\zeta(3)$ is irrational, but very little else is known about $\zeta(s)$ when s is odd. However, with a little extra work we can use (9.6) to evaluate $\zeta(s)$ for all even integers $s = 2k \geq 2$. Some of the techniques we use require rather careful analytic justification, using such concepts as uniform convergence, but for simplicity we will omit these details.

By taking logarithms in (9.6), we have

$$\ln \sin z = \ln z + \sum_{n \geq 1} \ln\left(1 - \frac{z^2}{n^2 \pi^2}\right),$$

and differentiating term by term we have

$$\cot z = \frac{1}{z} - \sum_{n \geq 1} \frac{2z}{n^2 \pi^2}\left(1 - \frac{z^2}{n^2 \pi^2}\right)^{-1}.$$

If we use the geometric series to write

$$\frac{2z}{n^2 \pi^2}\left(1 - \frac{z^2}{n^2 \pi^2}\right)^{-1} = \frac{2z}{n^2 \pi^2} \sum_{k \geq 0}\left(\frac{z^2}{n^2 \pi^2}\right)^k = 2\sum_{k \geq 0} \frac{z^{2k+1}}{n^{2k+2} \pi^{2k+2}} = 2\sum_{k \geq 1} \frac{z^{2k-1}}{n^{2k} \pi^{2k}},$$

and then collect powers of z, we get

$$\cot z = \frac{1}{z} - 2\sum_{k \geq 1}\sum_{n \geq 1} \frac{z^{2k-1}}{n^{2k} \pi^{2k}} = \frac{1}{z} - 2\sum_{k \geq 1} \frac{\zeta(2k) z^{2k-1}}{\pi^{2k}}, \tag{9.8}$$

which is the Laurent series for $\cot z$.

We will now compare (9.8) with a second expansion for $\cot z$. The exponential series

$$e^t = 1 + t + \frac{t^2}{2!} + \frac{t^3}{3!} + \cdots$$

implies that

$$\frac{e^t - 1}{t} = 1 + \frac{t}{2!} + \frac{t^2}{3!} + \cdots,$$

and the reciprocal of this has a Taylor series expansion which can be written in the form

$$\frac{t}{e^t - 1} = \left(1 + \frac{t}{2!} + \frac{t^2}{3!} + \cdots\right)^{-1} = \sum_{m \geq 0} \frac{B_m}{m!} t^m \qquad (9.9)$$

for certain constants $B_0, B_1, \ldots$, known as the *Bernoulli numbers*. Now

$$\begin{aligned}
\frac{t}{e^t - 1} &= \frac{t}{2}\left(\frac{e^t + 1}{e^t - 1} - 1\right) \\
&= \frac{t}{2}\left(\frac{e^{t/2} + e^{-t/2}}{e^{t/2} - e^{-t/2}} - 1\right) \\
&= \frac{t}{2}\left(\coth\frac{t}{2} - 1\right) \\
&= \frac{t}{2}\left(\mathrm{i}\cot\frac{\mathrm{i}t}{2} - 1\right)
\end{aligned}$$

where $\mathrm{i} = \sqrt{-1}$. Putting $z = \mathrm{i}t/2$ we get

$$\frac{t}{e^t - 1} = z\cot z - \frac{z}{\mathrm{i}} = z\cot z + \mathrm{i}z,$$

so dividing by z and using (9.9) we have

$$\cot z = -\mathrm{i} + \frac{1}{z}\sum_{m \geq 0}\frac{B_m}{m!}t^m = -\mathrm{i} + \sum_{m \geq 0}\frac{B_m}{m!}\left(\frac{2}{\mathrm{i}}\right)^m z^{m-1}.$$

By comparing coefficients with those in (9.8), we see that if $m = 2k \geq 2$ then

$$-2\frac{\zeta(2k)}{\pi^{2k}} = \frac{B_{2k}}{(2k)!}\left(\frac{2}{\mathrm{i}}\right)^{2k},$$

so that

$$\zeta(2k) = \frac{(-1)^{k-1}2^{2k-1}\pi^{2k}B_{2k}}{(2k)!}. \qquad (9.10)$$

Thus

$$\zeta(2) = \pi^2 B_2, \qquad \zeta(4) = -\frac{\pi^4 B_4}{3}, \qquad \zeta(6) = \frac{2\pi^6 B_6}{45},$$

and so on.

To evaluate the Bernoulli numbers, we write (9.9) in the form

$$t = \sum_{m \geq 0}\frac{B_m}{m!}t^m \cdot (e^t - 1) = \sum_{m \geq 0}\frac{B_m}{m!}t^m \cdot \sum_{n \geq 1}\frac{1}{n!}t^n. \qquad (9.11)$$

If we put $m + n = r$, we find that the coefficient of t^r in the right-hand side of (9.11) is

$$\sum_{m+n=r} \frac{B_m}{m! \, n!} = \sum_{m=0}^{r-1} \frac{B_m}{m! \, (m-r)!} = \frac{1}{r!} \sum_{m=0}^{r-1} \binom{r}{m} B_m .$$

Comparing this with the left-hand side of (9.11), we see that

$$\frac{1}{r!} \sum_{m=0}^{r-1} \binom{r}{m} B_m = \begin{cases} 1 & \text{if } r = 1, \\ 0 & \text{if } r > 1. \end{cases} \tag{9.12}$$

For $r = 1, 2, \ldots$, this is an infinite sequence of linear equations

$$\begin{aligned} B_0 &= 1, \\ B_0 + 2B_1 &= 0, \\ B_0 + 3B_1 + 3B_2 &= 0, \end{aligned}$$

and so on, which we can solve in succession to find each B_m. (A more efficient but less elementary method for evaluating Bernoulli numbers is given in Graham *et al.*, 1989, Chapter 6, Section 5.) The first few values are

$$B_0 = 1, \quad B_1 = -\frac{1}{2}, \quad B_2 = \frac{1}{6}, \quad B_3 = 0, \quad B_4 = -\frac{1}{30}, \quad B_5 = 0, \quad B_6 = \frac{1}{42},$$

and so on. (In particular, $B_m = 0$ for all odd $m > 1$, reflecting the fact that $\cot z$ is an odd function.) Substituting these values for even m in (9.10), we get

$$\zeta(2) = \frac{\pi^2}{6} = 1.64493406 \ldots ,$$

$$\zeta(4) = \frac{\pi^4}{90} = 1.08232323 \ldots ,$$

$$\zeta(6) = \frac{\pi^6}{945} = 1.01734306 \ldots ,$$

so that in the notation of Exercises 9.6 and 9.8 we have

$$P(2) = Q(2) = \frac{6}{\pi^2} = 0.607927101 \ldots ,$$

$$P(4) = Q(4) = \frac{90}{\pi^4} = 0.923938402 \ldots ,$$

$$P(6) = Q(6) = \frac{945}{\pi^6} = 0.982952592 \ldots ,$$

and so on.

The coefficients in the linear equations (9.12) are all rational numbers, so it follows by induction on r that the Bernoulli numbers are all rational. It then follows from (9.10) that $\zeta(2k)$ is a rational multiple of π^{2k}, so $P(2k)$ is a rational multiple of π^{-2k}. Now a complex number is said to be *algebraic* if it is a root of some non-trivial polynomial with integer coefficients (for instance, $\sqrt{2}$ is a root of $x^2 - 2$); all other complex numbers are called *transcendental*. In 1882 Lindemann proved that π is transcendental; it follows easily that π^{2k}, and hence $\zeta(2k)$ and $P(2k)$, are also transcendental, and are therefore irrational.

Exercise 9.9

Assuming Lindemann's result, prove the remarks in the last sentence.

9.7 Dirichlet series

We defined the Riemann zeta function as $\zeta(s) = \sum_{n=1}^{\infty} 1/n^s$, and then saw that $1/\zeta(s) = \sum_{n=1}^{\infty} \mu(n)/n^s$. These are just two examples of an important class of series of this general form.

Definition

If f is an arithmetic function, then its *Dirichlet series* is the series

$$F(s) = \sum_{n=1}^{\infty} \frac{f(n)}{n^s}.$$

For convenience, we will often abbreviate this to $F(s) = \sum f(n)/n^s$, with the convention that $\sum$ without limits denotes $\sum_{n=1}^{\infty}$. Just as generating functions $A(x) = \sum a_n x^n$ are useful for studying sequences (a_n) defined by recurrence relations, Dirichlet series $F(s)$ are useful for studying arithmetic functions f, especially those associated with primes and divisors. For instance, in 1837 Dirichlet used series of this type, called *L-series*, to prove Theorem 2.10, that if a and b are coprime then there are infinitely many primes $p \equiv b \bmod (a)$. The arithmetic functions u, N and μ have particularly simple Dirichlet series:

Example 9.1

If $f = u$ then $F(s) = \sum u(n)/n^s = \sum 1/n^s = \zeta(s)$.

Example 9.2

If $f = N$ then $F(s) = \sum N(n)/n^s = \sum n/n^s = \sum 1/n^{s-1} = \zeta(s - 1)$.

Example 9.3

If $f = \mu$ then $F(s) = \sum \mu(n)/n^s = 1/\zeta(s)$ by Theorem 9.4.

The next result helps to explain the importance of Dirichlet series: multiplication of Dirichlet series corresponds to the Dirichlet product of arithmetic functions.

Theorem 9.6

Suppose that

$$F(s) = \sum_{n=1}^{\infty} \frac{f(n)}{n^s}, \qquad G(s) = \sum_{n=1}^{\infty} \frac{g(n)}{n^s} \qquad \text{and} \qquad H(s) = \sum_{n=1}^{\infty} \frac{h(n)}{n^s},$$

where $h = f * g$. Then

$$H(s) = F(s)G(s)$$

for all s such that $F(s)$ and $G(s)$ both converge absolutely.

Proof

If $F(s)$ and $G(s)$ both converge absolutely, then we can multiply these series and rearrange their terms to give

$$
\begin{aligned}
F(s)G(s) &= \sum_{n=1}^{\infty} \frac{f(n)}{n^s} \cdot \sum_{n=1}^{\infty} \frac{g(n)}{n^s} \\
&= \sum_{m=1}^{\infty} \sum_{n=1}^{\infty} \frac{f(m)g(n)}{(mn)^s} \\
&= \sum_{k=1}^{\infty} \sum_{mn=k} \frac{f(m)g(n)}{k^s} \\
&= \sum_{k=1}^{\infty} \frac{(f * g)(k)}{k^s} \\
&= \sum_{k=1}^{\infty} \frac{h(k)}{k^s} \\
&= H(s).
\end{aligned}
$$

$\square$

Example 9.4

If we take $f = \mu$ and $g = u$, then $h = f * g = \mu * u = I$ by our definition of μ (Chapter 8, Sections 3 and 6). Now $I(1) = 1$ and $I(n) = 0$ for all $n > 1$, so $H(s) = \sum I(n)/n^s = 1$ for all s. We have $F(s) = \sum \mu(n)/n^s$, and $G(s) = \sum u(n)/n^s = \sum 1/n^s = \zeta(s)$, both absolutely convergent for $s > 1$; hence Theorem 9.6 gives

$$\sum_{n=1}^{\infty} \frac{\mu(n)}{n^s} \cdot \zeta(s) = 1 \,,$$

so that

$$\sum_{n=1}^{\infty} \frac{\mu(n)}{n^s} = \frac{1}{\zeta(s)}$$

for all $s > 1$, proving part of Theorem 9.4.

Example 9.5

Let $f = \phi$ and $g = u$. As before, $G(s) = \zeta(s)$ is absolutely convergent for $s > 1$. Now $1 \le \phi(n) \le n$ for all n, so $F(s) = \sum \phi(n)/n^s$ is absolutely convergent by comparison with $\sum n/n^s = \zeta(s-1)$ for $s - 1 > 1$, that is, for $s > 2$. Thus Theorem 9.6 is valid for $s > 2$. Now Theorem 5.8 gives $\phi * u = N$, so

$$\sum_{n=1}^{\infty} \frac{\phi(n)}{n^s} \cdot \zeta(s) = \sum_{n=1}^{\infty} \frac{N(n)}{n^s} = \sum_{n=1}^{\infty} \frac{n}{n^s} = \zeta(s-1) \,,$$

and hence

$$\sum_{n=1}^{\infty} \frac{\phi(n)}{n^s} = \frac{\zeta(s-1)}{\zeta(s)}$$

for all $s > 2$.

Exercise 9.10

Show that

$$\sum_{n=1}^{\infty} \frac{\tau(n)}{n^s} = \zeta(s)^2$$

for all $s > 1$.

Exercise 9.11

Express $\sum_{n=1}^{\infty} \sigma_k(n)/n^s$ in terms of the Riemann zeta function, where $\sigma_k(n) = \sum_{d|n} d^k$.

Exercise 9.12

Liouville's function λ is defined by

$$\lambda(p_1^{e_1} \cdots p_k^{e_k}) = (-1)^{e_1 + \cdots + e_k}$$

where $p_1, \ldots, p_k$ are distinct primes. Show that

$$\sum_{d|n} \lambda(d) = \begin{cases} 1 & \text{if } n \text{ is a perfect square,} \\ 0 & \text{otherwise,} \end{cases}$$

and hence show that

$$\sum_{n=1}^{\infty} \frac{\lambda(n)}{n^s} = \frac{\zeta(2s)}{\zeta(s)}$$

for all $s > 1$.

Exercise 9.13

Let $\nu(n)$ be the number of distinct primes dividing n (so that $\nu(60) = 3$, for instance). Show that

$$\sum_{n=1}^{\infty} \frac{\nu(n)}{n^s} = \zeta(s) \sum_{p} \frac{1}{p^s},$$

where p ranges over the set of primes. For which real s is this valid?

9.8 Euler products

Many Dirichlet series have product expansions analogous to that in Theorem 9.3, in which the factors are indexed by the primes. These are called *Euler products*. First we need to consider a stronger form of multiplicativity.

Definition

An arithmetic function f is *completely multiplicative* if $f(mn) = f(m)f(n)$ for all positive integers m and n.

Example 9.6

The functions N, u and I are completely multiplicative, whereas the multiplicative functions μ and ϕ are not.

Theorem 9.7

(a) If f is multiplicative, and $\sum_{n=1}^{\infty} f(n)$ is absolutely convergent, then

$$\sum_{n=1}^{\infty} f(n) = \prod_{p} \left(1 + f(p) + f(p^2) + \cdots\right).$$

(b) If f is completely multiplicative, and $\sum_{n=1}^{\infty} f(n)$ is absolutely convergent, then

$$\sum_{n=1}^{\infty} f(n) = \prod_{p} \left(\frac{1}{1 - f(p)}\right).$$

In each case, p ranges over all the primes.

Proof

(a) The proof follows that used for Theorem 9.3. Let $p_1, \ldots, p_k$ be the first k primes, and let

$$P_k = \prod_{i=1}^{k} \left(1 + f(p_i) + f(p_i^2) + \cdots\right).$$

The general term in the expansion of P_k is $f(p_1^{e_1}) \ldots f(p_k^{e_k}) = f(p_1^{e_1} \ldots p_k^{e_k})$, because f is multiplicative. Thus

$$P_k = \sum_{n \in A_k} f(n)$$

where $A_k = \{ n \mid n = p_1^{e_1} \ldots p_k^{e_k}, \ e_i \geq 0 \}$. We have

$$\left| P_k - \sum_{n=1}^{\infty} f(n) \right| = \left| \sum_{n \notin A_k} f(n) \right| \leq \sum_{n \notin A_k} |f(n)| \leq \sum_{n > p_k} |f(n)|,$$

since $n > p_k$ for each $n \notin A_k$. Now $\sum_{n=1}^{\infty} |f(n)|$ converges, so as $k \to \infty$ we have $\sum_{n > p_k} |f(n)| \to 0$ and hence $|P_k - \sum_{n=1}^{\infty} f(n)| \to 0$; thus $P_k \to \sum_{n=1}^{\infty} f(n)$ as $k \to \infty$, as required.

(b) If f is completely multiplicative, then $f(p^e) = f(p)^e$ for each prime-power p^e, so part (a) gives

$$
\begin{aligned}
\sum_{n=1}^{\infty} f(n) &= \prod_{p} \left(1 + f(p) + f(p^2) + \cdots\right) \\
&= \prod_{p} \left(1 + f(p) + f(p)^2 + \cdots\right) \\
&= \prod_{p} \left(\frac{1}{1 - f(p)}\right).
\end{aligned}
$$

$\square$

We can apply this result to Dirichlet series:

Corollary 9.8

Suppose that $\sum_{n=1}^{\infty} f(n)/n^s$ converges absolutely. If f is multiplicative, then

$$\sum_{n=1}^{\infty} \frac{f(n)}{n^s} = \prod_p \left(1 + \frac{f(p)}{p^s} + \frac{f(p^2)}{p^{2s}} + \cdots\right),$$

and if f is completely multiplicative, then

$$\sum_{n=1}^{\infty} \frac{f(n)}{n^s} = \prod_p \left(\frac{1}{1 - f(p)p^{-s}}\right).$$

Proof

In Theorem 9.7, we simply replace $f(n)$ with $f(n)n^{-s}$, which is multiplicative (or completely multiplicative) if and only if $f(n)$ is. $\square$

Example 9.7

The function u is completely multiplicative, so as a special case we get Theorem 9.3, that

$$\zeta(s) = \sum_{n=1}^{\infty} \frac{1}{n^s} = \prod_p \left(\frac{1}{1 - p^{-s}}\right)$$

for all $s > 1$.

Example 9.8

The Möbius function $\mu(n)$ is multiplicative, with $\mu(p) = -1$ and $\mu(p^e) = 0$ for all $e \geq 2$, so

$$\sum_{n=1}^{\infty} \frac{\mu(n)}{n^s} = \prod_p \left(1 + \frac{\mu(p)}{p^s} + \frac{\mu(p^2)}{p^{2s}} + \cdots\right) = \prod_p \left(1 - p^{-s}\right)$$

for all $s > 1$. Inverting the factors in this product we obtain $1/\zeta(s)$ by the previous example, so this completes the proof of Theorem 9.4 which we promised earlier.

Exercise 9.14

Find the Euler product expansion for the Dirichlet series $\sum_{n=1}^{\infty} |\mu(n)|/n^s$, and hence show that $\sum_{n=1}^{\infty} |\mu(n)|/n^s = \zeta(s)/\zeta(2s)$ for $s > 1$. Deduce that if $n > 1$ then $\sum_d \lambda(d) = 0$, where d ranges over the divisors of n such that n/d is square-free. (Here λ is Liouville's function, defined in Exercise 9.12.)

Exercise 9.15

Show that $\prod_{p \leq x}(1 - p^{-1}) \to 0$ as $x \to +\infty$.

9.9 Complex variables

In considering Dirichlet series $F(s) = \sum f(n)/n^s$, such as the Riemann zeta function $\sum 1/n^s$, we have assumed (often implicitly) that the variable s is real. For many purposes, this is adequate, but for some more advanced applications it is necessary to allow s to be complex. The advantage of this is that functions of a complex variable are often easier to deal with than those of a real variable: in particular, their domains of definition can often be extended by analytic continuation, and they can be integrated by the calculus of residues, techniques which are not available if we restrict to real variables.

Our earlier results on Dirichlet series and Euler products all extend to the case where s is a complex variable, provided we have absolute convergence. We therefore need to consider the subset of the complex plane $\mathbb{C}$ on which a Dirichlet series converges absolutely. We will see that, just as a power series converges absolutely on a disc (which may be the whole plane or a single point), a Dirichlet series has a half-plane of absolute convergence, which may be the whole plane or the empty set.

Following the traditional (if slightly bizarre) notation we put

$$s = \sigma + it \in \mathbb{C} \qquad \text{where} \qquad \sigma, t \in \mathbb{R}.$$

Then $n^s = n^{\sigma+it} = n^\sigma.n^{it} = n^\sigma.e^{it\ln(n)}$ with $n^\sigma > 0$ and $|e^{it\ln(n)}| = 1$, so $|n^s| = n^\sigma$. Now suppose that $F(s)$ converges absolutely (that is, $\sum |f(n)/n^s|$ converges) at some point $s = a + ib \in \mathbb{C}$; if $\sigma \geq a$ then

$$\left| \frac{f(n)}{n^{\sigma+it}} \right| = \left| \frac{f(n)}{n^\sigma} \right| \leq \left| \frac{f(n)}{n^a} \right| = \left| \frac{f(n)}{n^{a+ib}} \right|,$$

so $\sum f(n)/n^{\sigma+it}$ converges absolutely by the Comparison Test. This implies:

Theorem 9.9

Suppose that $\sum_{n=1}^{\infty} |f(n)/n^s|$ neither converges for all $s \in \mathbb{C}$, nor diverges for all $s \in \mathbb{C}$. Then there exists $\sigma_a \in \mathbb{R}$ such that $\sum_{n=1}^{\infty} |f(n)/n^s|$ converges for all $s = \sigma + it$ with $\sigma > \sigma_a$, and diverges for all $s = \sigma + it$ with $\sigma < \sigma_a$.

Proof

We take σ_a to be the least upper bound of all $a \in \mathbb{R}$ such that $\sum |f(n)/n^s|$ diverges at $s = a + ib$; by the preceding argument this coincides with the greatest lower bound of all $a \in \mathbb{R}$ such that $\sum |f(n)/n^s|$ converges at $s = a + ib$. $\square$

Definition

We call σ_a the *abscissa of absolute convergence* of $F(s)$, and $\{s = \sigma + it \in \mathbb{C} \mid \sigma > \sigma_a\}$ its *half-plane of absolute convergence*.

Note that the theorem says nothing about the behaviour of $F(s)$ when $\sigma = \sigma_a$. Note also that there are two other extreme possibilities, not covered by the theorem: $F(s)$ may converge absolutely for all $s \in \mathbb{C}$, or for no $s \in \mathbb{C}$; we then write $\sigma_a = -\infty$ or $+\infty$ respectively. A similar but more complicated argument shows that there exists $\sigma_c \leq \sigma_a$, called the *abscissa of convergence*, such that $F(s)$ converges for $\sigma > \sigma_c$ and diverges for $\sigma < \sigma_c$; if $\sigma_c < \sigma_a$, then convergence is conditional for $\sigma_c < \sigma < \sigma_a$.

Exercise 9.16

Find examples of Dirichlet series for which $\sigma_a = -\infty$ and $\sigma_a = +\infty$.

Example 9.9

Theorem 9.1 states that $\sum 1/n^s$ converges (absolutely) for all real $s > 1$ and diverges for $s \leq 1$. This series therefore has $\sigma_a = \sigma_c = 1$, so it converges absolutely for all $s = \sigma + it \in \mathbb{C}$ with $\sigma > 1$, and diverges for $\sigma < 1$. Similarly, $\sum (-1)^n/n^s$ has $\sigma_a = 1$, but in this case $\sigma_c = 0$ since the series converges for all $s > 0$ by the Alternating Test (Appendix C), but diverges for $s \leq 0$.

Example 9.10

If f is bounded, say $|f(n)| \leq M$ for all n, then $|f(n)/n^s| \leq M/n^\sigma$ where $s = \sigma + it$, so $\sum f(n)/n^s$ converges absolutely whenever $\sigma > 1$, by comparison with $\sum M/n^\sigma$. (It may converge absolutely for smaller σ, depending on the particular function f.) This applies to $f = \mu$ for example, with $M = 1$. More generally, if there are constants M and k such that $|f(n)| \leq Mn^k$ for all n, then $\sum f(n)/n^s$ converges absolutely for $\sigma > 1 + k$ by comparison with $\sum Mn^k/n^\sigma$. Now $|\phi(n)| \leq n$ for all n, so taking $k = 1$ we see that $\sum \phi(n)/n^s$ converges absolutely for $\sigma > 2$.

A complex function $F(s)$ is said to be *analytic* if it is differentiable with respect to s.

Theorem 9.10

A Dirichlet series $\sum_{n=1}^{\infty} f(n)/n^s$ represents an analytic function $F(s)$ for $\sigma > \sigma_c$, with derivative $F'(s) = -\sum_{n=1}^{\infty} f(n)\ln(n)/n^s$.

Proof (Outline proof)

For each $n \geq 1$, the function $f(n)/n^s = f(n)e^{-s\ln(n)}$ is analytic for all s (since the exponential function e^s is analytic), with derivative $-f(n)\ln(n)/n^s$. One now shows that $\sum f(n)/n^s$ converges uniformly on all compact (closed, bounded) subsets of the half-plane $\sigma > \sigma_c$, and then quotes the theorem that a uniformly convergent series of analytic functions has an analytic sum, which may be differentiated term by term. For full details, see Apostol (1976, Chapter 11, Section 7). $\qquad\square$

For example, the series $\sum 1/n^s$ defines an analytic function $\zeta(s)$ on the half-plane $\sigma > 1$. Riemann used analytic continuation to extend the domain of $\zeta(s)$: the resulting function, also denoted by $\zeta(s)$, is analytic on $\mathbb{C} \setminus \{1\}$, with a simple pole at $s = 1$ (this means that $(s - 1)\zeta(s)$ is analytic at $s = 1$, so that $\zeta(s)$ diverges there like $1/(s - 1)$). Note that we do *not* claim that the series $\sum 1/n^s$ converges outside the half-plane $\sigma > 1$: what Riemann showed is that there is a function $\zeta(s)$ which is analytic for all $s \neq 1$, and which agrees with $\sum 1/n^s$ for $\sigma > 1$. This is analogous to the situation with power series: for instance the geometric series $1 + z + z^2 + \cdots$ converges (absolutely) for all z with $|z| < 1$, and within this disc of convergence its sum is given by $(1 - z)^{-1}$; however, this function $(1 - z)^{-1}$ is analytic for *all* $z \neq 1$, even though the series diverges for $|z| \geq 1$.

Exercise 9.17

Show that $\zeta'(s) = -\sum \ln(n)/n^s$ and $-\zeta'(s)/\zeta(s) = \sum \Lambda(n)/n^s$ for all s with $\sigma > 1$, where Λ is Mangoldt's function (see Exercise 8.24).

Riemann showed that the extended function $\zeta(s)$ has zeros at $s = -2$, -4, -6, $\ldots$; these are called the *trivial zeros*, and he showed that the remaining *non-trivial zeros* all lie in the *critical strip* $0 \leq \sigma \leq 1$. The celebrated *Riemann Hypothesis* is the conjecture that these non-trivial zeros all lie on the line $\sigma = 1/2$. A great deal is now known about the location of the zeros of $\zeta(s)$: for instance, Hardy showed in 1914 that there are infinitely many on the line $\sigma = 1/2$. Despite strong evidence in its favour, the Riemann Hypothesis is still unproved; since many conjectures about the distribution of prime numbers depend on this result, the resolution of this problem remains one of the greatest challenges of number theory.

9.10 Supplementary exercises

Exercise 9.18

Let $\tau_k(n)$ be the number of k-tuples $(d_1, \ldots, d_k)$ of positive integers d_i such that $d_1 \ldots d_k = n$ (so that $\tau_2 = \tau$, for instance). Show that $\sum_{n=1}^{\infty} \tau_k(n)/n^s = \zeta(s)^k$ for all $s = \sigma + it$ with $\sigma > 1$.

Exercise 9.19

Show that $\sum_{n=1}^{\infty} \tau(n)^2/n^s = \zeta(s)^4/\zeta(2s)$ for $\sigma > 1$.

Exercise 9.20

Recall that $\pi(x)$ is the number of primes $p \leq x$. Show that if $q(x)$ denotes the number of square-free integers $m \leq x$, then

$$2^{\pi(x)} \geq q(x) \geq x\left(2 - \frac{\pi^2}{6}\right),$$

and hence

$$\pi(x) \;\geq\; \log_2 x + \log_2\left(2 - \frac{\pi^2}{6}\right)$$
$$= \;\frac{\ln x}{\ln 2} + \log_2\left(2 - \frac{\pi^2}{6}\right).$$

(This estimate for $\pi(x)$ is very weak: for instance it gives $\pi(10^9) \geq 28$, whereas in fact $\pi(10^9) \approx 5 \times 10^7$.)

Exercise 9.21

Let $f_k(n)$ denote the number of subgroups of finite index n in the group $\mathbb{Z}^k$ (see Exercise 8.26). Express the Dirichlet series $F_k(s) = \sum_{n=1}^{\infty} f_k(n)/n^s$ of f_k in terms of the Riemann zeta function. For which $s \in \mathbb{C}$ is your expression valid?

10

Sums of Squares

Our main aim in this chapter is to determine which integers can be expressed as the sum of a given number of squares, that is, which have the form $x_1^2 + \cdots + x_k^2$, where each $x_i \in \mathbb{Z}$, for a given k. We shall concentrate mainly on the two most important cases, characterising the sums of two squares, and showing that every non-negative integer is a sum of four squares. We shall adopt two completely different approaches to this problem: the first is mainly algebraic, making use of two number systems, the Gaussian integers and the quaternions; the second approach is geometric, based on the fact that the expression $x_1^2 + \cdots + x_k^2$ represents the square of the length of the vector $(x_1, \ldots, x_k)$ in $\mathbb{R}^k$. We shall therefore give two different proofs for several of the main theorems in this chapter. In mathematics, it is often useful to have more than one proof of a result, not because this adds anything to its validity (a single correct proof is enough for this), but rather because the extra proofs may add to our understanding of the result, and may enable us to extend it in different directions.

10.1 Sums of two squares

Definition

For each integer $k \geq 1$, let $S_k = \{n \mid n = x_1^2 + \cdots + x_k^2 \text{ for some } x_1, \ldots, x_k \in \mathbb{Z}\}$, the set of all sums of k squares.

Example 10.1

$S_1 = \{0, 1, 4, 9, \ldots\}$ is the set of all squares. By inspection, S_2, the set of sums of two squares, contains $0, 1, 2, 4, 5$ and 8, but not $3, 6$ or 7.

Lemma 10.1

The set S_2, consisting of the sums of two squares, is closed under multiplication, that is, if $s, t \in S_2$ then $st \in S_2$.

Proof

Let $s = a_1^2 + b_1^2$ and $t = a_2^2 + b_2^2$ be elements of S_2, where $a_1, b_1, a_2, b_2 \in \mathbb{Z}$. Then the identity

$$(a_1^2 + b_1^2)(a_2^2 + b_2^2) = (a_1 a_2 - b_1 b_2)^2 + (a_1 b_2 + b_1 a_2)^2 \tag{10.1}$$

shows that $st \in S_2$, since $a_1 a_2 - b_1 b_2, a_1 b_2 + b_1 a_2 \in \mathbb{Z}$. □

Example 10.2

We have $8 = 2^2 + 2^2$ and $10 = 3^2 + 1^2$, so $80 = 8.10 = (2.3 - 2.1)^2 + (2.1 + 2.3)^2 = 4^2 + 8^2$.

Comments

1. It follows immediately that the product of any finite set of elements of S_2 is also in S_2.

2. The identity (10.1) can be verified directly by expanding each side. However, it is more useful to define a pair of complex numbers $z_i = a_i + ib_i$ for $i = 1, 2$ (where $i = \sqrt{-1}$), so that $a_i^2 + b_i^2 = z_i \overline{z_i} = |z_i|^2$. Now the rules for multiplying complex numbers (with $i^2 = -1$) give $z_1 z_2 = (a_1 a_2 - b_1 b_2) + i(a_1 b_2 + b_1 a_2)$, so that $|z_1 z_2|^2 = (a_1 a_2 - b_1 b_2)^2 + (a_1 b_2 + b_1 a_2)^2$. One can therefore prove the identity by arguing that

$$|z_1 z_2|^2 = (z_1 z_2).\overline{(z_1 z_2)} = z_1 \overline{z_1}.z_2 \overline{z_2} = |z_1^2|.|z_2^2|$$

for all $z_1, z_2 \in \mathbb{C}$. We will see later that a similar identity holds for S_4, but not for S_3, so that sums of four squares are easier to deal with than sums of three squares.

3 By replacing b_1 with $-b_1$ in (10.1) we obtain the equivalent identity

$$(a_1^2 + b_1^2)(a_2^2 + b_2^2) = (a_1a_2 + b_1b_2)^2 + (a_1b_2 - b_1a_2)^2, \qquad (10.2)$$

which we will need later.

Lemma 10.1 suggests that, in determining the elements of S_2, we should first consider prime numbers: each integer $n \geq 2$ is a product of primes, and if these prime factors are all in S_2 then so is n. However, not all primes are sums of two squares, the prime 3 being the first counterexample.

Exercise 10.1

Which of the primes $p < 100$ are elements of S_2? Do you notice a pattern emerging?

The following *Two Squares Theorem* was stated by Fermat in 1640, and proved by Euler in 1754.

Theorem 10.2

Each prime $p \equiv 1$ mod (4) is a sum of two squares.

Proof

Since $p \equiv 1$ mod (4), Corollary 7.7 implies that $-1 \in Q_p$; thus $-1 \equiv u^2$ mod (p) for some u, so $u^2 + 1 = rp$ for some integer r. We can choose u so that $0 \leq u \leq p - 1$, giving $0 < r < p$. Now $rp = u^2 + 1^2 \in S_2$, so it follows that there is a smallest integer m such that $mp \in S_2$ and $0 < m < p$. If $m = 1$ then $p \in S_2$ and we are done, so assume that $m > 1$.

Since $mp \in S_2$, we have $mp = a_1^2 + b_1^2$ for some integers a_1 and b_1. Let a_2 and b_2 be the least absolute residues of a_1 and b_1 mod (m), so that $a_2 \equiv a_1$ and $b_2 \equiv b_1$ mod (m) and $|a_2|, |b_2| \leq m/2$. Then $a_2^2 + b_2^2 \equiv a_1^2 + b_1^2 \equiv 0$ mod (m), so $a_2^2 + b_2^2 = sm$ for some integer s; since $|a_2|, |b_2| \leq m/2$ we have $a_2^2 + b_2^2 \leq 2(m/2)^2 = m^2/2$, so $s \leq m/2$ and hence $s < m$.

We also have $s > 0$: if $s = 0$ then $a_2^2 + b_2^2 = 0$, so $a_2 = b_2 = 0$, giving $a_1 \equiv b_1 \equiv 0$ mod (m); then m divides a_1 and b_1, so m^2 divides $a_1^2 + b_1^2 = mp$ and hence m divides p, which is impossible since p is prime and $1 < m < p$. Thus $0 < s < m$.

Now $(a_1^2 + b_1^2)(a_2^2 + b_2^2) = mp.sm = m^2sp$, and identity (10.2) following Lemma 10.1 shows that

$$(a_1^2 + b_1^2)(a_2^2 + b_2^2) = (a_1a_2 + b_1b_2)^2 + (a_1b_2 - b_1a_2)^2,$$

so that

$$(a_1a_2 + b_1b_2)^2 + (a_1b_2 - b_1a_2)^2 = m^2sp.$$

Since $a_1a_2 + b_1b_2 \equiv a_2^2 + b_2^2 \equiv 0 \mod (m)$ and $a_1b_2 - b_1a_2 \equiv a_2b_2 - b_2a_2 \equiv 0$ $\mod (m)$, we can divide this equation through by m^2 to give

$$\left(\frac{a_1a_2 + b_1b_2}{m}\right)^2 + \left(\frac{a_1b_2 - b_1a_2}{m}\right)^2 = sp$$

where both $(a_1a_2 + b_1b_2)/m$ and $(a_1b_2 - b_1a_2)/m$ are integers. Thus $sp \in S_2$ with $0 < s < m$, contradicting the minimality of m. Hence $m = 1$ and the proof is complete.   □

We will give an alternative geometric proof of Theorem 10.2 in Section 10.6. We can now give a complete description of the elements of S_2 in terms of their prime-power factorisations.

Theorem 10.3

A positive integer n is a sum of two squares if and only if every prime $q \equiv 3$ $\mod (4)$ divides n to an even power (which may, of course, be 0 if $q \nmid n$).

Proof

($\Leftarrow$) By assumption, we can write

$$n = 2^e p_1^{e_1} \ldots p_k^{e_k} q_1^{2f_1} \ldots q_l^{2f_l} = 2^e p_1^{e_1} \ldots p_k^{e_k} (q_1^2)^{f_1} \ldots (q_l^2)^{f_l},$$

for some set of primes $p_i \equiv 1 \mod (4)$ and $q_j \equiv 3 \mod (4)$, where the exponents are integers $e \geq 0, e_i > 0$ and $f_j > 0$. Now $2 = 1^2 + 1^2 \in S_2$, Theorem 10.2 shows that each $p_i \in S_2$, and also each $q_j^2 = q_j^2 + 0^2 \in S_2$. Thus n is a product of elements of S_2, so Lemma 10.1 implies that $n \in S_2$.

($\Rightarrow$) Let $n \in S_2$, say $n = x^2 + y^2$. Let q be any prime such that $q \equiv 3$ $\mod (4)$, let $q^f \| n$, and suppose (for a contradiction) that f is odd. If d denotes the greatest common divisor of x and y, then $x = ad$ and $y = bd$ where $\gcd(a, b) = 1$, so $n = (a^2 + b^2)d^2$ and hence $nd^{-2} = a^2 + b^2$. If $q^e \| d$ then $q^{f-2e}|nd^{-2}$; now $f - 2e$ is odd and hence non-zero, so $q|nd^{-2} = a^2 + b^2$ and hence $a^2 \equiv -b^2 \mod (q)$. Now b cannot be divisible by q (for then q would divide both a and b, whereas they are coprime), so b is a unit mod (q). If c is the inverse of b in U_q, then multiplying through by c^2 we have $(ac)^2 \equiv -1$ $\mod (q)$, so that $-1 \in Q_q$. This is impossible for a prime $q \equiv 3 \mod (4)$ by Corollary 7.7, so f must be even.   □

Example 10.3

The integer $60\ (= 2^2.3.5)$ is not a sum of two squares, since the exponent of 3 dividing it is odd. However, $180\ (= 2^2.3^2.5)$ is a sum of two squares. To find them, first write 5 as a sum of two squares: $5 = 2^2 + 1^2$. Now multiplying through by $2^2.3^2$ we get $180 = 2^2.3^2.5 = (2.3.2)^2 + (2.3.1)^2 = 12^2 + 6^2$.

Example 10.4

The integer $221\ (= 13.17)$ is a sum of two squares, since $13 \equiv 17 \equiv 1 \bmod (4)$. To find these squares, imitate the proof of Lemma 10.1, with $13 = 3^2 + 2^2$ and $17 = 4^2 + 1^2$ corresponding to the equations $s = a_1^2 + b_1^2$ and $t = a_2^2 + b_2^2$. Then

$$221 = st = (a_1 a_2 - b_1 b_2)^2 + (a_1 b_2 + b_1 a_2)^2 = (3.4 - 2.1)^2 + (3.1 + 2.4)^2 = 10^2 + 11^2.$$

(Note that equation (10.2) sometimes gives a different expression, e.g. $221 = 14^2 + 5^2$ in this case.) Similarly, one can express $6409\ (= 221.29)$ as a sum of two squares by repeating this process: $221 = 10^2 + 11^2$ and $29 = 5^2 + 2^2$, so $6409 = (10.5 - 11.2)^2 + (10.2 + 11.5)^2 = 28^2 + 75^2$.

Exercise 10.2

Write each of the following integers as a sum of two squares, or show that this is impossible: 130, 260, 847, 980, 1073.

Exercise 10.3

Find all the pairs $(x, y) \in \mathbb{Z}^2$ satisfying $x^2 + y^2 = 50$.

As a special case of Theorem 10.3, we have the following stronger form of Theorem 10.2:

Corollary 10.4

A prime p is a sum of two squares if and only if $p = 2$ or $p \equiv 1 \bmod (4)$.

10.2 The Gaussian integers

A representation of an integer n as a *difference* of two squares, say $n = x^2 - y^2$, gives rise to a factorisation of n as a product of two integers,

$$n = x^2 - y^2 = (x + y)(x - y),$$

where $x + y \equiv x - y \bmod (2)$. Conversely, if $n = rs$ where $r \equiv s \bmod (2)$, then by writing $x = (r + s)/2$ and $y = (r - s)/2$ we get $n = x^2 - y^2$ with $x, y \in \mathbb{Z}$. This link with factorisations can be extended to *sums* of two squares if we write

$$n = x^2 + y^2 = (x + yi)(x - yi),$$

where $i = \sqrt{-1} \in \mathbb{C}$: given any factorisation $n = rs$ of this form, we now write $x = (r + s)/2$ and $y = (r - s)/2i$, provided these are integers. This suggests that we should study the complex numbers of the form $x + yi$ ($x, y \in \mathbb{Z}$), known as the *Gaussian integers*, and in particular their factorisations.

The set

$$\mathbb{Z}[i] = \{x + yi \mid x, y \in \mathbb{Z}\}$$

of all Gaussian integers is closed under addition, subtraction and multiplication: for instance, $(a + bi)(c + di) = (ac - bd) + (ad + bc)i$, and if a, b, c, d are integers then so are $ac - bd$ and $ad + bc$. The usual axioms (associativity, distributivity, etc.) are satisfied, so $\mathbb{Z}[i]$ is a ring; however, it is not a field, since not every non-zero element of $\mathbb{Z}[i]$ is a unit (see Exercise 10.4). Thus $\mathbb{Z}[i]$ shares many of the basic properties of $\mathbb{Z}$, so it is not surprising that many of our earlier results about divisibility and factorisation of integers extend in a natural way to Gaussian integers.

There are two other important properties of $\mathbb{Z}$ shared by $\mathbb{Z}[i]$. The first is that if $r, s \neq 0$ then $rs \neq 0$, or equivalently, if $rs = 0$ then $r = 0$ or $s = 0$ (this is true in $\mathbb{Z}[i]$ since it is true in $\mathbb{C}$, which contains $\mathbb{Z}[i]$). A ring with this property is called an *integral domain*. This property is useful since it allows one to cancel non-zero factors: if $r's = r''s$ with $s \neq 0$, then $(r' - r'')s = 0$ so that $r' - r'' = 0$ and hence $r' = r''$.

The second important property of $\mathbb{Z}$ is the Division Algorithm (Corollary 1.2), which allows one to divide an integer a by a non-zero integer b, with a remainder which is small compared with b. An integral domain R is a *Euclidean domain* if, for each $a \in R \setminus \{0\}$ there is an integer $d(a) \geq 0$ such that

(1) $d(ab) \geq d(b)$ for all $a, b \neq 0$, with equality if and only if a is a unit;

(2) for all $a, b \in R$ with $b \neq 0$, there exist $q, r \in R$ such that $a = qb + r$, with $r = 0$ or $d(r) < d(b)$.

The function d assigns a measure of size to each non-zero element of R. Condition (1) simply states that this function behaves reasonably with respect to products, while condition (2) is the analogue of Corollary 1.2, giving a remainder r which is small in comparison with b.

Example 10.5

The ring $\mathbb{Z}$ is a Euclidean domain, with $d(r) = |r|$: condition (1) is clear, using the fact that the units of $\mathbb{Z}$ are ± 1, while condition (2) is simply Corollary 1.2.

Example 10.6

If F is any field, then the ring $F[x]$ of polynomials in one variable x, with coefficients in F, is a Euclidean domain. For each non-zero $f = f(x) \in F[x]$ we define $d(f) = \deg(f)$, the degree of the polynomial $f(x)$. Then condition (1) follows from the facts that $\deg(fg) = \deg(f) + \deg(g)$, and that f is a unit if and only if it is a non-zero constant polynomial, that is, $\deg(f) = 0$; condition (2) follows from polynomial division.

Example 10.7

The ring $\mathbb{Z}[i]$ of Gaussian integers is a Euclidean domain, with $d(z) = z\bar{z} = |z|^2 = x^2 + y^2$ for each $z = x + yi \in \mathbb{Z}[i]$. (This is sometimes called the *norm* of z, written $N(z)$.) Condition (1) is straightforward, using the facts that $d(zw) = d(z)d(w)$ for all z, w (see Comment 2 of Section 1), and that the units in $\mathbb{Z}[i]$ are ± 1 and $\pm i$ (see Exercises 10.4 and 10.5).

Exercise 10.4

Show that if $u \in \mathbb{Z}[i]$ then the following are equivalent:

(a) u is a unit in $\mathbb{Z}[i]$;

(b) $d(u) = 1$;

(c) $u = +1$ or $\pm i$.

Exercise 10.5

Verify condition (1) for $\mathbb{Z}[i]$, that $d(ab) \geq d(b)$ for all non-zero $a, b \in \mathbb{Z}[i]$, with equality if and only if a is a unit.

We will give a geometric proof of condition (2) for $\mathbb{Z}[i]$, though an algebraic proof is also possible. The Gaussian integers $q = x+yi \in \mathbb{Z}[i]$ can be regarded as the points $(x, y) \in \mathbb{R}^2$ with integer coordinates; as such they are the vertices of a tessellation (tiling) of the plane by squares with side-length 1. If b is a non-zero element of $\mathbb{Z}[i]$, then the multiples qb of b, with $q \in \mathbb{Z}[i]$, are also the vertices of a tessellation by squares, obtained by multiplying the original tessellation by b; equivalently, we rotate the original tessellation about the origin by an angle $\arg(b)$, and expand it by a factor $|b|$, so these squares have side-length $|b|$. Every complex number (and hence every Gaussian integer) a is in at least one of these squares, and its distance from one of the vertices qb is at most $|b|/\sqrt{2}$ (attained if a is the centre of a square). If we define $r = a - qb$, then $r \in \mathbb{Z}[i]$, $a = qb + r$, and $|r| \leq |b|/\sqrt{2}$, so $d(r) = |r|^2 \leq |b|^2/2 < |b|^2 = d(b)$, as required.

The main results from Chapters 1 and 2 concerning divisors and factorisations all extend to any Euclidean domain R, with only minor changes of terminology. It is a useful exercise to adapt their proofs to establish the following more general results, and to see what they mean for $\mathbb{Z}[i]$. First we need some terminology.

Two elements $a, b \in R$ are *associates* if $a = ub$ for some unit u of R; this is an equivalence relation, since the units form a group under multiplication. An element $a \in R$, which is not 0 or a unit, is *irreducible* if its only divisors are units or its associates; otherwise, a is *reducible*. In $\mathbb{Z}$, for instance, the associates of a are $\pm a$, and the irreducible elements are those of the form $\pm p$ where p is prime.

The main result we need here states that each element of a Euclidean domain R, other than 0 or a unit, is a product of powers of irreducible elements; moreover, this representation is unique, apart from permuting factors and replacing them with their associates. In the case $R = \mathbb{Z}$ this is the Fundamental Theorem of Arithmetic (Theorem 2.3). The proof for general Euclidean domains is very similar, and we will omit the details, since they can be found in many algebra textbooks.

In order to apply this to the Gaussian integers, we need to determine the irreducible elements of $\mathbb{Z}[i]$.

Each prime $q \equiv 3 \bmod (4)$ in $\mathbb{Z}$ is irreducible in $\mathbb{Z}[i]$. For if $q = zw$ with $z, w \in \mathbb{Z}[i]$, then $d(z)d(w) = d(q) = q^2$ in $\mathbb{Z}$, so either $d(z) = d(w) = q$ or $\{d(z), d(w)\} = \{1, q^2\}$. In the first case, putting $z = x+yi$ we get $q = x^2+y^2 \in S_2$, which is impossible by Corollary 10.4; hence either $d(z) = 1$ or $d(w) = 1$, so z or w is a unit, and q is irreducible. For instance, the primes $q = 3, 7, 11, 19, \ldots$ are all irreducible as Gaussian integers. Each prime $q \equiv 3 \bmod (4)$ gives rise to four associates $\pm q$ and $\pm qi$, all irreducible in $\mathbb{Z}[i]$; by the uniqueness of factorisation in $\mathbb{Z}[i]$, these are the only irreducible elements dividing q.

On the other hand, any prime $p = 2$ or $p \equiv 1 \bmod (4)$ is in S_2, so $p = x^2+y^2$

for some integers x and y, giving a factorisation $p = (x + yi)(x - yi)$ of p in $\mathbb{Z}[i]$. These factors $\pi = x + yi$ and $\overline{\pi} = x - yi$ must be irreducible: if $x \pm yi = st$ in $\mathbb{Z}[i]$ then $d(s)d(t) = d(x \pm yi) = p$ in $\mathbb{Z}$, so $d(s) = 1$ or $d(t) = 1$ and hence s or t must be a unit. For instance, $2 = (1 + i)(1 - i)$, so $1 + i$ and $1 - i$ are irreducible, and multiplying by units we obtain four associate irreducible elements: $1+i$, $i(1+i) = -1+i$, $-(1+i) = -1-i$ and $-i(1+i) = 1-i$. However, if $p \equiv 1 \bmod (4)$ we obtain eight irreducible elements $\pm x \pm yi$ and $\pm y \pm xi$, consisting of four associates of each of π and $\overline{\pi}$; for instance $5 = 1^2 + 2^2 = (1 + 2i)(1 - 2i)$, giving the irreducible elements $\pm 1 \pm 2i$ and $\pm 2 \pm i$. In either case, the uniqueness of factorisation implies that these are the only irreducible divisors of p in $\mathbb{Z}[i]$.

These irreducible elements $\pi, \overline{\pi}$ and q, together with their associates, are in fact the only irreducible elements of $\mathbb{Z}[i]$. For suppose that z is irreducible in $\mathbb{Z}[i]$. Now z divides the positive integer $z\overline{z} = d(z)$, so there is a least positive integer n such that z divides n in $\mathbb{Z}[i]$; since z is not a unit, $n > 1$. Now n must be prime, for if $n = ab$ in $\mathbb{Z}$ with $a, b > 0$, then z divides a or b in $\mathbb{Z}[i]$ (by the analogue of Lemma 2.1(b) for $\mathbb{Z}[i]$), so $a = n$ or $b = n$ by the minimality of n. We have already determined which irreducible Gaussian integers divide the various primes $n = 2$, $n = p \equiv 1$ and $n = q \equiv 3 \bmod (4)$, so z must be an associate of some $\pi, \overline{\pi}$ or q, as required.

The uniqueness of factorisation in $\mathbb{Z}[i]$ implies that the representation of a prime $p \in S_2$ as $x^2 + y^2$ is essentially unique, apart from the obvious changes of transposing x and y, and multiplying either or both of them by -1. More precisely, if $r(n)$ denotes the number of pairs $(x, y) \in \mathbb{Z}^2$ such that $x^2 + y^2 = n$, then $r(2) = 4$, from the representations $2 = (\pm 1)^2 + (\pm 1)^2$, and $r(p) = 8$ for primes $p \equiv 1 \bmod (4)$, from $p = (\pm x)^2 + (\pm y)^2 = (\pm y)^2 + (\pm x)^2$; of course, Corollary 10.4 gives $r(q) = 0$ for primes $q \equiv 3 \bmod (4)$.

Using our knowledge of the irreducible elements of $\mathbb{Z}[i]$, we can in fact evaluate $r(n)$ for all n. Suppose that n factorises in $\mathbb{Z}$ as

$$n = 2^e p_1^{e_1} \ldots p_k^{e_k} q_1^{f_1} \ldots q_l^{f_l}$$

for primes $p_i \equiv 1$ and $q_j \equiv 3 \bmod (4)$, and integers $e \geq 0$ and $e_i, f_j > 0$. By factorising each prime $2, p_i$ and q_j in $\mathbb{Z}[i]$, we find that n has factorisation

$$
\begin{aligned}
n &= (1+i)^e (1-i)^e \prod_i \pi_i^{e_i} (\overline{\pi_i})^{e_i} \prod_j q_j^{f_j} \\
&= i^e (1-i)^{2e} \prod_i \pi_i^{e_i} (\overline{\pi_i})^{e_i} \prod_j q_j^{f_j}
\end{aligned}
$$

in $\mathbb{Z}[i]$, where $1-i, \pi_i, \overline{\pi_i}$ and q_j are all irreducible, and $\pi_i \overline{\pi_i} = p_i$ for $i = 1, \ldots, k$. Now $n = x^2 + y^2$ if and only if $n = (x + yi)(x - yi)$ in $\mathbb{Z}[i]$, so $r(n)$ is the number of distinct factors z of n in $\mathbb{Z}[i]$ such that $n = z\bar{z}$. Any factor z of n must be a product of a unit u and various irreducible factors of n, that is,

$$z = u(1-i)^a \prod_i \pi_i^{a_i}(\bar{\pi}_i)^{b_i} \prod_j q_j^{c_j}$$

where $0 \le a \le 2e$, $0 \le a_i, b_i \le e_i$, and $0 \le c_j \le f_j$. Then

$$\begin{aligned} \bar{z} &= \bar{u}(1+i)^a \prod_i (\bar{\pi}_i)^{a_i} \pi_i^{b_i} \prod_j q_j^{c_j} \\ &= \bar{u}i^a(1-i)^a \prod_i \pi_i^{b_i}(\bar{\pi}_i)^{a_i} \prod_j q_j^{c_j}, \end{aligned}$$

and by using $u\bar{u} = 1$ we can combine these factorisations of z and $\bar{z}$ to obtain the factorisation

$$z\bar{z} = i^a(1-i)^{2a} \prod_i \pi_i^{a_i+b_i}(\bar{\pi}_i)^{a_i+b_i} \prod_j q_j^{2c_j}.$$

By comparing this with the factorisation of n, and using the uniqueness of factorisation of Gaussian integers, we see that $z\bar{z} = n$ if and only if $a = e$, $a_i + b_i = e_i$ and $2c_j = f_j$ for all i and j. Now $r(n)$ is the number of such factors z of n, so $r(n) = 0$ unless each f_j is even, confirming Theorem 10.3; when this happens, the values of a $(= e)$ and c_j $(= f_j/2)$ are uniquely determined by n, while there are four choices for the unit u, and $e_i + 1$ choices for each pair a_i, b_i (since $a_i = e_i - b_i = 0, 1, \ldots, e_i$). Thus

$$r(n) = \begin{cases} 4\prod_{i=1}^k (e_i + 1) & \text{if } f_1, \ldots, f_l \text{ are all even,} \\ 0 & \text{otherwise.} \end{cases}$$

Note that Theorem 8.3 implies that $\prod_{i=1}^k (e_i + 1) = \tau(n_1)$, the number of divisors of $n_1 = \prod_{i=1}^k p_i^{e_i}$.

Example 10.8

Let $n = 30420 = 2^2.5.13^2.3^2$. The only prime $q_j \equiv 3 \bmod (4)$ dividing n is $q_1 = 3$, so $l = 1$ with $f_1 = 2$ (the exponent of 3), which is even. The primes $p_i \equiv 1 \bmod (4)$ dividing n are $p_1 = 5$ and $p_2 = 13$, so $k = 2$ with $e_1 = 1$ and $e_2 = 2$ (the exponents of 5 and 13). Thus $r(30420) = 4(1+1)(2+1) = 24$. Since $5 = (2+i)(2-i)$ and $13 = (3+2i)(3-2i)$, and since $e = 2$, the factors z such that $z\bar{z} = n$ have the form

$$z = u(1-i)^2.(2+i)^{a_1}(2-i)^{1-a_1}(3+2i)^{a_2}(3-2i)^{2-a_2}.3$$

where u is a unit, $0 \le a_1 \le 1$ and $0 \le a_2 \le 2$. Now $u(1 - i)^2.3 = 6v$, where $v = -iu$ is a unit, so if we take $a_1 = 0, 1$ and $a_2 = 0, 1, 2$ in turn then we get 24 factors

$$z = x + yi = 6v(-2 \pm 29i), \quad 6v(26 \pm 13i), \quad 6v(22 \pm 19i).$$

These give 24 representations of $n = 30420$ as $x^2 + y^2$, each of which is obtained from one of
$$12^2 + 174^2, \ 156^2 + 78^2, \ 132^2 + 114^4$$

by transposing x and y, or multiplying them by ± 1, or both.

Exercise 10.6

Calculate $r(221)$, and find all representations of 221 as a sum of two squares. (See Example 10.4.)

Exercise 10.7

Calculate $r(16660)$, and find all representations of 16660 as a sum of two squares.

Exercise 10.8

Show that $r(n) = 4(\tau_1(n) - \tau_3(n))$ for all n, where $\tau_1(n)$ and $\tau_3(n)$ denote the number of divisors d of n such that $d \equiv 1$ or $3 \bmod (4)$ respectively (see Chapter 8, Exercise 8.21).

10.3 Sums of three squares

Gauss proved that $n \in S_3$ if and only if $n \ne 4^e(8k + 7)$; thus $7, 15, 23, 28, \ldots$ are not sums of three squares. It is a simple exercise (see below) to prove that no integer $n = 4^e(8k + 7)$ can be a sum of three squares. The converse, which we will omit for lack of space, is rather harder, mainly because the set S_3 is not closed under multiplication: for instance 3 and 5 are sums of three squares, but 15 is not.

Exercise 10.9

Show that if $n \in S_3$ then $n \not\equiv 7 \bmod (8)$.

Exercise 10.10

Show that if $n \in S_3$ and n is divisible by 4, then $n/4 \in S_3$.

Exercise 10.11

Deduce that if $n = 4^e(8k + 7)$ then $n \notin S_3$.

Exercise 10.12

In how many ways can 14 and 11 be written as sums of three squares?

10.4 Sums of four squares

Perhaps surprisingly, it is easier to deal with sums of four squares: first we need the following result.

Lemma 10.5

The set S_4, consisting of the sums of four squares, is closed under multiplication, that is, if $s, t \in S_4$ then $st \in S_4$.

Proof

This follows immediately from the identity

$$
\begin{aligned}
(a_1^2 + b_1^2 + c_1^2 + d_1^2)(a_2^2 + b_2^2 + c_2^2 + d_2^2) = \ &(a_1a_2 - b_1b_2 - c_1c_2 - d_1d_2)^2 \\
&+ (a_1b_2 + b_1a_2 + c_1d_2 - d_1c_2)^2 \\
&+ (a_1c_2 - b_1d_2 + c_1a_2 + d_1b_2)^2 \\
&+ (a_1d_2 + b_1c_2 - c_1b_2 + d_1a_2)^2,
\end{aligned}
$$

$$(10.3)$$

which can be proved directly (at some length) by expanding each side. (We will shortly give an alternative explanation of this identity in terms of quaternions.)

$\square$

Comment

By replacing b_1, c_1 and d_1 with $-b_1, -c_1$ and $-d_1$, we obtain the identity

$$
\begin{aligned}
(a_1^2 + b_1^2 + c_1^2 + d_1^2)(a_2^2 + b_2^2 + c_2^2 + d_2^2) \;=\; & (a_1 a_2 + b_1 b_2 + c_1 c_2 + d_1 d_2)^2 \\
& + (a_1 b_2 - b_1 a_2 - c_1 d_2 + d_1 c_2)^2 \\
& + (a_1 c_2 + b_1 d_2 - c_1 a_2 - d_1 b_2)^2 \\
& + (a_1 d_2 - b_1 c_2 + c_1 b_2 - d_1 a_2)^2,
\end{aligned}
$$

$$(10.4)$$

which we will need later.

The following *Four Squares Theorem* was proved by Lagrange in 1770.

Theorem 10.6

Every non-negative integer is a sum of four squares.

Proof

Clearly $0, 1, 2 \in S_4$, so by Lemma 10.5 it is sufficient to prove that every odd prime p is in S_4. We do this by following the method of proof of Theorem 10.2 as far as possible. First we show that some positive multiple of p is in S_4.

Of the elements of $\mathbb{Z}_p$, exactly $(p+1)/2$ are squares, namely 0 and the $(p-1)/2$ quadratic residues in Q_p. Thus the set $K = \{z \in \mathbb{Z}_p \mid z = k^2, \; k \in \mathbb{Z}_p\}$ contains $(p+1)/2$ elements, and a similar argument shows that the set $L = \{z \in \mathbb{Z}_p \mid z = -1 - l^2, \; l \in \mathbb{Z}_p\}$ also has $(p+1)/2$ elements. Each of these two subsets thus accounts for more than half of the elements of $\mathbb{Z}_p$, so their intersection $K \cap L$ is non-empty. This means that there exist integers $u, v \in \mathbb{Z}$ such that $u^2 \equiv -1 - v^2 \bmod (p)$, that is, $u^2 + v^2 + 1 = rp$ for some integer $r > 0$. (As an example, if $p = 7$ then $K = \{0, 1, 2, 4\}$ and $L = \{2, 4, 5, 6\}$ so we can take $u^2 \equiv -1 - v^2 \equiv 2$ (or 4) $\bmod (7)$, say $u = 3$ and $v = 2$ with $u^2 + v^2 + 1 = 14$.) Since $u^2 + v^2 + 1 = u^2 + v^2 + 1^2 + 0^2$, we have shown that some positive multiple rp of p is in S_4. By replacing u and v with their least absolute residues $\bmod (p)$ we may assume that $|u|, |v| \le p/2$, so that $r < p$. It follows that there exists a least positive integer $m < p$ such that $mp \in S_4$, say

$$
mp = a_1^2 + b_1^2 + c_1^2 + d_1^2. \tag{10.5}
$$

If $m = 1$ then $p \in S_4$ and we are home, so assume that $m > 1$.

Imitating the proof of Theorem 10.2, we take the least absolute residues a_2, b_2, c_2, d_2 of $a_1, b_1, c_1, d_1 \bmod (m)$, so $a_2 \equiv a_1 \bmod (m)$, etc., and $|a_2|, |b_2|,$

$|c_2|, |d_2| \leq m/2$. We have $a_2^2 + b_2^2 + c_2^2 + d_2^2 \equiv a_1^2 + b_1^2 + c_1^2 + d_1^2 \equiv 0 \bmod (m)$, so

$$a_2^2 + b_2^2 + c_2^2 + d_2^2 = sm \tag{10.6}$$

for some integer s. Now we would like to be able to assert that $s < m$, so that the proof could be completed as in Theorem 10.2; unfortunately, our bound on a_2, b_2, c_2 and d_2 merely implies that $sm \leq 4.(m/2)^2 = m^2$, so that $s \leq m$, which is not strong enough. However, if m is odd then since least absolute residues are integers we actually have $|a_2|, |b_2|, |c_2|, |d_2| < m/2$, so $s < m$ as required. We therefore need to eliminate the possibility that m is even. If m is even, then equation (10.5) implies that all, or two, or none of a_1, b_1, c_1 and d_1 are odd; by renaming these variables we may assume that a_1 and b_1 have the same parity, as do c_1 and d_1, that is, $a_1 \pm b_1$ and $c_1 \pm d_1$ are all even. But then

$$\left(\frac{a_1 + b_1}{2}\right)^2 + \left(\frac{a_1 - b_1}{2}\right)^2 + \left(\frac{c_1 + d_1}{2}\right)^2 + \left(\frac{c_1 - d_1}{2}\right)^2 = \frac{a_1^2 + b_1^2 + c_1^2 + d_1^2}{2} = \frac{mp}{2}$$

is an element of S_4 which is a positive multiple of p, contradicting the minimality of m. Thus m is odd, so $s < m$ as shown above.

We now show that $s > 0$. If $s = 0$, then equation (10.6) implies that $a_2 = b_2 = c_2 = d_2 = 0$, so a_1, b_1, c_1 and d_1 are all divisible by m, and equation (10.5) implies that p is divisible by m. This is impossible, since p is prime and $1 < m < p$, so we must have $s > 0$.

Equations (10.5) and (10.6) show that

$$(a_1^2 + b_1^2 + c_1^2 + d_1^2)(a_2^2 + b_2^2 + c_2^2 + d_2^2) = mp.sm = m^2 sp,$$

and we can use identity (10.4) to write this as

$$\begin{aligned} m^2 sp = \ & (a_1 a_2 + b_1 b_2 + c_1 c_2 + d_1 d_2)^2 + (a_1 b_2 - b_1 a_2 - c_1 d_2 + d_1 c_2)^2 \\ & + (a_1 c_2 + b_1 d_2 - c_1 a_2 - d_1 b_2)^2 + (a_1 d_2 - b_1 c_2 + c_1 b_2 - d_1 a_2)^2. \end{aligned} \tag{10.7}$$

The congruences $a_1 \equiv a_2 \bmod (m)$, etc., together with (10.6), show that

$$a_1 a_2 + b_1 b_2 + c_1 c_2 + d_1 d_2 \equiv a_2^2 + b_2^2 + c_2^2 + d_2^2 \equiv 0 \bmod (m),$$

$$a_1 b_2 - b_1 a_2 - c_1 d_2 + d_1 c_2 \equiv a_2 b_2 - b_2 a_2 - c_2 d_2 + d_2 c_2 \equiv 0 \bmod (m),$$

etc., so each bracketed term on the right-hand side of (10.7) is divisible by m. We can therefore rewrite (10.7) as

$$\begin{aligned} sp = \ & \left(\frac{a_1 a_2 + b_1 b_2 + c_1 c_2 + d_1 d_2}{m}\right)^2 + \left(\frac{a_1 b_2 - b_1 a_2 - c_1 d_2 + d_1 c_2}{m}\right)^2 \\ & + \left(\frac{a_1 c_2 + b_1 d_2 - c_1 a_2 - d_1 b_2}{m}\right)^2 + \left(\frac{a_1 d_2 - b_1 c_2 + c_1 b_2 - d_1 a_2}{m}\right)^2, \end{aligned}$$

so that sp is a positive multiple of p contained in S_4. Since $s < m$ this contradicts the minimality of m, so $m = 1$ and the result is proved. (We will give an alternative geometric proof later in this chapter.) □

Exercise 10.13

Express the following integers as sums of four squares: 247, 308, 465.

Exercise 10.14

In how many ways can 28 be written as a sum of four squares?

10.5 Digression on quaternions

Just as the two-squares identity (10.1) in Lemma 10.1 can be explained in terms of complex numbers, the four-squares identity (10.3) in the proof of Lemma 10.5 can be derived from a generalisation of complex numbers known as the quaternions. In the first half of the 19th century, Hamilton tried to find a 3-dimensional number system which would model the real world $\mathbb{R}^3$ in the same way as the 2-dimensional system $\mathbb{C}$ of complex numbers is an algebraic model of the plane $\mathbb{R}^2$. He wanted this system to retain as many as possible of the basic properties of $\mathbb{C}$, and in particular he wanted the length of a product to be equal to the product of the lengths of its factors (see Comment 2 following Lemma 10.1). After several years without success, he eventually realised in 1843 that this property would require a 4-dimensional system, rather than one based on $\mathbb{R}^3$. Its elements, which Hamilton called *quaternions*, are the points $q = (a, b, c, d) \in \mathbb{R}^4$, and addition and subtraction are performed by the usual method for vectors. To define multiplication, it is useful to write each quaternion in the form

$$q = a + bi + cj + dk = a1 + bi + cj + dk,$$

where $a, b, c, d \in \mathbb{R}$ and $1, i, j, k$ denote the standard basis vectors of $\mathbb{R}^4$. (This is analogous to identifying each point $(a, b) \in \mathbb{R}^2$ with the complex number $a + bi = a1 + bi$.) Hamilton defined the products of the basis vectors by

$$i^2 = j^2 = k^2 = -1, \quad ij = k = -ji, \quad jk = i = -kj, \quad ki = j = -ik,$$

together with the rule that $1^2 = 1, 1i = i = i1$ and so on. (Notice that multiplication is *not* commutative, since $ij \neq ji$ for example.) By assuming distributivity (that is, $q(q' + q'') = qq' + qq''$ and $(q' + q'')q = q'q + q''q$ for all q, q', q''), we find that the product of any pair of quaternions

$$q_1 = a_1 + b_1i + c_1j + d_1k \quad \text{and} \quad q_2 = a_2 + b_2i + c_2j + d_2k$$

is given by

$$
\begin{aligned}
q_1 q_2 &= (a_1 a_2 - b_1 b_2 - c_1 c_2 - d_1 d_2) + (a_1 b_2 + b_1 a_2 + c_1 d_2 - d_1 c_2)\mathrm{i} \\
&\quad + (a_1 c_2 - b_1 d_2 + c_1 a_2 + d_1 b_2)\mathrm{j} + (a_1 d_2 + b_1 c_2 - c_1 b_2 + d_1 a_2)\mathrm{k}.
\end{aligned}
$$

$$(10.8)$$

The *conjugate* of a quaternion $q = a + b\mathrm{i} + c\mathrm{j} + d\mathrm{k}$ is the quaternion $\bar{q} = a - b\mathrm{i} - c\mathrm{j} - d\mathrm{k}$, and the *length* $|q|$ of q is given by

$$
|q| = \sqrt{a^2 + b^2 + c^2 + d^2}.
$$

Exercise 10.15

Verify that $|q|^2 = q\bar{q}$ for all quaternions q, and that $\overline{q_1 q_2} = \bar{q_2}.\bar{q_1}$ for all quaternions q_1, q_2.

Exercise 10.16

Use Exercise 10.15 to prove that $|q_1 q_2|^2 = |q_1|^2.|q_2|^2$, and deduce identity (10.3).

The quaternion number system is usually denoted by $\mathbb{H}$, in honour of Hamilton. He wrote two books and numerous papers on quaternions, exploiting their 4-dimensional nature to study space and time simultaneously. Soon after Hamilton's discovery of the quaternions, Cayley and Graves independently discovered a non-associative 8-dimensional system $\mathbb{O}$, the *octonions*, which leads to an eight-squares identity, analogous to those we have seen for two and four squares. Hamilton's failure to find a 3-dimensional number system was not due to any lack of effort or ability on his part: in 1878 Frobenius proved that $\mathbb{R}, \mathbb{C}$ and $\mathbb{H}$ are the only systems with the required properties (to be precise, these are the only finite-dimensional associative division algebras over $\mathbb{R}$), and in 1898 Hurwitz showed there is a k-squares identity of the required form only for $k = 1, 2, 4$ and 8. These facts help to explain why sums of three squares are harder to study than sums of two or four squares. For more on quaternions and related number systems, see Ebbinghaus *et al.* (1991).

10.6 Minkowski's Theorem

We will now reconsider some of the preceding results from a geometric point of view. The proof of Theorem 7.11 (quadratic reciprocity) involves the counting of lattice-points in a subset of Euclidean space. This idea is useful elsewhere, for instance in studying the function $r(n)$ considered in Section 2 (see Exercise 10.26). Before applying lattices to sums of squares, we first need to study their properties a little more formally.

Definition

A *lattice* in $\mathbb{R}^n$ is a set of the form

$$\Lambda = \{\alpha_1 v_1 + \cdots + \alpha_n v_n \mid \alpha_i \in \mathbb{Z}\}$$

where $v_1, \ldots, v_n$ form a basis for the vector space $\mathbb{R}^n$. We then call $v_1, \ldots, v_n$ a *basis* for Λ.

Example 10.9

If $n = 1$ then $\mathbb{R}^n = \mathbb{R}$ and $\Lambda = \{\alpha_1 v_1 \mid \alpha_1 \in \mathbb{Z}\}$ for some non-zero $v_1 \in \mathbb{R}$, so Λ is the subgroup of $\mathbb{R}$ generated by v_1. If $v_1 = 1$ or -1, for instance, we get $\Lambda = \mathbb{Z}$.

Example 10.10

If $n = 2$ and we choose v_1 and v_2 to be the standard basis vectors $(1, 0)$ and $(0, 1)$ of $\mathbb{R}^2$, then $\Lambda = \{(\alpha_1, \alpha_2) \mid \alpha_1, \alpha_2 \in \mathbb{Z}\}$ is the *square lattice*, or *integer lattice* $\mathbb{Z}^2 \subset \mathbb{R}^2$.

Example 10.11

Similarly, if we choose v_1, v_2 and v_3 to be the standard basis vectors for $\mathbb{R}^3$ then Λ is the *simple cubic lattice* $\mathbb{Z}^3 \subset \mathbb{R}^3$, which plays a major role in crystallography.

Lemma 10.7

If Λ is a lattice in $\mathbb{R}^n$, then Λ is a subgroup of $\mathbb{R}^n$ under addition.

Proof

Let $v_1, \ldots, v_n$ be a basis for Λ. Clearly the zero vector $0 = \sum 0.v_i$ is in Λ. If $v = \sum \alpha_i v_i$ and $w = \sum \beta_i v_i$ are in Λ then $\alpha_i, \beta_i \in \mathbb{Z}$ for all i, so $\alpha_i - \beta_i \in \mathbb{Z}$ and hence $v - w = \sum (\alpha_i - \beta_i) v_i \in \Lambda$. $\qquad \square$

Definition

If Λ is a lattice in $\mathbb{R}^n$, then vectors $v, w \in \mathbb{R}^n$ are *equivalent* (modulo Λ), written $v \sim w$, if $v - w \in \Lambda$. It follows from Lemma 10.7 that $\sim$ is an equivalence relation; the equivalence classes are simply the cosets $\Lambda + v = v + \Lambda$ of the subgroup Λ in the group $\mathbb{R}^n$. If $v_1, \ldots, v_n$ is a basis for Λ, we call the set

$$F = \{\alpha_1 v_1 + \cdots + \alpha_n v_n \mid 0 \leq \alpha_i < 1\}$$

a *fundamental region* for Λ; the sets $F + l$ $(l \in \Lambda)$ tessellate $\mathbb{R}^n$, that is, they cover $\mathbb{R}^n$ without overlapping. This is equivalent to the following property:

Lemma 10.8

For each $v \in \mathbb{R}^n$ there is a unique $w \in F$ with $v \sim w$.

Proof

Let $v = \sum \alpha_i v_i \in \mathbb{R}^n$, so each $\alpha_i \in \mathbb{R}$. If we define $\beta_i = \alpha_i - \lfloor \alpha_i \rfloor$, the fractional part of α_i, and put $w = \sum \beta_i v_i$, then $w \in F$ since $0 \leq \beta_i < 1$ for all i, and $w = \sum \alpha_i v_i - \sum \lfloor \alpha_i \rfloor v_i = v - l$ with $l = \sum \lfloor \alpha_i \rfloor v_i \in \Lambda$, so $v \sim w$. For the uniqueness of w, suppose that we also have $v \sim w' \in F$. We have $w' = \sum \beta_i' v_i$ where $0 \leq \beta_i' < 1$ for each i, so $|\beta_i - \beta_i'| < 1$. Since v is equivalent to both w and w', we have $w \sim w'$, so $w - w' \in \Lambda$ and hence $\beta_i - \beta_i' \in \mathbb{Z}$. It follows that $\beta_i = \beta_i'$ for all i, so $w = w'$ as required. $\qquad\square$

Comment

An alternative interpretation of this result is that each $v \in \mathbb{R}^n$ lies in a set $F + l = \{f + l \mid f \in F\}$ for a unique $l \in \Lambda$ (namely, $l = v - w$, so that $v = w + l \in F + l$). These sets $F + l$ are called *translates* of F, since they are obtained from F by translating F by l. Lemma 10.8 then asserts that these translates tessellate $\mathbb{R}^n$, that is, they cover $\mathbb{R}^n$ without overlapping.

We can use Lemma 10.8 to define a function $\phi : \mathbb{R}^n \to F$ by $\phi(v) = w$, where $v \sim w \in F$; thus w is the unique coset representative for $v + \Lambda$ contained in F. We can apply ϕ to a subset $X \subset \mathbb{R}^n$ by dividing X into regions $X \cap (F+l)$, one in each translate of F, and then translating each region by $-l$ to $(X - l) \cap F \subseteq F$. Note that the images $(X - l) \cap F$ of different regions may overlap if X is sufficiently large (see Lemma 10.9).

To make this last remark more precise we define the n-dimensional volume of a set $X \subset \mathbf{R}^n$ to be

$$\mathrm{vol}(X) = \int\!\!\int \cdots \int 1 \, dx_1 \, dx_2 \, \ldots \, dx_n \,,$$

provided this exists and is finite, where the integration is over all $(x_1, x_2, \ldots, x_n) \in X$. When $n = 1, 2$ or 3 this represents the length, area or volume of X.

Example 10.12

If X is the n-dimensional unit cube, defined by $0 \leq x_i \leq 1$ for $i = 1, \ldots, n$, then $\mathrm{vol}(X) = 1$. Similarly the subset $C \subset X$ defined by $0 \leq x_i < 1$, which is a fundamental region for the integer lattice $\Lambda = \mathbf{Z}^n$, also has $\mathrm{vol}(C) = 1$.

An important example we will need is the n-dimensional open ball $B_n(r)$ of radius r, the subset of $\mathbf{R}^n$ defined by $x_1^2 + \cdots + x_n^2 < r^2$.

Exercise 10.17

Let $V_n = \mathrm{vol}\big(B_n(1)\big)$, the volume of the n-dimensional unit ball. By considering cross-sections $x_n = x$ for $-1 \leq x \leq 1$, show that

$$V_n = \int_{-1}^{1} V_{n-1}(1 - x^2)^{(n-1)/2} dx = 2V_{n-1}I_n$$

for all $n \geq 2$, where $I_n = \int_0^{\pi/2} \sin^n \theta \, d\theta$ satisfies the reduction formula $I_n = (n-1)I_{n-2}/n$. By evaluating V_1, I_0 and I_1 directly, show that

$$V_n = \begin{cases} (2\pi)^m/n(n-2)\ldots 4.2 & \text{if } n = 2m \text{ is even,} \\[2mm] 2^{m+1}\pi^m/n(n-2)\ldots 3.1 & \text{if } n = 2m+1 \text{ is odd.} \end{cases}$$

(For those familiar with the gamma function, one can write this more concisely as $V_n = \pi^{n/2}/\Gamma(\frac{n}{2} + 1)$.)

Exercise 10.18

Deduce that the n-dimensional open ball $B_n(r)$ of radius r has volume $2r, \pi r^2, 4\pi r^3/3$ or $\pi^2 r^4/2$ for $n = 1, 2, 3$ or 4.

Exercise 10.19

By inscribing an octagon in a disc, show that $\pi > 2\sqrt{2}$. (We will need this inequality later.)

Exercise 10.20

Prove that the ellipse $x^2/a^2 + y^2/b^2 = 1$ encloses a set $X \subset \mathbb{R}^2$ (defined by $x^2/a^2 + y^2/b^2 < 1$) satisfying $\mathrm{vol}(X) = \pi ab$.

Lemma 10.9

If $\mathrm{vol}(X) > \mathrm{vol}(F)$ then the restriction $\phi|_X$ of ϕ to X is not one-to-one.

   (In other words, if X is sufficiently large then there are at least two distinct equivalent points in X.)

Proof

Because the translates $F+l$ $(l \in \Lambda)$ tessellate $\mathbb{R}^n$, it follows that X is the disjoint union of the subsets $X_l = X \cap (F + l)$ $(l \in \Lambda)$. If $v \in X_l$ then $\phi(v) = v - l$, so ϕ translates X_l to a congruent subset $\phi(X_l) = X_l - l$ of F. Since translations preserve volumes, we have $\mathrm{vol}(\phi(X_l)) = \mathrm{vol}(X_l)$. Now

$$\mathrm{vol}(X) = \sum_{l \in \Lambda} \mathrm{vol}(X_l) = \sum_{l \in \Lambda} \mathrm{vol}(\phi(X_l)).$$

If $\phi|_X$ is one-to-one then the translates $\phi(X_l)$ cannot overlap, so

$$\sum_{l \in \Lambda} \mathrm{vol}(\phi(X_l)) \leq \mathrm{vol}(F)$$

and hence $\mathrm{vol}(X) \leq \mathrm{vol}(F)$, against our assumption.   □

Definition

A subset X of $\mathbb{R}^n$ is *centrally symmetric* if, whenever $v \in X$, we also have $-v \in X$; it is *convex* if, whenever $v, w \in X$, the line-segment vw also lies in X, that is, $tv + (1 - t)w \in X$ for all t such that $0 \leq t \leq 1$.

Example 10.13

$B_n(r)$ is centrally symmetric, since if $x_1^2 + \cdots + x_n^2 < r^2$ then $(-x_1)^2 + \cdots + (-x_n)^2 < r^2$. Similarly, the region $x^2/a^2 + y^2/b^2 < 1$ bounded by an ellipse is centrally symmetric.

Exercise 10.21

Show that $B_n(r)$ is convex.

The following theorem, proved by Minkowski around 1890, has some far-reaching consequences.

Theorem 10.10

Let Λ be a lattice in $\mathbb{R}^n$ with fundamental region F, and let X be a centrally symmetric convex set in $\mathbb{R}^n$ with $\mathrm{vol}(X) > 2^n \mathrm{vol}(F)$. Then X contains a non-zero lattice-point of Λ.

Proof

The lattice $2\Lambda = \{2v \mid v \in \Lambda\}$ has a fundamental region $2F = \{2v \mid v \in F\}$ with $\mathrm{vol}\,(2F) = 2^n \mathrm{vol}\,(F)$. Thus $\mathrm{vol}\,(X) > \mathrm{vol}\,(2F)$, so by applying Lemma 10.9 to X and 2Λ we see that there exist $v \neq w$ in X with $v - w \in 2\Lambda$. Since $w \in X$ and X is centrally symmetric, we have $-w \in X$. Since X is convex and $v, -w \in X$, the midpoint $\frac{1}{2}(v - w)$ of the line-segment from v to $-w$ is also in X. Now $v - w \in 2\Lambda$, so $\frac{1}{2}(v - w) \in \Lambda$, giving the required non-zero lattice-point in X. $\square$

Example 10.14

The lattice $\Lambda = \mathbb{Z}^n$ has a fundamental region F (such as the unit cube in $\mathbb{R}^n$) with $\mathrm{vol}(F) = 1$. The set $X = \{\sum \alpha_i v_i \mid |\alpha_i| < 1\}$ is centrally symmetric and convex, with $\mathrm{vol}(X) = 2^n = 2^n \mathrm{vol}(F)$, but X contains no non-zero lattice-points. This shows that Minkowski's Theorem fails if we relax the lower bound on $\mathrm{vol}(X)$.

Exercise 10.22

By finding suitable counterexamples, show that Minkowski's Theorem fails if either of the conditions 'centrally symmetric' or 'convex' is omitted.

In order to apply Minkowski's Theorem one needs to be able to calculate volumes of fundamental regions. This is easily done using determinants. Suppose that Λ has a basis $\{v_1, \ldots, v_n\}$, where each $v_i = (\alpha_{i1}, \ldots, \alpha_{in}) \in \mathbb{R}^n$. If A is the $n \times n$ matrix (α_{ij}) formed from these vectors v_i (as row or column vectors), then

$$\mathrm{vol}(F) = |\det(A)|\,.$$

This is because the linear transformation $\mathbb{R}^n \to \mathbb{R}^n$ induced by A sends the standard basis vectors e_i of $\mathbb{R}^n$ to the basis vectors v_i of Λ, and hence sends the set $C = \{\alpha_1 e_1 + \cdots + \alpha_n e_n \mid 0 \le \alpha_i < 1\}$ to the fundamental region $F = \{\alpha_1 v_1 + \cdots + \alpha_n v_n \mid 0 \le \alpha_i < 1\}$ for Λ. Since $\mathrm{vol}(C) = 1$, and since any linear transformation A multiplies volumes by $|\det(A)|$, it follows that $\mathrm{vol}(F) = |\det(A)|$.

We now apply these ideas to give another proof of Theorem 10.2, the *Two Squares Theorem*, which we state again:

Theorem 10.2

Each prime $p \equiv 1 \bmod (4)$ is a sum of two squares.

Proof

Since $p \equiv 1 \bmod (4)$, Corollary 7.7 implies that $u^2 \equiv -1 \bmod (p)$ for some integer u. Now suppose that the following condition is true:

there exist $x, y \in \mathbb{Z}$ with $y \equiv ux \bmod (p)$ and $0 < x^2 + y^2 < 2p$.
$$(10.9)$$
Then $x^2 + y^2 \equiv x^2 + u^2 x^2 \equiv x^2 - x^2 \equiv 0 \bmod (p)$, so $x^2 + y^2 = kp$ for some integer k. The inequalities in (10.9) become $0 < kp < 2p$, so $k = 1$ and $x^2 + y^2 = p$, as required. It is therefore sufficient to prove (10.9). We can do this using Minkowski's Theorem, since the first condition $y \equiv ux \bmod (p)$ in (10.9) defines a lattice Λ in $\mathbb{R}^2$ (as we shall prove below), the condition $x^2 + y^2 < 2p$ defines a centrally symmetric convex set X, namely the disc $B_2(\sqrt{2p})$, and the condition $0 < x^2 + y^2$ specifies a non-zero point (x, y); Minkowski's Theorem then guarantees the existence of a point (x, y) satisfying all three of these conditions, so (10.9) is proved.

To justify this, we must verify all the hypotheses in Minkowski's Theorem. First let

$$\Lambda = \{(x, y) \in \mathbb{Z}^2 \mid y \equiv ux \bmod (p)\}.$$

It is easily checked that this is a subgroup of $\mathbb{R}^2$ containing the linearly independent vectors $v_1 = (1, u)$ and $v_2 = (0, p)$. If $(x, y) \in \Lambda$ then let $\alpha_1 = x$ and $\alpha_2 = (y - ux)/p$; these are integers, with $\alpha_1 v_1 + \alpha_2 v_2 = (x, y)$, so Λ is generated by v_1 and v_2. Thus Λ is a lattice with v_1 and v_2 forming a basis, so it has a fundamental region F with 2-dimensional volume (or area)

$$\mathrm{vol}(F) = \left| \det \begin{pmatrix} 1 & u \\ 0 & p \end{pmatrix} \right| = p.$$

Now let $X = B_2(\sqrt{2p}) = \{(x, y) \in \mathbb{R}^2 \mid x^2 + y^2 < 2p\}$, an open disc of radius $r = \sqrt{2p}$ centred at the origin. This is centrally symmetric and convex, and Exercise 10.18 gives $\mathrm{vol}(X) = \pi r^2 = 2\pi p$. Now $\pi > 2\sqrt{2} > 2$ by Exercise

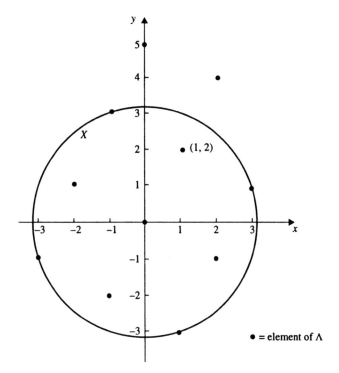

Figure 10.1. The proof of Theorem 10.2 for $p = 5$, with $u = 2$.

10.19, so $\mathrm{vol}(X) > 2^2\mathrm{vol}(F)$, and hence Minkowski's Theorem gives a non-zero lattice point $(x, y) \in X \cap \Lambda$. (Figure 10.1 illustrates this in the case $p = 5$, with $u = 2$ and $(x, y) = (1, 2)$.) We now have a pair of integers x and y satisfying (10.9), so the proof is complete. $\qquad\qquad\qquad\qquad\qquad\qquad\qquad\qquad\qquad\quad$ $\square$

We can also use this method to prove Theorem 10.6, the Four Squares Theorem.

Theorem 10.6

Every non-negative integer is a sum of four squares.

Proof

As in our earlier proof of Theorem 10.6, in Section 4, it is sufficient to prove that every odd prime p is a sum of four squares. First we show (as before) that there exist integers u, v satisfying $u^2 + v^2 \equiv -1 \bmod (p)$. For such a pair u, v

let

$$\Lambda = \{(x,y,z,t) \in \mathbb{Z}^4 \mid z \equiv ux + vy \text{ and } t \equiv vx - uy \mod (p)\}.$$

Exercise 10.23

Show that Λ is a lattice in $\mathbb{R}^4$ with basis

$$v_1 = (1,0,u,v), \quad v_2 = (0,1,v,-u), \quad v_3 = (0,0,p,0), \quad v_4 = (0,0,0,p).$$

Continuing the proof, we deduce from Exercise 10.23 that a fundamental region F for Λ has volume

$$\text{vol}(F) = \left| \det \begin{pmatrix} 1 & 0 & u & v \\ 0 & 1 & v & -u \\ 0 & 0 & p & 0 \\ 0 & 0 & 0 & p \end{pmatrix} \right| = p^2.$$

Now let $X = B_4(\sqrt{2p}) = \{(x,y,z,t) \in \mathbb{R}^4 \mid x^2 + y^2 + z^2 + t^2 < 2p\}$, an open ball of radius $r = \sqrt{2p}$. This is centrally symmetric and convex, with 4-dimensional volume

$$\text{vol}(X) = \frac{\pi^2 r^4}{2} = 2\pi^2 p^2$$

by Exercise 10.18. Now Exercise 10.19 gives $\pi^2 > 8$, so $\text{vol}(X) > 2^4 \text{vol}(F)$ and hence Minkowski's Theorem implies that X contains a non-zero lattice point. Thus there exist integers x, y, z, t such that $0 < x^2 + y^2 + z^2 + t^2 < 2p$ and $z \equiv ux + vy$, $t \equiv vx - uy \mod (p)$. Then

$$\begin{aligned}
x^2 + y^2 + z^2 + t^2 &\equiv x^2 + y^2 + u^2 x^2 + 2uvxy + v^2 y^2 + v^2 x^2 - 2uvxy + u^2 y^2 \\
&\equiv (1 + u^2 + v^2)(x^2 + y^2) \\
&\equiv 0 \mod (p)
\end{aligned}$$

since $u^2 + v^2 \equiv -1 \mod (p)$, so $x^2 + y^2 + z^2 + t^2 = p$ and the proof is complete.
□

10.7 Supplementary exercises

Exercise 10.24

Show that an odd prime p can be written in the form $2x^2 + y^2$ if and only if $-2 \in Q_p$, or equivalently $p \equiv 1$ or $3 \mod (8)$. (Hint: apply Minkowski's Theorem, with X the interior of an ellipse.)

Exercise 10.25

Show that a prime p can be written in the form $x^2 + xy + y^2$ if and only if $p = 3$ or $p \equiv 1 \bmod (3)$.

Exercise 10.26

Show that the number $r(n)$ of representions of n as a sum of two squares has average value π, that is,

$$\frac{1}{n} \sum_{m=1}^{n} r(m) \to \pi \quad \text{as} \quad n \to \infty.$$

(Hint: representations $m = x^2 + y^2$ of integers $m \leq n$ correspond to integer lattice points (x, y) within distance $\sqrt{n}$ of the origin.) What can be said about the average number of representations of n as a sum of k squares?

Exercise 10.27

Show that $\sum_{n=1}^{\infty} r(n)/n^s = 4L(s)\zeta(s)$ for all $s > 1$, where $L(s) = 1^{-s} - 3^{-s} + 5^{-s} - \cdots$.

11
Fermat's Last Theorem

In this final chapter, we will discuss one of the classic problems of number theory, whose solution in 1993 by Andrew Wiles must be considered one of the greatest achievements of modern mathematics. Although the problem was first posed in the 17th century, its roots can be traced back, through the Greek mathematicians Diophantos and Pythagoras, to the unknown Babylonian mathematicians who recorded their results on clay tablets nearly four thousand years ago.

11.1 The problem

Pierre de Fermat (1601–1665) was a judge, living in the French city of Toulouse. Although mathematics was not his profession, and although he published virtually nothing during his life (preferring to communicate his results in letters to colleagues throughout Europe), he made fundamental contributions in areas such as calculus, probability theory and number theory, and he is generally regarded as one of the greatest of all mathematicians. *Fermat's Last Theorem* (which we will abbreviate to FLT) is the following assertion, which he wrote in the margin of his copy of Bachet's Latin translation of the *Arithmetica* of Diophantos around 1637:

Theorem 11.1

There are no positive integer solutions a, b, c of the equation

$$a^n + b^n = c^n \tag{11.1}$$

for integers $n \geq 3$.

He gave no proof, claiming that the margin was too small (a phrase which has since become a classic excuse for failing to justify a mathematical statement). Indeed it is not at all clear whether Fermat had a valid proof (most experts think not), though for a while he clearly thought so; it might therefore be more correct to call FLT a conjecture, rather than a theorem. Fermat made many similar number-theoretic assertions, and most of them were later shown to be correct, while a few (such as his conjecture about Fermat primes) were disproved; this one remained the last to be settled, and hence both its name (it was far from the last work Fermat did) and also its status as one of the classic problems of number theory. For over 350 years, some of the greatest mathematicians worked on FLT, occasionally making significant progress without ever achieving a complete proof: several times, proofs were claimed, but none of them survived serious scrutiny. Eventually, in 1993, amid great publicity, a proof was announced by Andrew Wiles, a British mathematician working in Princeton, USA. The full details were published two years later, and although only a handful of mathematicians have had the time and the expertise to check the very lengthy and difficult proof, the general verdict is that this great problem has at last been solved.

For the background to FLT, and an explanation of the condition $n \geq 3$, we must first go back several thousand years, and do some geometry.

11.2 Pythagoras's Theorem

This is one of the most famous results of elementary geometry. It states that if a right-angled triangle has sides a, b and c (the hypotenuse, or longest side), then

$$a^2 + b^2 = c^2 . \tag{11.2}$$

There are many proofs of this. Perhaps the most attractive (and the simplest) is shown in Figure 11.1.

On the left we have a square S with sides of length $a + b$, containing four copies of the right-angled triangle, one in each corner; the region of S not

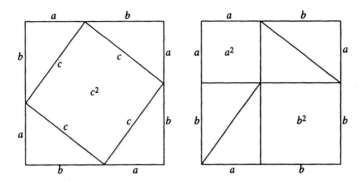

Figure 11.1. The proof of Pythagoras's Theorem.

covered by the triangles is another square, of area c^2. On the right, the four triangles have been moved around within S to form two rectangles; the uncovered region of S now consists of two squares of areas a^2 and b^2. Since moving the triangles leaves their areas unchanged, the two uncovered regions have equal areas, so $a^2 + b^2 = c^2$.

The converse is also true: if a, b and c are positive real numbers satisfying (11.2), then there is a right-angled triangle with sides a, b and c.

Although the theorem is usually associated with the Greek philosopher and mathematician Pythagoras, who lived in the 6th century BC, it is in fact at least a thousand years older; Pythagoras probably learnt of it in his travels in Egypt and the Middle East.

11.3 Pythagorean triples

From the point of view of number theory, there is considerable interest in finding *integer* solutions of equation (11.2). A *Pythagorean triple* is a triple (a, b, c) of positive integers satisfying $a^2 + b^2 = c^2$; these triples correspond to the *Pythagorean triangles*, right-angled triangles whose sides all have integer lengths. It is a classic problem, combining geometry and number theory, to find all such triples. The best-known example is $(3, 4, 5)$, arising from the equation

$$3^2 + 4^2 = 5^2.$$

Multiplying through by $2, 3, \ldots$ we obtain further Pythagorean triples $(6, 8, 10)$, $(9, 12, 15)$, and so on. Similarly the equation $5^2 + 12^2 = 13^2$ gives the triple $(5, 12, 13)$, together with $(10, 24, 26)$, $(15, 36, 39)$, etc.

A Pythagorean triple (a, b, c) and its associated Pythagorean triangle are said to be *primitive* if the integers a, b and c are coprime, that is, $\gcd(a, b, c) = 1$.

Thus $(3, 4, 5)$ and $(5, 12, 13)$ are primitive, whereas the other triples given above are not. It follows easily from equation (11.2) that (a, b, c) is primitive if and only if any two of a, b and c are mutually coprime: a common factor of two of them would also have to divide the third. It is also clear that every Pythagorean triple is a multiple (ma, mb, mc) of a primitive triple (a, b, c) for some integer $m \geq 1$, so to classify the Pythagorean triples it is sufficient to find all the primitive triples.

The importance of the primitive triples seems to have been known to the Babylonians: clay tablet number 322 in the Plimpton collection at the University of Columbia contains a list of primitive Pythagorean triples, including such non-obvious examples as $(4961, 6480, 8161)$. This tablet is believed to date from the period 1900–1600 BC. (Babylonian mathematics at that time was considerably more sophisticated than most people realise: for instance, the Babylonians had an approximation to $\sqrt{2}$ which is correct to six decimal places.)

Exercise 11.1

Verify that $(4961, 6480, 8161)$ is a primitive Pythagorean triple.

Exercise 11.2

Show that neither 1 nor 2 can appear in any Pythagorean triple, but that every integer $k \geq 3$ can appear.

Exercise 11.3

Prove that for each integer k there are only finitely many Pythagorean triples containing k.

Exercise 11.4

Find all the Pythagorean triples containing an integer $k \leq 7$.

Exercise 11.5

Show that if (a, b, c) is a primitive Pythagorean triple, then exactly one of a and b is even, and exactly one of them is divisible by 3; how many of a, b and c can be divisible by 5?

11.4 Isosceles triangles and irrationality

As a slight digression, let us consider whether there can be an 'isosceles' Pythagorean triple (a, b, c), meaning one in which $a = b$, so that the corresponding Pythagorean triangle is isosceles.

Theorem 11.2

There is no Pythagorean triple (a, b, c) with $a = b$.

Proof

The proof is by contradiction. If such a triple (a, a, c) exists, then $c^2 = 2a^2$, so c^2 is even and hence so is c. Putting $c = 2c_1$ (where c_1 is an integer) we get $4c_1^2 = 2a^2$, so $a^2 = 2c_1^2$, showing that a^2 is even and hence so is a. Putting $a = 2a_1$ we see that $c_1^2 = 2a_1^2$, which gives us another isosceles Pythagorean triple (a_1, a_1, c_1) with strictly smaller terms than the first one. Applying this process again to our new triple, we can get a third triple (a_2, a_2, c_2) with yet smaller terms, and by repeating the process we get an infinite sequence of such triples. Their first entries then form a strictly decreasing infinite sequence

$$a > a_1 > a_2 > \ldots$$

of positive integers, which is impossible: any such sequence of integers must sooner or later contain negative terms. (The corresponding sequence of Pythagorean triangles is shown in Figure 11.2; clearly they cannot all have sides of integer lengths.) Thus there can be no isosceles Pythagorean triple. $\square$

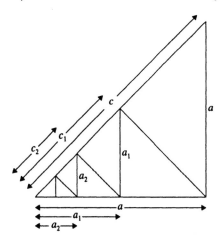

Figure 11.2. A sequence of Pythagorean triangles.

This type of argument has become known as *Fermat's method of descent*: to show that a given equation has no positive integer solutions, we show that any such solution gives rise to a smaller one, and hence (by iteration) to an infinite decreasing sequence of positive integer solutions, which is impossible. Fermat used this technique many times, and we shall see another example of it in Theorem 11.5. In this particular case, the argument also proves that $\sqrt{2}$ must be irrational: if $\sqrt{2} = c/a$ for integers a and c then $c^2 = 2a^2$, which we have shown to be impossible. The discovery of irrational numbers was a great shock to the Pythagoreans, who tried to base their science and philosophy on the properties of the integers and their ratios (the rational numbers). In several areas, such as music, this proved very successful; however, the discovery of the irrationality of one of the most important constants in geometry, the ratio of the diagonal and the side of a square, was so serious that, according to legend, a follower of Pythagoras called Hippasus of Metapontum was deliberately drowned in the Mediterranean either for making the discovery or possibly for publicising the terrible news. The irrationality of $\sqrt{2}$ showed that number theory (as it was then understood) was inadequate to explain geometry, and as a consequence Greek mathematics subsequently split into two fairly separate areas of geometry and number theory. It was only in the late 19th century that the relationship between rational and irrational numbers was satisfactorily explained, when Weierstrass and Dedekind showed how to construct the real numbers from the rationals.

Exercise 11.6

Use Fermat's method of descent to show that there is no Pythagorean triple (a, b, c) in which $c = 2b$, and deduce that $\sqrt{3}$ is irrational. Draw the sequence of Pythagorean triangles which play the role of Figure 11.2 in this situation.

Exercise 11.7

Use Fermat's method of descent to show that there is no Pythagorean triple (a, b, c) in which $a = 2b$, and deduce that $\sqrt{5}$ is irrational.

11.5 The classification of Pythagorean triples

Let us return to our aim of classifying the primitive Pythagorean triples. The solution to this problem was given in the 3rd century AD by Diophantos of Alexandria, in Book II of his *Arithmetica*, and a more geometric version can also be found in Book X of Euclid's *Elements*.

Theorem 11.3

If u and v are coprime positive integers of opposite parity, with $u > v$, then the numbers
$$a = u^2 - v^2, \quad b = 2uv, \quad c = u^2 + v^2 \tag{11.3}$$
form a primitive Pythagorean triple. Conversely, every primitive Pythagorean triple (a, b, c) is given by (11.3) (possibly with a and b transposed) for such a pair u, v.

(This may be the way the Babylonians created their list of triples; for example, if $u = 81$ and $v = 40$ we get the triple $(a, b, c) = (4961, 6480, 8161)$.)

Proof

The numbers a, b and c in (11.3) are positive integers, and one can easily verify that
$$(u^2 - v^2)^2 + (2uv)^2 = (u^2 + v^2)^2$$
for all u and v, so (a, b, c) is a Pythagorean triple. Suppose that (a, b, c) is not primitive, so a, b and c are all divisible by some prime p. If $p = 2$ then a is even; since $a = u^2 - v^2$ it follows that u and v have the same parity, which is false. If p is odd then p divides $(a + c)/2 = u^2$, and hence divides u; it therefore divides $u^2 - a = v^2$ and hence divides v, contradicting the fact that u and v are coprime. In either case we have a contradiction, so (a, b, c) must be primitive.

For the converse, suppose that (a, b, c) is a primitive Pythagorean triple. Since $a^2 + b^2 = c^2$ we have $a^2 + b^2 \equiv c^2 \bmod (4)$. Now $x^2 \equiv 0$ or $1 \bmod (4)$ as x is even or odd, and the only solutions of the equation $[x] + [y] = [z]$ with $[x], [y], [z] = [0]$ or $[1]$ in $\mathbb{Z}_4$ are $[0] + [0] = [0]$, $[0] + [1] = [1]$ and $[1] + [0] = [1]$. Since a and b are not both even, it follows that one is odd and the other is even. Transposing a and b if necessary, we can assume that a is odd and b is even, say $b = 2d$ for some integer d. Then
$$4d^2 = b^2 = c^2 - a^2 = (c + a)(c - a),$$
so at least one of the factors $c \pm a$ is even, and since they differ by $2a$ they are both even. Thus
$$d^2 = \left(\frac{c + a}{2}\right)\left(\frac{c - a}{2}\right),$$

with both factors $(c \pm a)/2$ integers. These factors are coprime, since any common factor would also divide their sum (which is c) and their difference (which is a), and would therefore divide $\gcd(a, c)$, which is 1. Since their product is a perfect square, both of these factors must be perfect squares by Lemma 2.4, say

$$\frac{c+a}{2} = u^2 \quad \text{and} \quad \frac{c-a}{2} = v^2$$

for some positive integers u and v. Adding and subtracting these two equations, we then have $c = u^2 + v^2$ and $a = u^2 - v^2$, while the equations $b = 2d$ and $d^2 = u^2v^2$ imply that $b = 2uv$. Thus equations (11.3) are satisfied, and these show that u and v must be coprime, since any common factor would divide a, b and c. Since a is odd and positive, u and v must have opposite parity with $u > v$. $\square$

This gives us a complete description of the primitive Pythagorean triples, and by taking integer multiples of these we immediately get a description of all the Pythagorean triples:

Corollary 11.4

The general form for a Pythagorean triple (a, b, c) is given by

$$a = m(u^2 - v^2), \quad b = 2muv, \quad c = m(u^2 + v^2)$$

(or possibly with a and b transposed), where u and v are coprime positive integers of opposite parity with $u > v$, and m is a positive integer.

There is an alternative approach, which classifies all *rational* solutions (a, b, c) of equation (11.2), including, of course, the Pythagorean triples. To avoid trivial solutions, let us assume that $b \neq 0$. This implies that $c \neq 0$, so dividing (11.2) by c^2 we get

$$x^2 + y^2 = 1, \tag{11.4}$$

where

$$x = \frac{a}{c} \quad \text{and} \quad y = \frac{b}{c}$$

are both rational numbers. Now (11.4) is the equation of a circle C of radius 1 in the xy-plane, centred at the origin $O = (0, 0)$. If $P = (x, y)$ is any point on C, other than the point $Q = (-1, 0)$, then the line PQ has gradient

$$t = \frac{y}{1 + x},$$

which is a rational number if x and y are both rational. (Note that $t = \tan\theta$, where θ is the angle PQ makes with the x-axis.) Now $y^2 = 1 - x^2 = (1-x)(1+x)$, and so dividing by $(1 + x)^2$ we get

$$t^2 = \frac{1-x}{1+x}.$$

Solving for x we have $(1 + x)t^2 = 1 - x$ and hence

$$x = \frac{1-t^2}{1+t^2},$$

and then the equation $y = (1 + x)t$ gives

$$y = \frac{2t}{1+t^2}.$$

(These are the formulae for $\cos 2\theta$ and $\sin 2\theta$ in terms of $t = \tan\theta$, corresponding to the fact that the radius OP makes an angle 2θ with the x-axis.) If t is any rational number, then this pair of equations defines a rational solution of (11.4), and conversely every rational solution of (11.4) is obtained in this way from a unique rational number t. (Strictly speaking, we have to include $t = \infty$ here, to account for the solution $x = -1, y = 0$; normally, it is dangerous to treat ∞ as if it were a number, but in this particular case it can be justified quite rigorously by taking limits as $t \to \infty$.) We have therefore classified the rational solutions (x, y) of equation (11.4) in terms of a single rational parameter t, and from these we can obtain all rational solutions (a, b, c) of equation (11.2) in the form (cx, cy, c) where c is rational.

We can now deduce the representation (11.3) of a primitive Pythagorean triple (a, b, c), where we assume (as usual) that b is even. Since t is rational we can put $t = v/u$ where u and v are coprime integers, both positive since $y = b/c > 0$ implies that $t > 0$. Then

$$a = cx = \frac{c(1 - t^2)}{1 + t^2} = (u^2 - v^2) \cdot \frac{c}{u^2 + v^2},$$

and similarly

$$b = 2uv \cdot \frac{c}{u^2 + v^2}.$$

Since u and v are coprime, so are uv and $u^2 + v^2$ (see Exercise 11.8); since $b/2$ is an integer, this last equation implies that $u^2 + v^2$ divides c. Then the positive integer $c/(u^2 + v^2)$ is a common factor of a and b, so it must be 1, giving $a = u^2 - v^2$, $b = 2uv$ and $c = u^2 + v^2$.

Exercise 11.8

Show that if u and v are coprime, then uv and $u^2 + v^2$ are also coprime.

11.6 Fermat

The *Arithmetica* contained many problems and solutions similar to this. Diophantos was mainly interested in finding all the *rational* solutions of a given equation, though his name is now attached to the subject of *Diophantine equations*, where the problem is to find all the *integer* solutions. Diophantos wrote in Greek, but in 1621 Claude Gaspard de Bachet published a Latin translation and commentary on the *Arithmetica*, thus making it much more accessible. Fermat, having read how Diophantos solved equation (11.2), was drawn to consider the analogous equation (11.1), where the exponent 2 is replaced with a larger integer n. In his copy of Bachet's book he wrote (in Latin):

"On the other hand, it is impossible to separate a cube into two cubes, or a biquadrate [fourth power] into two biquadrates, or generally any power except a square into two powers with the same exponent. I have discovered a truly marvellous proof of this, which however the margin is not large enough to contain."

In modern terminology, this becomes FLT as stated in Theorem 11.1. In his later correspondence, Fermat stated this result only for the case $n = 3$, where he may or may not have had a proof. He certainly had a proof for $n = 4$ (using his method of descent), and he may at one stage have felt that his method would work for all $n \geq 3$. It seems likely that he soon realised how difficult such an extension would be, and consequently did not repeat his general assertion. After his death, his son Samuel published an edition of Bachet's translation of the *Arithmetica* containing Fermat's comments, including the famous statement of FLT.

In proving FLT, it is not in fact necessary to consider all integers $n \geq 3$. Suppose that FLT is true for some exponent m (so that $x^m + y^m = z^m$ has no positive integer solutions), and that m divides n, say $n = lm$. If a triple (a, b, c) satisfies (11.1) then

$$(a^l)^m + (b^l)^m = (c^l)^m \,,$$

so that by putting $x = a^l$, $y = b^l$ and $z = c^l$ we get a positive integer solution of $x^m + y^m = z^m$, which is impossible; thus FLT is true for exponent n. By the Fundamental Theorem of Arithmetic, every integer $n \geq 3$ is divisible by $m = 4$ or by an odd prime $m = p$, so this argument shows that it is sufficient to prove FLT for exponent 4 and for all odd prime exponents. This is still a major task (partly because there are infinitely many odd primes), but it is somewhat easier than the original problem.

11.7 The case $n = 4$

This is the easiest case of FLT. It is an immediate corollary of the following result (proved by Fermat):

Theorem 11.5

There are no positive integer solutions x, y and z of

$$x^4 + y^4 = z^2 . \tag{11.5}$$

Before proving this, we state its more important consequences:

Corollary 11.6

There are no positive integer solutions a, b and c of the equation $a^4 + b^4 = c^4$.

Proof

If there were a solution, then by putting $x = a$, $y = b$ and $z = c^2$ we would get a positive integer solution of (11.5), which is impossible by Theorem 11.5. $\square$

As shown in the preceding section, this immediately implies FLT for all exponents divisible by 4:

Corollary 11.7

If n is divisible by 4 then there are no positive integer solutions a, b and c of the equation $a^n + b^n = c^n$.

Proof of Theorem 11.5.

We use Fermat's method of descent, as in the proof of Theorem 11.2. If there is a positive integer solution of (11.5), then by dividing through by any common factors we can find a primitive solution (x, y, z), with x, y and z mutually coprime. It follows that (x^2, y^2, z) is a primitive Pythagorean triple, so (transposing x and y if necessary to make y^2 even) we see from Theorem 11.3 that

$$x^2 = u^2 - v^2, \quad y^2 = 2uv, \quad z = u^2 + v^2$$

where u and v are coprime positive integers of opposite parity. The first of these equations can be written in the form $x^2 + v^2 = u^2$, so (x, v, u) is a Pythagorean triple, primitive since u and v are coprime. Since x is odd, Theorem 11.3 therefore gives

$$x = u_1^2 - v_1^2, \quad v = 2u_1v_1, \quad u = u_1^2 + v_1^2$$

for some coprime positive integers u_1 and v_1, so that

$$y^2 = 4u_1v_1(u_1^2 + v_1^2).$$

Now u_1, v_1 and $u_1^2 + v_1^2$ are mutually coprime, so by Lemma 2.4 this equation shows that they must be perfect squares, say

$$u_1 = x_1^2, \quad v_1 = y_1^2, \quad u_1^2 + v_1^2 = z_1^2,$$

and so

$$x_1^4 + y_1^4 = z_1^2.$$

Thus the triple (x_1, y_1, z_1) is another integer solution of (11.5), and since $z_1 > 1$ we have

$$z_1 < z_1^4 = (u_1^2 + v_1^2)^2 = u^2 < u^2 + v^2 = z.$$

By iterating this process, using each solution to create a smaller solution, we get an infinite sequence of integer solutions (x_n, y_n, z_n) of (11.5), whose third terms z_n form an infinite decreasing sequence of positive integers. This is impossible, so no positive integer solutions of (11.5) can exist.   □

Exercise 11.9

Prove that there are no positive integer solutions of the Diophantine equation $x^4 - y^4 = z^2$.

Exercise 11.10

Deduce from the previous exercise that the area of a Pythagorean triangle cannot be a perfect square. (These results are both due to Fermat.)

11.8 Odd prime exponents

We have now reduced FLT to the cases where the exponent is an odd prime p, the problem being to show that the equation

$$a^p + b^p = c^p \tag{11.6}$$

has no positive integer solutions. Here progress is much more difficult, and we will merely outline some of the methods used and the main results obtained.

We saw in Chapter 3 that an effective way of proving that an equation has no integer solutions is to prove that for some integer n, the corresponding congruence mod (n) has no solutions. While this method is not powerful enough to prove FLT on its own, it can at least give us some helpful information about possible solutions of (11.6).

Perhaps the most obvious choice for n is to take $n = p$, so that (11.6) implies

$$a^p + b^p \equiv c^p \bmod (p).$$

Now we saw in Chapter 4 that $x^p \equiv x \bmod (p)$ for all integers x, so this congruence reduces to

$$a + b \equiv c \bmod (p),$$

which has rather too many solutions to be very helpful: for each of the p^2 pairs of classes $[a]$ and $[b]$ in $\mathbb{Z}_p$ there is a unique class $[c]$ ($= [a + b]$) satisfying the congruence. The one useful fact we can obtain is that if any two of a, b and c are congruent to 0 then so is the third. If we restrict attention to primitive triples (a, b, c) then there are only two possibilities:

(I) p divides none of a, b and c;

(II) p divides exactly one of a, b and c.

These are traditionally known as cases I and II of FLT. Even this trivial observation has proved useful, since it transpires that different techniques are effective in these two cases, with case I proving to be rather easier to deal with.

At this point, it is useful to replace c with $-c$; since p is odd we have $(-c)^p = -c^p$, so now the problem is to show that

$$a^p + b^p + c^p = 0 \quad (a, b, c \in \mathbb{Z}) \quad \text{implies} \quad abc = 0. \tag{11.7}$$

(Note that $abc = 0$ is simply a quick way of saying that $a = 0$ or $b = 0$ or $c = 0$; the advantage of this reformulation of the problem is that we now have complete symmetry between a, b and c, and this more than compensates for the slight disadvantage of having to consider negative integers.)

Let us consider a specific example, say $p = 3$, so we have

$$a^3 + b^3 + c^3 = 0.$$

For simplicity we will restrict attention to case I, so a, b and c are all coprime to 3. Now

$$-a^3 = b^3 + c^3 = (b + c)(b^2 - bc + c^2), \tag{11.8}$$

and we claim that the two factors on the right are mutually coprime. To see this, suppose that some prime m divides them both. Then $c \equiv -b \bmod (m)$ from the first factor, so the second factor gives $3b^2 \equiv 0 \bmod (m)$; since m is prime, we must have $m = 3$ or $m|b$. We cannot have $m = 3$, since m divides $-a^3$ whereas a is coprime to 3; if m divides b then it also divides c (since it divides $b + c$), contradicting the primitivity of (a, b, c). Thus the two factors in equation (11.8) are coprime, and since their product is a cube Lemma 2.4 implies that they are both cubes, say

$$b + c = r^3 \quad \text{and} \quad b^2 - bc + c^2 = u^3, \quad \text{so} \quad a = -ru$$

for some integers r and u. By the symmetry in a, b and c, we also have

$$c + a = s^3 \quad \text{and} \quad c^2 - ca + a^2 = v^3, \quad \text{so that} \quad b = -sv,$$
$$a + b = t^3 \quad \text{and} \quad a^2 - ab + b^2 = w^3, \quad \text{so that} \quad c = -tw,$$

where $s, t, v, w \in \mathbb{Z}$.

We now consider congruences mod (7): this may seem a rather arbitrary choice, and we will justify it more fully later, but a simple explanation is that the only cubes in $\mathbb{Z}_7$ are the classes $[0], [1]$ and $[-1]$, which are easy to add. For instance, this observation implies that any solution of

$$a^3 + b^3 + c^3 \equiv 0 \bmod (7)$$

has at least one of the classes $[a], [b]$ or $[c]$ equal to $[0]$, so 7 must divide at least one of a, b and c. Without any loss of generality, we can assume that 7 divides c. Then

$$r^3 + s^3 + (-t)^3 = r^3 + s^3 - t^3 = (b + c) + (c + a) - (a + b) = 2c \equiv 0 \bmod (7),$$

so the same observation about cubes in $\mathbb{Z}_7$ implies that 7 divides at least one of r, s and t. If 7 divides r then it divides $r^3 = b + c$; but 7 divides c and so it also divides b, which is impossible since (a, b, c) is primitive. Thus 7 cannot divide r, and a similar argument shows that it cannot divide s, so 7 must divide t. Thus 7 divides $a + b$, so $a \equiv -b \bmod (7)$ and hence

$$w^3 = a^2 - ab + b^2 \equiv 3b^2 \bmod (7).$$

Since 7 divides c we also have

$$u^3 = b^2 - bc + c^2 \equiv b^2 \bmod (7).$$

Now u is coprime to 7 (for otherwise 7 would divide both c and $a = -ru$, contradicting primitivity); thus u is a unit mod (7), so $ui \equiv 1 \bmod (7)$ for some integer i. Then

$$(wi)^3 = w^3 i^3 \equiv 3b^2 i^3 \equiv 3u^3 i^3 = 3(ui)^3 \equiv 3 \bmod (7),$$

so [3] is a cube in $\mathbb{Z}_7$. By inspection, this is not true, so this contradiction has proved case I of FLT for the prime $p = 3$.

In this argument, the only special properties of the primes 3 and 7 we have used are:

(1) if $x^3 + y^3 + z^3 \equiv 0$ mod (7) then x, y or $z \equiv 0$ mod (7) ;

(2) the class [3] is not a cube in $\mathbb{Z}_7$.

It follows that the above argument proves case I of FLT for any odd prime p, provided we can find a prime q such that p and q satisfy these two conditions. More precisely, the argument establishes the following theorem, proved by Sophie Germain in the early 19th century:

Theorem 11.8

Let p and q be odd primes such that

(1) if $x^p + y^p + z^p \equiv 0$ mod (q) then x, y or $z \equiv 0$ mod (q) ;

(2) the class $[p]$ is not a p-th power in $\mathbb{Z}_q$;

then case I of FLT is true for exponent p, that is, there are no positive integer solutions of $a^p + b^p = c^p$ with a, b and c all coprime to p.

(Sophie Germain is one of the few women to have made a substantial contribution to number theory. She had to fight strong social prejudice against women doing mathematics, initially using the masculine pseudonym Antoine Le Blanc to gain acceptance for her work. She also obtained significant results in applied mathematics.)

Exercise 11.11

Convert our preceding argument for the primes $p = 3$ and $q = 7$ into a proof of Theorem 11.8. In place of equation (11.8) you will need to show that

$$b^p + c^p = (b + c)(b^{p-1} - b^{p-2}c + b^{p-3}c^2 - \cdots + c^{p-1}),$$

with the two factors on the right mutually coprime.

In order to apply Theorem 11.8 to a particular exponent p, we need to find a suitable prime q, so that conditions (1) and (2) are satisfied. Our best chance of doing this will be to choose q so that there are relatively few p-th powers in $\mathbb{Z}_q$. If p does not divide $q - 1$ then every element of $\mathbb{Z}_q$ is a p-th power, so in particular condition (2) must fail; if $q = kp + 1$, on the other hand, then there are just k distinct p-th powers in U_q, and hence just $k + 1$ (including [0]) in $\mathbb{Z}_q$.

Exercise 11.12

Prove the statements about p-th powers in the preceding sentence.

Theorem 2.10 (Dirichlet's Theorem) guarantees that for each p there are infinitely many primes of the form $q = kp + 1$, so by trying some small values of k we can look for primes q satisfying (1) and (2). Since q is odd, k must be even, and the best situation is when $k = 2$, that is, when the integer $q = 2p+1$ is prime:

Exercise 11.13

Show that if p and $q = 2p+1$ are both primes, then the only p-th powers in $\mathbb{Z}_q$ are the classes $[0], [1]$ and $[-1]$. Hence show that conditions (1) and (2) of Theorem 11.8 are satisfied.

This exercise, together with Theorem 11.8, immediately proves case I of FLT for all odd primes p such that $2p + 1$ is prime. Many small primes p, such as $3, 5, 11, \ldots$, satisfy this condition, but it is not known whether there are infinitely many of them.

Exercise 11.14

List all the odd primes $p < 100$ for which $2p + 1$ is prime.

If $2p + 1$ is not prime then we can try other values of k in order to find a suitable prime q. By this method, Sophie Germain and Legendre were able to prove case I of FLT for all primes $p < 100$.

Exercise 11.15

Show that if $p = 7$ then the prime $q = 29$ satisfies the conditions of Theorem 11.8. Find a suitable prime q when $p = 13$.

Of course, Theorem 11.8 is relevant only to case I of FLT, and it tells us nothing about the harder case II, where p divides one of a, b and c. Nevertheless, complete proofs of FLT were found, initially for small primes p, and then for larger classes of primes. In 1753 Euler proved FLT for $p = 3$; his proof was essentially correct, though it contained a minor gap which was noticed and corrected much later; Gauss also proved the case $p = 3$, though as so often with this prolific genius, his proof was not published until after his death. In 1825 Dirichlet and Legendre proved FLT for $p = 5$, and in 1839 Lamé proved it for $p = 7$, Dirichlet having already dealt with the slightly easier case of exponent 14 in 1832. However, the steadily increasing difficulty of these proofs made it clear that some new general method was required, which would deal with whole classes of primes rather than individual cases.

11.9 Lamé and Kummer

In 1847 Lamé announced what he thought was a proof for all odd primes p, based on the factorisation

$$a^p + b^p = (a + b)(a + \zeta b)(a + \zeta^2 b) \ldots (a + \zeta^{p-1} b) \qquad (11.9)$$

where ζ is a complex number such that $\zeta^p = 1 \neq \zeta$; for instance, we could take $\zeta = \cos(2\pi/p) + \mathrm{i}\sin(2\pi/p)$ where $\mathrm{i} = \sqrt{-1}$. The factors on the right are all examples of *cyclotomic integers*, complex numbers of the form

$$a_0 + a_1\zeta + a_2\zeta^2 + \cdots + a_{p-1}\zeta^{p-1} \qquad (a_r \in \mathbb{Z}).$$

(The word *cyclotomic* means circle-dividing: in the usual geometric representation of complex numbers $z = x + \mathrm{i}y$ as points (x, y) in the plane, the points $\zeta, \zeta^2, \ldots, \zeta^p = 1$ divide the unit circle $x^2 + y^2 = 1$ into p equal segments.) The set $\mathbb{Z}[\zeta]$ of cyclotomic integers is closed under addition, subtraction and multiplication, since we can use the equation $\zeta^p = 1$ to express any powers ζ^r $(r \geq p)$ in terms of lower powers of ζ. Thus $\mathbb{Z}[\zeta]$, like $\mathbb{Z}$ and $\mathbb{Z}[\mathrm{i}]$, is a ring. Lamé argued that if the left-hand side of equation (11.9) is a p-th power c^p, then a result similar to Lemma 2.4 would show that each factor on the right-hand side would have to be a p-th power in $\mathbb{Z}[\zeta]$, and from this he could obtain the required contradiction by Fermat's method of descent. Unfortunately, Lemma 2.4 depends on the uniqueness of prime power factorisations (Theorem 2.3), and while this is valid for ordinary integers, it is not generally valid for cyclotomic integers (the smallest prime for which it fails is $p = 23$). Thus Lemma 2.4 cannot be extended from $\mathbb{Z}$ to $\mathbb{Z}[\zeta]$, so Lamé's 'proof' is incorrect, as he soon discovered.

Exercise 11.16

Show that the roots of the polynomial $x^p + 1$ are $-1, -\zeta, -\zeta^2, \ldots, -\zeta^{p-1}$, and hence prove equation (11.9).

Exercise 11.17

Show that ζ is a root of the cyclotomic polynomial $\Phi_p(x) = 1 + x + x^2 + \cdots + x^{p-1}$, and deduce that every cyclotomic integer z can be written in the form

$$z = b_0 + b_1\zeta + b_2\zeta^2 + \cdots + b_{p-2}\zeta^{p-2} \qquad (b_i \in \mathbb{Z}).$$

Use the irreducibility of $\Phi_p(x)$ (Chapter 2, Example 2.2) to show that this representation of z is unique.

In the early 1840s Kummer, investigating generalisations of the law of quadratic reciprocity to higher powers, had already discovered this difficulty concerning unique factorisation. In order to overcome it he introduced concepts such as *ideals*, which are substitutes for the missing primes in $\mathbb{Z}[\zeta]$, and the *class number* h_p, an integer which measures how badly unique factorisation fails in $\mathbb{Z}[\zeta]$; these concepts played a crucial role in the subsequent development of ring theory and algebraic number theory. Kummer also devised a general method which proved FLT for the *regular* primes p, those which do not divide h_p (a condition which means that unique factorisation either holds in $\mathbb{Z}[\zeta]$, or at least does not fail too badly). He showed that an odd prime p is regular if and only if it does not divide the numerators of the Bernoulli numbers $B_2, B_4, B_6, \ldots, B_{p-3}$. We recall from Chapter 9, Section 6 that the Bernoulli numbers are a sequence of rational numbers B_n given by expanding $t/(e^t - 1)$ as a power series

$$\frac{t}{e^t - 1} = \sum_{n=0}^{\infty} \frac{B_n}{n!} t^n \,,$$

and since they are not difficult to compute (at least, for small n), it is straightforward to determine which small primes are regular. All the odd primes $p < 100$ except $37, 59$ and 67 are regular, so they satisfy FLT; 37 is not regular since it divides the numerator of

$$B_{32} = -\frac{37 \times 683 \times 305065927}{510} \,.$$

Kummer conjectured that there are infinitely many regular primes; this is still an open problem, though it is known that there are infinitely many irregular primes.

Exercise 11.18

Show that B_n is a rational number for all $n \geq 0$. Calculate enough Bernoulli numbers to show that the odd primes $p \leq 13$ are all regular.

11.10 Modern developments

We saw in Theorem 4.3 that if a is coprime to p then $a^{p-1} \equiv 1 \mod (p)$, so in particular every odd prime p divides $2^{p-1} - 1$. In many cases the integer $(2^{p-1} - 1)/p$ is not divisible by p, that is, $2^{p-1} \not\equiv 1 \mod (p^2)$, and in 1909 Wieferich proved case I of FLT for all primes p satisfying this condition. With patience or a computer, this condition is straightforward to check, for instance by repeatedly multiplying and reducing mod (p^2), and the only primes $p <$ 3×10^9 to fail the condition are 1093 and 3511. Soon afterwards Mirimanoff proved case I of FLT for all primes p such that $3^{p-1} \not\equiv 1 \bmod (p^2)$; this includes the primes 1093 and 3511, so case I was now proved for all $p < 3 \times 10^9$. These results of Wieferich and Mirimanoff also implied case I of FLT for all primes of the form $2^a.3^b \pm 1$ or $\pm 2^a \pm 3^b$ where $a, b \geq 0$, and hence in particular for all Fermat and Mersenne primes. In recent times, computers have been used to show that any counterexample to FLT would have to involve huge numbers: for instance by 1992 FLT was known to be true for all primes $p < 4000000$.

In parallel with these computational attacks on FLT, mathematicians have also recently used geometric ideas, in a sense returning us to the roots of the problem discussed at the beginning of this chapter. If $P(x, y)$ is a polynomial in two variables, then the real solutions (x, y) of the equation $P(x, y) = 0$ form a 1-dimensional geometric structure in the xy-plane, namely the graph of the equation. If we allow x and y to be complex numbers, then the solutions of $P(x, y) = 0$ form a 2-dimensional structure (in a 4-dimensional space), that is, a surface: this is because the complex numbers themselves form a surface rather than a line. The surface given by $P(x, y) = 0$ looks like a sphere with finitely many handles attached, and we define the genus g of this equation to be the number of handles; for instance, it can be shown that the equation

$$P_n(x, y) = x^n + y^n - 1 = 0$$

has genus

$$g = \frac{(n-1)(n-2)}{2}.$$

In 1922 Mordell conjectured that if a polynomial $P(x, y)$ has rational coefficients, and if $g \geq 2$, then the equation $P(x, y) = 0$ has only finitely many pairs x, y of rational solutions. This conjecture was proved in 1983 by Faltings. Now the polynomial $P_n(x, y)$ visibly has rational coefficients, and if $n \geq 4$ it has genus $g \geq 2$, so by Faltings's proof of the Mordell Conjecture it follows that for each $n \geq 4$ there are only finitely many rational solutions of $x^n + y^n = 1$ (including the 'obvious' solutions where $x = 0$ or $y = 0$). This is easily seen to be equivalent to the result that there are only finitely many primitive solutions $a, b, c \in \mathbb{Z}$ of $a^n + b^n = c^n$ for each $n \geq 4$ (FLT asserts that there are none). When $n = 3$ we have $g = 1$, so we cannot use the Mordell Conjecture to prove this, but in this case the result follows from the solution of FLT for exponent 3. When $n = 2$, however, we have already seen that there are infinitely many rational solutions x and y of $x^2 + y^2 = 1$, corresponding to the infinite number of primitive Pythagorean triples.

The most important modern development has been to connect FLT with the theory of elliptic curves; this is a central area of current research in pure mathematics, where number theory, algebra, geometry and topology all interact in a particularly interesting way. An *elliptic curve* E is the surface corresponding to a polynomial equation $P(x, y) = 0$ of genus $g = 1$; this surface looks like a sphere with one handle attached, that is, a doughnut (the technical name is a *torus*). For example, if $Q(x)$ is a cubic polynomial with distinct roots, then the equation

$$P(x, y) = y^2 - Q(x) = 0$$

defines an elliptic curve. (The tradition of calling these surfaces 'elliptic curves' is very inappropriate, since they are neither ellipses nor curves: they are 'elliptic' only in the sense that equations of genus 1 occur when one performs the integration required to determine the circumference of an ellipse; they are 'curves' only in the sense that they are generalisations to complex numbers of the curves $P(x, y) = 0$ given by restricting x and y to real numbers.)

In the 1950s, Taniyama and Shimura developed a conjecture that if an elliptic curve E is defined by a polynomial $P(x, y)$ with rational coefficients, then it must be *modular*: this is a difficult condition to define precisely, but essentially it means that E can be constructed in a particular way using hyperbolic geometry, matrices and congruences. In the 1980s Frey, Ribet and Serre showed that the Taniyama–Shimura Conjecture implies FLT. The argument is as follows. If FLT is false then there must be a primitive solution (a, b, c) of $a^p + b^p = c^p$ for some prime $p > 3$, since we know that FLT is true for $p = 3$. By permuting terms and changing signs if necessary, we can assume that b is even and $a \equiv -1$ mod (4). Now the equation

$$y^2 = x(x - a^p)(x + b^p)$$

defines an elliptic curve E (called a *Frey curve*), since the right-hand side is a cubic polynomial $Q(x)$ with distinct roots. This equation clearly has rational coefficients, so if the Taniyama–Shimura Conjecture is true then E must be modular. However, some very difficult work by Frey, Ribet and Serre showed that the Frey curves E, if they exist, are not modular, so this contradiction proves FLT. Thus a proof of the Taniyama–Shimura Conjecture would immediately imply FLT.

This argument added enormously to the interest in this conjecture, which had already achieved considerable importance in its own right. On 23rd June 1993, at a conference at the Isaac Newton Institute in Cambridge, Andrew Wiles outlined a proof of the conjecture, or rather, enough of the conjecture to imply FLT. There was considerable excitement, both within and outside the mathematical community, that one of the classic problems of mathematics had apparently been solved. There followed an anxious delay of more than a year while he filled in the details: they were very complicated, and a number of previous 'proofs' of FLT, some from very respectable mathematicians, had subsequently turned out to be incorrect. For a while, there was a gap in the proof which proved very difficult to fill, but eventually Wiles succeeded in overcoming this obstacle, and the full proof (about 200 pages long, part of it written jointly with Richard Taylor) was published in 1995 (Taylor and Wiles, 1995; Wiles, 1995). At first sight, this great achievement appears to close a long chapter in the development of mathematics, since there are few interesting corollaries one can deduce from FLT. However, the methods developed for the proof have much wider applications, both in number theory and in related topics such as Galois theory, so one can expect interest to remain high for many years. As has happened several times in the history of FLT, it is the proofs rather than the theorems which have the important consequences.

11.11 Further reading

The exercises in this chapter are quite hard, so instead of setting any supplementary exercises we will close the chapter by suggesting some further reading on FLT.

Expository papers by Cox (1994) and Gouvêa (1994), and a short note by Ribet (1993), written soon after Wiles announced his proof, offer concise summaries of FLT and its background, and a paper by Mazur (1991), written a little earlier, describes some of the methods subsequently used to prove FLT and related results. All of these are intended for a general mathematical readership. Wiles's proof of FLT appears in Wiles (1995), with some important subsidiary results in a joint paper with Taylor (Taylor and Wiles, 1995); both of these papers are written for specialists, and they are *very* difficult. The best sources for the history of FLT and related mathematical developments are the books by Edwards (1977) and Ribenboim (1979). Aczel (1996) and Singh (1997) have written two very readable and non-technical accounts of FLT and its solution, with Singh's book covering rather more ground. At a more technical level, van der Poorten (1996) presents many of the ideas underlying the proof of FLT in a very clear and often light-hearted way.

Appendix A
Induction and Well-ordering

Throughout this book, we are mainly interested in the properties of the set $\mathbb{N} = \{1, 2, 3, \ldots\}$ of natural numbers (some authors also include 0 in $\mathbb{N}$). There is one very important principle, or method of proof, which applies to this number system, but not to any of the other standard number systems, such as the set $\mathbb{Z}$ of all integers, or the sets $\mathbb{Q}, \mathbb{R}$ or $\mathbb{C}$ of rational, real or complex numbers. There are three versions of this principle, known as the principle of induction, the principle of strong induction, and the well-ordering principle; they are logically equivalent, in the sense that each implies the other, but in different contexts one of them may be more convenient to use than the others.

The Principle of Induction. The most familiar version of this principle concerns statements $P(n)$ about integers n :

(1) If $P(1)$ is true, and $P(n)$ implies $P(n + 1)$ for all $n \in \mathbb{N}$, then $P(n)$ is true for all $n \in \mathbb{N}$.

For an example, see the proof of Corollary 2.2 (with k in place of n). The justification for this principle is that $P(1)$ is true, and $P(1)$ implies $P(2)$, so $P(2)$ is true; since $P(2)$ implies $P(3)$, $P(3)$ is also true, and by continuing we can prove $P(n)$ for each $n \in \mathbb{N}$. (In some applications we may need to start at some other integer n_0, such as 0, and prove $P(n)$ for all integers $n \geq n_0$, but this makes no significant difference.) An equivalent form of this principle concerns sets of integers, rather than statements about integers:

(1') Suppose that $A \subseteq \mathbb{N}$, that $1 \in A$, and that $n \in A$ implies $n + 1 \in A$ for all $n \in \mathbb{N}$; then $A = \mathbb{N}$.

To see that (1) implies (1'), assume (1), and suppose that A satisfies the hypotheses of (1'). Let $P(n)$ be the statement '$n \in A$', so $P(1)$ is true since $1 \in A$; if $P(n)$ is true then $n \in A$, so $n + 1 \in A$, and hence $P(n+1)$ is true; thus $P(n)$ implies $P(n+1)$, so $P(n)$ is true for all $n \in \mathbb{N}$ by (1); thus $n \in A$ for all $n \in \mathbb{N}$, so $A = \mathbb{N}$. For the converse (that (1') implies (1)), given $P(n)$ take $A = \{n \in \mathbb{N} \mid P(n) \text{ is true}\}$; then $1 \in A$ (since $P(1)$ is true), and if $n \in A$ then $P(n)$ is true, so $P(n+1)$ is true, giving $n + 1 \in A$; hence $A = \mathbb{N}$ by (1'), so $P(n)$ is true for all $n \in \mathbb{N}$.

The Principle of Strong Induction. This also has two equivalent forms. The conclusions are the same as those for induction, but the hypotheses are stronger:

- **(2)** If $P(1)$ is true, and $P(1), P(2), \ldots, P(n)$ together imply $P(n+1)$, then $P(n)$ is true for all $n \in \mathbb{N}$.

- **(2')** Suppose that $B \subseteq \mathbb{N}$, that $1 \in B$, and that if $1, 2, \ldots, n \in B$ then $n + 1 \in B$; then $B = \mathbb{N}$.

For an example of (2), see the proof of Theorem 2.3. The proof that (2) and (2') are equivalent is similar to that for (1) and (1'). This form of induction is used when the hypothesis $P(n)$ alone is not strong enough to prove $P(n+1)$.

The Well-ordering Principle. This refers to the order relation $\le$ on $\mathbb{N}$:

- **(3)** If $C \subseteq \mathbb{N}$ and C is non-empty, then C has a least element.

By a least element, we mean some $c \in C$ such that $c \le d$ for all $d \in C$. This principle is used in the proof of Theorem 1.1. The corresponding statement is easily seen to be false if we replace $\mathbb{N}$ with any of the other standard number systems: for instance, the set of positive rational numbers has no least element.

To show that these principles are equivalent, we show that $(1') \Rightarrow (2') \Rightarrow (3) \Rightarrow (1')$.

$(1') \Rightarrow (2')$. Suppose that B satisfies the hypotheses of (2'). Let $A = \{n \in \mathbb{N} \mid 1, 2, \ldots, n \in B\}$. Then $A \subseteq \mathbb{N}$, and $1 \in A$ (since $1 \in B$). If $n \in A$ then $1, 2, \ldots, n \in B$ (by definition of A), so $n + 1 \in B$ (by one of the hypotheses in (2')), so $1, 2, \ldots, n + 1 \in B$ and hence $n + 1 \in A$ (by definition of A); thus $n \in A$ implies $n + 1 \in A$, so $A = \mathbb{N}$ by (1'). This means that for each $n \in \mathbb{N}$ we have $1, 2, \ldots, n \in B$, so in particular $n \in B$; thus $B = \mathbb{N}$, as required.

$(2') \Rightarrow (3)$. We show that if $C \subseteq \mathbb{N}$, and C has no least element, then C is empty. Let $B = \mathbb{N} \setminus C$, the complement of C in $\mathbb{N}$. Then $1 \in B$, for otherwise $1 \in C$ and so 1 is a least element of C (since it is a least element of $\mathbb{N}$). If $1, 2, \ldots, n \in B$ then $1, 2, \ldots, n \notin C$; it follows that $n + 1 \notin C$ (for otherwise

$n + 1$ would be a least element of C), so $n + 1 \in B$. Thus B satisfies the hypotheses of $(2')$, so $B = \mathbf{N}$ and C is empty.

$(3) \Rightarrow (1')$. Suppose that A satisfies the hypotheses of $(1')$, and let $C = \mathbf{N} \setminus A$. If C is non-empty, then it has a least element c. Since $1 \in A$ and $c \in C$, we have $c \neq 1$, so $c - 1 \in \mathbf{N}$. Now $c - 1 < c$, so $c - 1 \notin C$ (for otherwise c could not be a least element of C), and hence $c - 1 \in A$. But $n \in A$ implies $n + 1 \in A$, so putting $n = c - 1$ we see that $c \in A$, contradicting the fact that $c \in C$. Thus C is empty, so $A = \mathbf{N}$.

Appendix B
Groups, Rings and Fields

A *group* consists of a set G together with a binary operation $*$ satisfying the following axioms:

- *Closure:* if $g, h \in G$ then $g * h \in G$;

- *Associativity:* $f * (g * h) = (f * g) * h$ for all $f, g, h \in G$;

- *Identity:* there is an element $e \in G$ such that $g * e = g = e * g$ for all $g \in G$;

- *Inverses:* for each $g \in G$ there is an element $h \in G$ such that $g*h = e = h*g$.

In many cases (such as here), the symbol $*$ is omitted, and we write simply gh for $g * h$, and fgh for the product in the associativity axiom. A product $g * g * \cdots * g$, with i factors, is written g^i. The element e is called the *identity element* of G, often denoted by the symbol 1. The element h in the last axiom is called the *inverse* of g, often written $h = g^{-1}$, so this axiom becomes $gg^{-1} = 1 = g^{-1}g$. The inverse of g^i is written g^{-i}, and g^0 denotes 1. The *order* $|G|$ of a group G is the number of elements of the set G; if this is finite, we say that G is a *finite* group.

A group G is *abelian*, or *commutative*, if it satisfies the additional axiom:

- *Commutativity:* $gh = hg$ for all $g, h \in G$.

In an abelian group, the binary operation is often denoted by $+$, the identity by 0 (usually called the *zero element*), and the inverse of g by $-g$, so for instance $g + 0 = g = 0 + g$ and $g + (-g) = 0 = (-g) + g$ for all g.

A *subgroup* of a group G is a subset H of G which is also a group with respect to the same binary operation as G; this is equivalent to the conditions:

- if $g, h \in H$ then $gh \in H$;

- $1 \in H$;

- if $g \in H$ then $g^{-1} \in H$.

We write $H \leq G$ to denote that H is a subgroup of G.

If $H \leq G$ and $g \in G$ then the *right coset* of H containing g is the subset $Hg = \{hg \mid h \in H\}$ of G. Each right coset of H contains $|H|$ elements. Right cosets Hg_1 and Hg_2 are either equal or disjoint, so they partition G into disjoint subsets. The number of distinct right cosets of H in G is called the *index* $|G : H|$ of H in G. If G is finite, then $|G| = |G : H|.|H|$, which proves Lagrange's Theorem, that $|H|$ divides $|G|$. Similar results hold for left cosets $gH = \{gh \mid h \in H\}$.

The *order* of an element $g \in G$ is the least integer $n > 0$ such that $g^n = 1$, provided such an integer exists; if it does not, g has *infinite order*. If G is finite, then every element g has finite order n for some n; the powers $g, g^2, \ldots, g^{n-1}, g^n \,(= 1)$ of g then form a subgroup of G, so n divides $|G|$ by Lagrange's Theorem.

A group G is *cyclic* if there exists an element $c \in G$, called a *generator* for G, such that every $g \in G$ has the form $g = c^i$ for some integer i. If c has finite order n, then $|G| = n$, and G is denoted by C_n. Such a group G has one subgroup H of order m for each m dividing n, and no other subgroups; H is a cyclic group of order m, with generator $c^{n/m}$.

A *homomorphism* between groups G and G' is a function $\theta : G \to G'$ such that $\theta(gh) = \theta(g)\theta(h)$ for all $g, h \in G$; if θ is a bijection, it is called an *isomorphism*. If such an isomorphism exists, we say that G and G' are *isomorphic*, written $G \cong G'$. This means that G and G' have the same algebraic structure, and differ only in the notation for their elements.

The *direct product* $G_1 \times G_2$ of groups G_1 and G_2 consists of all ordered pairs (g_1, g_2) with $g_i \in G_i$ for $i = 1, 2$. This is a group, with binary operation $(g_1, g_2)(h_1, h_2) = (g_1 h_1, g_2 h_2)$; the identity element is $(1_1, 1_2)$, where 1_i is the identity element in G_i, and the inverse of (g_1, g_2) is (g_1^{-1}, g_2^{-1}). There are subgroups $G_1' = \{(g_1, 1_2) \mid g_1 \in G_1\} \cong G_1$ and $G_2' = \{(1_1, g_2) \mid g_2 \in G_2\} \cong G_2$. Direct products $G_1 \times \cdots \times G_k$ are defined similarly for $k > 2$. If m and n are coprime, then $C_m \times C_n \cong C_{mn}$, since if c_1 and c_2 generate C_m and C_n then (c_1, c_2), which has order mn, generates $C_m \times C_n$.

By a *ring*, we mean a commutative ring with identity. This is a set R with two binary operations (addition $r + s$ and multiplication $r.s$, usually written rs), and with distinct elements 0 and 1 such that

- *Additive structure:* $(R, +)$ is an abelian group, with zero element 0;

- *Commutativity:* $rs = sr$ for all $r, s \in R$;

- *Associativity:* $r(st) = (rs)t$ for all $r, s, t \in R$;

- *Distributivity:* $r(s + t) = rs + rt$ for all $r, s, t \in R$;

- *Identity:* $r1 = r$ for all $r \in R$.

The number systems $\mathbb{Z}$ (integers), $\mathbb{Z}_n$ (integers mod (n)), $\mathbb{Q}$ (rational numbers), $\mathbb{R}$ (real numbers) and $\mathbb{C}$ (complex numbers) are all examples of rings.

The direct product $R_1 \times \cdots \times R_k$ of rings $R_1, \ldots, R_k$ is defined in much the same way as the direct product of groups: its elements are the k-tuples $(r_1, \ldots, r_k)$ such that $r_i \in R_i$ for all i, with componentwise operations. A *homomorphism* between rings R and R' is a function $\theta : R \rightarrow R'$ such that $\theta(r + s) = \theta(r) + \theta(s)$ and $\theta(rs) = \theta(r)\theta(s)$ for all $r, s \in R$, and $\theta(1) = 1$; if θ is a bijection, it is called an *isomorphism*. If such an isomorphism exists, we say that R and R' are *isomorphic*, written $R \cong R'$. If m and n are coprime, then $\mathbb{Z}_m \times \mathbb{Z}_n \cong \mathbb{Z}_{mn}$.

An element $r \in R$ is a *unit* if $rs = 1$ for some $s \in R$; the units form a group under multiplication, with 1 as the identity element. A *field* is a ring R in which every element $r \neq 0$ is a unit. The number systems $\mathbb{Q}, \mathbb{R}$ and $\mathbb{C}$ are fields, as is $\mathbb{Z}_n$ if n is prime.

Appendix C

Convergence

An infinite series $\sum_{n=1}^{\infty} a_n$ *converges* to l if its partial sums $s_n = a_1 + \cdots + a_n$ have limit l as $n \to \infty$; if there is no such l then the series *diverges*. For example, the *geometric series* $1 + a + a^2 + \cdots$ converges to $1/(1-a)$ if $|a| < 1$ (since $s_n = 1 + a + \cdots + a^{n-1} = (1-a^n)/(1-a) \to 1/(1-a)$), but it diverges if $|a| \geq 1$. An infinite product $\prod_{n=1}^{\infty} a_n$ *converges* to l if $a_1 a_2 \ldots a_n \to l$ as $n \to \infty$.

The *Comparison Test* states that if $a_n \geq b_n \geq 0$ for all n, and $\sum_{n=1}^{\infty} a_n$ converges, then $\sum_{n=1}^{\infty} b_n$ also converges, with $\sum_{n=1}^{\infty} b_n \leq \sum_{n=1}^{\infty} a_n$; equivalently, if $a_n \geq b_n \geq 0$ for all n, and $\sum_{n=1}^{\infty} b_n$ diverges, then $\sum_{n=1}^{\infty} a_n$ also diverges.

The *Integral Test* states that if f is a real-valued decreasing function such that $f(x) \geq 0$ for all $x \geq 1$ and $f(x) \to 0$ as $x \to +\infty$, then the series $\sum_{n=1}^{\infty} f(n)$ and the integral $\int_1^{+\infty} f(x)\, dx$ either both converge or both diverge. For instance, take $f(x) = x^{-s}$ for some constant $s > 0$; since $\int x^{-s}\, dx = x^{1-s}/(1-s)$ for $s \neq 1$, and $\int x^{-1}\, dx = \ln x$, we see that the integral $\int_1^{+\infty} x^{-s}\, dx$ converges for $s > 1$ and diverges otherwise; it follows that the series $\sum_{n=1}^{\infty} n^{-s}$ does likewise. In particular, the *harmonic series* $\sum_{n=1}^{\infty} 1/n$ diverges.

The *Alternating Test* states that if the terms a_n are real and alternating in sign, and if $a_n \to 0$ as $n \to \infty$, then $\sum_{n=1}^{\infty} a_n$ converges. For instance, $\sum_{n=1}^{\infty} (-1)^n/n$ converges.

An infinite series $\sum_{n=1}^{\infty} a_n$ is *absolutely convergent* if $\sum_{n=1}^{\infty} |a_n|$ converges; absolute convergence implies convergence, but the converse is false. A series which is convergent but not absolutely convergent, such as $\sum_{n=1}^{\infty} (-1)^n/n$, is called *conditionally convergent*. The terms of an absolutely convergent series can be rearranged or bracketed together without altering its sum; this fails for conditionally convergent series. If $\sum_{n=1}^{\infty} a_n$ and $\sum_{n=1}^{\infty} b_n$ converge absolutely to l and m, then their product $\sum_{n=1}^{\infty}(a_1 b_n + a_2 b_{n-1} + \cdots + a_n b_1)$ converges absolutely to lm.

A series of functions $\sum_{n=1}^{\infty} f_n(x)$ *converges* to $f(x)$ on a set X if, for each $x \in X$, the partial sums $s_n(x) = f_1(x) + \cdots + f_n(x)$ converge to $f(x)$ as $n \to \infty$; thus, for each $x \in X$ and each $\varepsilon > 0$, there exists N (which may depend on x and ε) such that $|s_n(x) - f(x)| < \varepsilon$ for all $n > N$. If N depends only on ε (and not on x), we say that the series is *uniformly convergent* on X. A uniformly convergent series of integrable functions can be integrated term by term; if the terms are differentiable, and the series of derivatives is uniformly convergent, then the series may be differentiated term by term.

If a complex function $f(z)$ is analytic (that is, differentiable) for all z close to some $a \in \mathbf{C}$, then near a it is represented by a *Taylor series* $f(z) = \sum_{n=0}^{\infty} a_n(z - a)^n$ where $a_n = f^{(n)}(a)/n!$. If $f(z)$ has a pole of order k at a (that is, $(z - a)^k f(z)$ is analytic and non-zero near a), then it is represented by a *Laurent series* $f(z) = \sum_{n=-k}^{\infty} b_n(z - a)^n$ near a.

Appendix D

Table of primes $p < 1000$.

$2, 3, 5, 7, 11, 13, 17, 19, 23, 29, 31, 37, 41, 43, 47, 53, 59, 61, 67, 71, 73, 79, 83, 89, 97,$

$101, 103, 107, 109, 113, 127, 131, 137, 139, 149, 151, 157, 163, 167, 173, 179, 181,$
$191, 193, 197, 199,$

$211, 223, 227, 229, 233, 239, 241, 251, 257, 263, 269, 271, 277, 281, 283, 293,$

$307, 311, 313, 317, 331, 337, 347, 349, 353, 359, 367, 373, 379, 383, 389, 397,$

$401, 409, 419, 421, 431, 433, 439, 443, 449, 457, 461, 463, 467, 479, 487, 491, 499,$

$503, 509, 521, 523, 541, 547, 557, 563, 569, 571, 577, 587, 593, 599,$

$601, 607, 613, 617, 619, 631, 641, 643, 647, 653, 659, 661, 673, 677, 683, 691,$

$701, 709, 719, 727, 733, 739, 743, 751, 757, 761, 769, 773, 787, 797,$

$809, 811, 821, 823, 827, 829, 839, 853, 857, 859, 863, 877, 881, 883, 887,$

$907, 911, 919, 929, 937, 941, 947, 953, 967, 971, 977, 983, 991, 997.$

Solutions to Exercises

Chapter 1

1.1 Put $a = 2q + r$ with $r = 0$ or 1, so $n = a^2 = (2q+r)^2 = 4(q^2+qr)+r^2$ with $r^2 = 0$ or 1.

1.2 0 or 1; 0, 1 or 4; 0, 1, 3 or 4. (Imitate Example 1.2, with $b = 3, 5$ and 6.)

1.3 (a) If $b = qa$ and $c = q'b$ then $c = (q'q)a$.

(b) If $b = qa$ and $d = q'c$ then $bd = (qq')ac$.

(c) $b = qa$ iff $mb = q(ma)$.

(d) If $a = qd$ then $|a| = |q|.|d| \geq |d|$ since $|q| \geq 1$.

1.4 No: $1|1$ and $1|2$, but $1 + 1$ does not divide $1 + 2$.

1.5 $1745 = 1.1485 + 260$, $1485 = 5.260 + 185$, $260 = 1.185 + 75$, $185 = 2.75 + 35$, $75 = 2.35 + 5$, $35 = 7.5 + 0$, so $\gcd(1485, 1745) = 5$.

1.6 We could take $u' = u + 1066 = 1061$ and $v' = v - 1492 = -1485$, so $1492u' + 1066v' = 1492u + 1492.1066 + 1066v - 1066.1492 = 1492u + 1066v = d$.

1.7 The solution of Exercise 1.5 gives $\gcd(1485, 1745) = 5 = 75 - 2.35 = 75 - 2.(185 - 2.75) = -2.185 + 5.75 = -2.185 + 5.(260 - 1.185) = 5.260 - 7.185 = 5.260 - 7.(1485 - 5.260) = -7.1485 + 40.260 = -7.1485 + 40.(1745 - 1.1485) = -47.1485 + 40.1745$, so take $u = -47, v = 40$.

1.8 $\gcd(a, b)$ divides a and b, and hence so does any factor c of $\gcd(a, b)$.

251

Conversely $\gcd(a, b) = au + bv$ by Theorem 1.7, so by Corollary 1.4 any common factor c divides $\gcd(a, b)$.

1.9 By Exercise 1.8, an integer c divides $a_1, \ldots, a_k$ iff it divides $\gcd(a_1, a_2)$, a_3, $\ldots, a_k$; the largest such c is both $\gcd(a_1, \ldots, a_k)$ and $\gcd(\gcd(a_1, a_2), a_3, \ldots, a_k)$.

1.10 $1155 = 1.1092 + 63$, $1092 = 17.63 + 21$, $63 = 3.21 + 0$, so $\gcd(1092, 1155) = 21$; then $2002 = 95.21 + 7$, $21 = 3.7 + 0$, so $d = \gcd(21, 2002) = 7$. Similarly, $\gcd(910, 780) = 130$ and $\gcd(130, 286) = 26$, so $d = \gcd(26, 195) = 13$.

1.11 Use the solution of Exercise 1.9. Let d_i denote $\gcd(a_1, \ldots, a_i)$. Theorem 1.7 gives $d_2 = a_1 u + a_2 v$ for some u, v, then $d_3 = d_2 u' + a_3 v' = a_1 uu' + a_2 vu' + a_3 v'$ for some u', v', and so on until eventually $d = d_k$ has the form $a_1 u_1 + \cdots + a_k u_k$. The solution of Exercise 1.10 gives $21 = \gcd(1092, 1155) = 18.1092 - 17.1155$ and $7 = \gcd(21, 2002) = -95.21 + 1.2002 = -95.(18.1092 - 17.1155) + 1.2002 = -1710.1092 + 1615.1155 + 1.2002$.

1.12 In (a) take $a = b = c = 2$, in (b) take $a = b = 2$ and $c = 1$.

1.13 $\operatorname{lcm}(1485, 1745) = (1485 \times 1745)/\gcd(1485, 1745) = 518265$ by Theorem 1.12 and Exercise 1.5.

1.14 If $l | c$ then since $a | l$ and $b | l$ we have $a | c$ and $b | c$. Conversely, Theorem 1.1 gives $c = ql + r$ with $0 \le r < l$; since a and b divide c and l, they divide r, so r is a common multiple; since l is the least positive common multiple, $r = 0$ and so $l | c$.

1.15 Exercise 1.5 gives $d = 5$, which divides $c = 15$, so solutions exist, with $e = 15/5 = 3$; Exercise 1.7 gives $u = -47, v = 40$, so $x_0 = 3u = -141, y_0 = 3v = 120$ and the general solution is $x = -141 + 349n$, $y = 120 + 297n$ $(n \in \mathbb{Z})$.

1.16 Solutions exist iff $d | c$, where $d = \gcd(a_1, \ldots, a_k)$: Exercise 1.11 gives $d = a_1 u_1 + \cdots + a_k u_k$ for some $u_i \in \mathbb{Z}$, so if $c = de$ then $c = a_1 x_1 + \cdots + a_k x_k$ with $x_i = u_i e \in \mathbb{Z}$; conversely, if $c = a_1 x_1 + \cdots + a_k x_k$ then $d | c$ since $d | a_i$ for each i.

1.17 $h(2) = 1$ since the only case is $b = 1$ with $2 = 2.1 + 0$, giving $\gcd(2, 1) = 1$ in one step. If $a > 2$ then taking $b = a - 1$ gives $a = 1.b + 1, b = b.1 + 0$, taking two steps, so $h(a) \ge 2$. Considering all $b < a$ individually gives $h(3) = h(4) = 2, h(5) = 3, h(6) = 2, h(7) = 3, h(8) = 4$ (attained by $b = 5$).

1.18 By induction, $0 \le f_n < f_{n+1}$ for all $n \ge 2$, so Euclid's algorithm gives $f_{n+2} = 1.f_{n+1} + f_n$, $f_{n+1} = 1.f_n + f_{n-1},\ldots,5 = 1.3+2$, $3 = 1.2+1$, $2 = 2.1+0$. Thus $r_i = f_n, f_{n-1},\ldots,f_3,f_2,0$ for $i = 1,2,\ldots,n$, so $\gcd(f_{n+2}, f_{n+1}) = f_2 = 1$. This takes n steps, so $h(f_{n+2}) \ge n$.

1.19 Among all n-step applications of Euclid's algorithm, we can minimise a by taking the least possible values of $q_1,\ldots,q_n$ and r_{n-1}, and then working back through the equations to find $r_{n-2},\ldots,r_1,b$ and a. Now $q_1,\ldots,q_{n-1} \ge 1$ (since $a > b > r_1 > r_2 > \ldots$), $q_n \ge 2$ (since $q_n r_{n-1} = r_{n-2} > r_{n-1}$), and $r_{n-1} \ge 1$ (since $r_{n-1} > r_n = 0$), so putting $q_1 = \cdots = q_{n-1} = 1$, $q_n = 2$ and $r_{n-1} = 1$ we find that the equations (in reverse order) become $r_{n-2} = 2r_{n-1} = 2$, $r_i = r_{i+1} + r_{i+2}$ $(n - 3 \ge i \ge 1)$, $b = r_1 + r_2$, $a = b + r_1$. Thus the sequence $r_{n-1}, r_{n-2},\ldots,r_1,b,a$ starts with $1,2,\ldots$, and then each term is the sum of its two predecessors, so it agrees with $f_2, f_3, f_4, \ldots$. In particular, its n-th and $(n+1)$-th terms b and a are f_{n+1} and f_{n+2}.

1.20 If Euclid's algorithm takes m steps to calculate $\gcd(a,b)$ for some $b < a \le f_{n+2}$, then $a \ge f_{m+2}$ by Exercise 1.19; thus $f_{m+2} \le f_{n+2}$ and hence $m \le n$ for all such a,b, so $h(a) \le n$. Taking $a = f_{n+2}$ we get $h(f_{n+2}) \le n$, so $h(f_{n+2}) = n$ by Exercise 1.18; taking $a < f_{n+2}$ we get $m < n$ for all $b < a$, so $h(a) < n$.

1.21 Exercise 1.19 implies that if $a > b > 0$ and Euclid's algorithm computes $\gcd(a,b)$ in n steps, then $a \ge f_{n+2}$. Induction on n shows that $f_n = (\phi^n - \psi^n)/\sqrt{5}$, where $\phi, \psi = (1 \pm \sqrt{5})/2$ are the roots of $\lambda^2 = \lambda + 1$. Now f_n is an integer, and $|\psi^n/\sqrt{5}| < 1/2$ for all n, so $f_n = \{\phi^n/\sqrt{5}\}$ and hence $a \ge \{\phi^{n+2}/\sqrt{5}\} \approx \phi^{n+2}/\sqrt{5}$. Hence n is bounded above by approximately $\log_\phi(a\sqrt{5}) - 2 = \log_\phi(a) + \frac{1}{2}\log_\phi(5) - 2 \approx 4.785 \log_{10}(a) - 0.328$. (Since $\log_{10}(a)$ is approximately the number of digits in the decimal expansion of a, this says that Euclid's algorithm needs at most about $5k$ steps to compute the greatest common divisor of two k-digit numbers; in fact, the average number of steps is less than $2k$.)

1.22 If q,r in Corollary 1.2 satisfy $r \le |b|/2$, use these. Otherwise $|b|/2 < r < |b|$, so if $b > 0$ write $a = (q+1)b + (r - b)$ and replace q,r with $q+1, r - b$, where $-|b|/2 < r - b < 0 \le |b|/2$; similarly, if $b < 0$ write $a = (q - 1)b + (r + b)$. The proof of uniqueness is similar to that in Theorem 1.1. Now iterate this, writing $a = q_1 b + r_1$, $b = q_2 r_1 + r_2$, $r_1 = q_3 r_2 + r_3$, $\ldots$ with $-|b|/2 < r_1 \le |b|/2$ and $-|r_i|/2 < r_{i+1} \le |r_i|/2$; as in Euclid's algorithm, eventually some $r_n = 0$, and then

$\gcd(a, b) = |r_{n-1}|$. (This algorithm is generally faster than Euclid's (involving fewer divisions on average), by a factor of $\log_2 \phi \approx 0.694$; it tends to halt sooner because the remainders approach 0 faster.)

1.23 $1492 = 1.1066 + 426$, $1066 = 3.426 - 212$, $426 = 2.212 + 2$, $212 = 106.2 + 0$, so $\gcd(1066, 1492) = 2$; this takes four steps, rather than five for Euclid's algorithm in Example 1.3. $1745 = 1.1485 + 260$, $1485 = 6.260 - 75$, $260 = 3.75 + 35$, $75 = 2.35 + 5$, $35 = 7.5 + 0$, so $\gcd(1485, 1745) = 5$; this takes five steps, compared with six for Euclid's algorithm in Exercise 1.5.

1.24 If $a = f_{n+2}$ and $b = f_{n+1}$, then successive steps are $a = 2b - f_{n-1}$, $b = 3f_{n-1} - f_{n-3}$, $f_{n-1} = 3f_{n-3} - f_{n-5}$, etc. Apart from the first and last steps, each $q_i = 3$ and the remainders are alternate decreasing Fibonacci numbers. For instance, $a = f_{n+2} = f_{n+1} + f_n = 2f_{n+1} - f_{n-1} = 2b - f_{n-1}$ (since $f_n = f_{n+1} - f_{n-1}$), with $|f_{n-1}| < |f_{n+1}|/2$. Then $b = f_{n+1} = f_{n-1} + f_n = 2f_{n-1} + f_{n-2} = 3f_{n-1} - f_{n-3}$ (since $f_n = f_{n-1} + f_{n-2}$ and $f_{n-2} = f_{n-1} - f_{n-3}$), with $|f_{n-3}| < |f_{n-1}|/2$, etc.

1.25 The line $ax + by = c$ cuts the x- and y-axes at c/a and c/b, so its length in the first quadrant $(x, y \geq 0)$ is $c\sqrt{a^{-2} + b^{-2}}$. By Theorem 1.13, successive integer points (x, y) on this line have x- and y-coordinates differing by b and a, so they are distance $\sqrt{a^2 + b^2}$ apart. Hence the first quadrant contains such a point provided $\sqrt{a^2 + b^2} \leq c\sqrt{a^{-2} + b^{-2}}$, that is, $c \geq \sqrt{(a^2 + b^2)/(a^{-2} + b^{-2})} = ab$. If $ab - a - b = ax + by$ with $x, y \geq 0$ then $a|b(1 + y)$, so $a|(1 + y)$ and hence $a \leq 1 + y$; similarly $b \leq 1 + x$, so $ax + by \geq a(b - 1) + b(a - 1) = 2ab - a - b > ab - a - b$. (It can be shown that every integer $c > ab - a - b$ has the form $ax + by$, with $x, y \geq 0$.)

Chapter 2

2.1 Corollary 2.2, with each $a_i = a$, gives $p|a$, say $a = pq$; then $a^k = p^k q^k$ is divisible by p^k. This can fail if p is composite, e.g. 4 divides 2^k but not 2, where $k \geq 2$.

2.2 $132 = 2^2.3.11$, $400 = 2^4.5^2$, $1995 = 3.5.7.19$, so $\gcd(132, 400) = 2^2 = 4$, $\gcd(132, 1995) = 3$, $\gcd(400, 1995) = 5$ and $\gcd(132, 400, 1995) = 1$.

2.3 (a) Yes: $a = pq$ with $\gcd(p, q) = 1$, so $a^2 = p^2 q^2$ and $\gcd(a^2, p^2) = p^2$. (b) No: take $a = p$ and $b = p^3$. (c) Yes: $a = pq_1$ and $b = pq_2$ with $\gcd(p, q_i) = 1$, so $ab = p^2 q_1 q_2$ and $\gcd(ab, p^4) = p^2$. (d) No: take $a = p = 2$.

2.4 $m^{1/n} \in \mathbb{Q}$ iff m is the n-th power of an integer: imitate Corollary 2.5, with n replacing 2.

2.5 $\mathrm{li}\, x$ and $x/\ln x$ have derivatives $1/\ln x$ and $(\ln x - 1)/(\ln x)^2$; these have ratio $\ln x/(\ln x - 1) \to 1$ as $x \to +\infty$, so l'Hôpital's rule gives the result.

2.6 Write $p = 3q + r$ where $r = 0, 1$ or 2. If $r = 0$ then $3|p$, so $p = 3$; hence if $p \neq 3$ then $r = 1$ or 2. Now imitate the proof of Theorem 2.9, with $m = 3p_1 \ldots p_k - 1$ coprime to 3, using $(3s + 1)(3t + 1) = 3(3st + s + t) + 1$.

2.7 $24, 25, 26, 27, 28$ are composite. $(k+1)! + 2, (k+1)! + 3, \ldots, (k+1)! + (k+1)$ are divisible by $2, 3, \ldots, k+1$ respectively, and are therefore composite.

2.8 $2^{32} = 2^4.2^{28} = (641 - 5^4).2^{28} = 641.2^{28} - (5.2^7)^4 = 641.2^{28} - (641 - 1)^4$ has the form $641q - 1$ by the Binomial Theorem, so $641|2^{32} + 1 = F_5$.

2.9 If a is odd then $a^m + 1$ is even and greater than 2, and hence composite; thus a is even. Now imitate the proof of Lemma 2.11, putting $x = a$ rather than $x = 2$.

2.10 If $a > 2$ then $a^m - 1 = (a - 1)(a^{m-1} + a^{m-2} + \cdots + 1)$ is composite; hence $a = 2$. If $m = rs$ with $r, s > 1$ then $a^m - 1 = (a^r)^s - 1 = (a^r - 1)((a^r)^{s-1} + (a^r)^{s-2} + \cdots + 1)$ is composite; hence m is prime.

2.11 No, since 3 does not divide $8 + 7 + \cdots + 3 = 50$. Yes, since 11 divides $-8 + 7 - \cdots - 7 + 3 = 0$.

2.12 $n = 157$ and 641 and 1103 are prime (check for prime factors $p \leq \sqrt{n}$), but $221 = 13.17$.

2.13 2, 3, 5, 7, 11, 13, 17, 19, 23, 29, 31, 37, 41, 43, 47, 53, 59, 61, 67, 71, 73, 79, 83, 89, 97.

2.14 $M_{13} = 2^{13} - 1 = 8191$ is prime, since no prime $p \leq \sqrt{M_{13}}$ (i.e. $p \leq 89$) divides it (use a calculator).

2.15 $247 = 13.19$ and $6887 = 71.97$.

2.16 $3992003 = 1997.1999$, with both 1997 and 1999 prime: write a program to test successive primes $p < \sqrt{3992003} < 2000$ as factors.

2.17 Only for $p = 3$. If $p \neq 3$ then $p = 3q \pm 1$ for some integer q, so $p^2 + 2 = 9q^2 \pm 6q + 3$ is divisible by 3, and is therefore composite.

2.18 If $a|p$ then $a|(p-1)! + 1$ since $p|(p-1)! + 1$; but if $a < p$ then a is also a factor of $(p-1)!$, so $a = 1$.

2.19 Allow the exponents e_i to be positive or negative integers; the uniqueness result is the same.

2.20 The multiples are $q, 2q, \ldots, iq$ where $iq \leq n < (i+1)q$, that is, $i \leq n/q < i + 1$, so there are $i = \lfloor n/q \rfloor$ of them. Hence each prime p divides $\lfloor n/p \rfloor$ of the factors $1, 2, \ldots, n$ in $n!$, each contributing a term p to the prime-power factorisation of $n!$; also p^2 divides $\lfloor n/p^2 \rfloor$ factors, each contributing an extra term p, and so on; the total number of terms p is $\lfloor n/p \rfloor + \lfloor n/p^2 \rfloor + \cdots$, a finite sum since $\lfloor n/p^i \rfloor = 0$ if $p^i > n$.

2.21 In decimal notation, the number of 0s is the greatest integer m such that $10^m|n$, so $m = \min(e, e')$ where $2^e \| n$ and $5^{e'} \| n$. In base b notation, where $b = p_1^{f_1} \ldots p_k^{f_k}$ and $p_i^{e_i} \| n$, it is the greatest m for which $b^m|n$, that is, $\min\{\lfloor e_1/f_1 \rfloor, \ldots, \lfloor e_k/f_k \rfloor\}$.

2.22 Use induction on n, and $F_n^2 = F_{n+1} + 2F_n - 2$.

2.23 $M_{17} = 131071$ is prime.

2.24 If $n < p \leq 2n$ then p divides $(2n)!$ but not $(n!)^2$, so $p|\binom{2n}{n}$. There are $\pi(2n) - \pi(n)$ such primes, each $> n$, and $\binom{2n}{n} < (1+1)^{2n} = 2^{2n}$ by the Binomial Theorem, so $n^{\pi(2n)-\pi(n)} < \prod_{n<p\leq 2n} p \leq \binom{2n}{n} < 2^{2n}$. Taking logarithms, $(\pi(2n) - \pi(n)) \lg n < 2n$, so $(\pi(2n) - \pi(n))/n < 2/\lg n$.

Chapter 3

3.1 If $m = 10q + r$ with $0 \leq r < 10$, then $m^2 = 100q^2 + 20qr + r^2$ has the same final digit as r^2; but $r^2 = 0, 1, 4, \ldots, 81$ never has final digit $2, 3, 7$ or 8 for $r = 0, 1, 2, \ldots, 9$. No, since the remainder on division by 4 is 3.

3.2 (a) 27, since $34 \times 17 \equiv 5 \times -12 = -60 \equiv 27$ with $0 \leq 27 < 29$.

   (b) -10, since $19 \times 14 \equiv -4 \times -9 = 36 \equiv -10$ with $|-10| \leq 23/2$.

   (c) 5, since $5^2 = 25 \equiv 6$, so $5^4 \equiv 6^2 = 36 \equiv -2$, giving $5^{10} = (5^4)^2.5^2 \equiv (-2)^2.6 = 24 \equiv 5$.

   (d) 3, since $1! + 2! + 3! + \cdots + 10! \equiv 1 + 2 + 6 + 24 = 33 \equiv 3 \bmod (10)$, using the fact that $10|n!$ if $n > 4$.

3.3 Since a and $a + 1$ are consecutive integers, one of them must be even, so $2|a(a + 1)(2a + 1)$. If 3 divides a or $a + 1$ then it divides $a(a + 1)(2a + 1)$; if 3 does not divide a or $a + 1$ then $a \equiv 1 \bmod (3)$, so $2a + 1 \equiv 2.1 + 1 = 3 \equiv 0 \bmod (3)$ and so $3|a(a + 1)(2a + 1)$. In either case $a(a + 1)(2a + 1)$ is divisible by 2 and by 3, and hence by $2.3 = 6$.

3.4 Imitate Example 3.7, using the moduli $n = 2, 3$ and 5 respectively.

3.5 $x \equiv 2, 7$ or $12 \bmod (15)$, that is, $x \equiv 2 \bmod (5)$.

3.6 $6x \equiv 2 \bmod (4)$ has general solution $x \equiv 1$ or $3 \bmod (4)$, but if we divide $a = 6$ and $b = 2$ by $m = 2$, the congruence $3x \equiv 1 \bmod (4)$ has general solution $x \equiv 3 \bmod (4)$.

3.7 (a) $x \equiv 4 \bmod (7)$.

   (b) No solutions.

   (c) $x \equiv 18 \bmod (50)$.

   (d) $x \equiv 19$ or $44 \bmod (50)$.

3.8 $x \equiv 53 \bmod (60)$.

3.9 $x \equiv 79 \bmod (252)$.

3.10 The given congruences are equivalent to $x \equiv 2 \bmod (4)$, $x \equiv 6 \bmod (7)$ and $x \equiv 2 \bmod (5)$, with solution $x \equiv 62 \bmod (140)$.

3.11 $x \equiv 169 \bmod (440)$.

3.12 $x \equiv 2$ or $46 \bmod (55)$.

3.13 (a) $2^4 = 16$ solutions, since $k = 3$ primes divide 168 and $168 \equiv 0$ mod (8).

(b) $1.2.0 = 0$ solutions, using $70 = 2.5.7$.

(c) $2.2 = 4$ solutions, using $91 = 7.13$.

(d) $1.1.3 = 3$ solutions, using $140 = 4.5.7$.

3.14 (a) No solutions, since $5 \not\equiv 4$ mod (7).

(b) $x \equiv 19$ mod (42).

(c) $x \equiv 1413$ mod (2200).

(d) The congruences are equivalent to $x \equiv 3$ or 7 mod (10), $x \equiv 13$ mod (24) and $x \equiv 22$ mod (45), with solution $x \equiv 157$ mod (360).

3.15 Imitate Example 3.14: the number is the least positive remainder of $x_0 = -35a_1 + 21a_2 + 15a_3$ mod (105), where a_1, a_2 and a_3 are the remainders mod (3), mod (5) and mod (7). These are small, so you can calculate x_0 in your head as fast as your friend can calculate the remainders, then reduce x_0 mod (105).

3.16 (a) $x \equiv 53$ mod (60).

(b) $x \equiv 2$ mod (4), $x \equiv 6$ mod (7), $x \equiv 2$ mod (5), equivalently $x \equiv 62$ mod (140).

(c) $x \equiv 3$ mod (6) and $x \equiv 2$ mod (5), equivalently $x \equiv -3$ mod (30).

3.17 $x \equiv 2$ mod (3) and $x \equiv 5, 8, 9$ mod (11), equivalently $x \equiv 5, 8, 20$ mod (33).

3.18 The general solution of $x \equiv 6$ mod (7), $x \equiv 2$ mod (6), $x \equiv 1$ mod (5) and $x \equiv 0$ mod (4) is $x \equiv 356$ mod (420), and the least non-negative solution is $x = 356$.

3.19 Choose k distinct primes $p_1, \ldots, p_k$. By Theorem 3.10 there is a solution x for the simultaneous congruences $x \equiv -i$ mod (p_i^2); then $x + i$ is not square-free for $i = 1, \ldots, k$.

3.20 (a) $\pm 1, \pm 3, \pm 5, 7$.

(b) $0, \pm 2, \pm 4, \pm 6$.

(c) $2, 3, 5, 7, 11, 13, 29$.

(d) No, since no integer x satisfies $x^2 \equiv 3$ mod (7).

3.21 If $r_1 + r_2 n_1 + r_3 n_1 n_2 + \cdots + r_k n_1 n_2 \ldots n_{k-1} \equiv s_1 + s_2 n_1 + s_3 n_1 n_2 + \cdots + s_k n_1 n_2 \ldots n_{k-1}$ mod (n), where $r_i, s_i \in R_i$, then $r_1 \equiv s_1$ mod (n_1), so $r_1 = s_1$; cancelling, and then dividing by n_1, we get $r_2 + r_3 n_2 + \cdots + r_k n_2 \ldots n_{k-1} \equiv s_2 + s_3 n_2 + \cdots + s_k n_2 \ldots n_{k-1} \bmod (n_2 \ldots n_k)$, so $r_2 \equiv s_2 \bmod (n_2)$ and hence $r_2 = s_2$; continuing, we get $r_i = s_i$ for all i. Thus the $n_1 \ldots n_k$ different choices for $r_1, \ldots, r_k$ give integers r in distinct classes mod (n); since $n = n_1 \ldots n_k$, these form a complete set of residues mod (n).

Chapter 4

4.1 [1] in $\mathbb{Z}_2$, none in $\mathbb{Z}_3$, $\pm[2]$ in $\mathbb{Z}_5$, none in $\mathbb{Z}_7$ or $\mathbb{Z}_{11}$, $\pm[5]$ in $\mathbb{Z}_{13}$, $\pm[4]$ in $\mathbb{Z}_{17}$. Conjecture: there are two roots if $p \equiv 1 \bmod (4)$, none if $p \equiv 3 \bmod (4)$, one if $p = 2$ (see the last paragraph of Section 4.1 for confirmation).

4.2 $3^{22} \equiv 1 \bmod (23)$ by Theorem 4.3, and $91 \equiv 3 \bmod (22)$, so $3^{91} \equiv 3^3 = 27 \equiv 4 \bmod (23)$.

4.3 Putting the fractions over a common denominator, we have $r = (n_1 + \cdots + n_{p-1})/(p-1)!$, where $n_i = (p-1)!/i$ for each i. Corollary 4.5 gives $(p-1)! \equiv -1 \bmod (p)$, so p does not divide the denominator. In the proof of Corollary 4.5, write $f(x) = a_0 + a_1 x + a_2 x^2 + \cdots$, so $a_1 = -n_1 - \cdots - n_{p-1}$; since p divides each a_i, it divides the numerator $n_1 + \cdots + n_{p-1}$ of r. After any cancelling, p still divides the numerator. Now $a_0 = (p-1)! + 1$ and $a_0 + a_1 p + a_2 p^2 + \cdots = f(p) = (p-1)! + 1 - p^{p-1}$, so $a_1 p + a_2 p^2 + \cdots = -p^{p-1}$, giving $a_1 = -p^{p-2} - a_2 p - a_3 p^2 - \cdots$; since p divides each a_i, p^2 divides a_1 provided $p > 3$.

4.4 511 fails: $2^9 = 512 \equiv 1 \bmod (511)$, and $511 \equiv 7 \bmod (9)$, so $2^{511} \equiv 2^7 \not\equiv 2 \bmod (511)$ and 511 is composite. However, 509 passes: $2^9 \equiv 3 \bmod (509)$, and $509 = 9.56 + 5$, so $2^{509} \equiv 3^{56}.2^5 \bmod (509)$; now $3^{10} = 59049 \equiv 5 \bmod (509)$ and $56 = 10.5 + 6$, so $2^{509} \equiv 5^5.3^6.2^5 = (2^2.5^3).(2.3^6).(2^2.5^2) = 500.1458.100 \equiv -9. - 69.100 = 900.69 \equiv -118.69 = -8142 = -16.509 + 2 \equiv 2 \bmod (509)$. Thus 509 could be prime or composite. (It is, in fact, prime.)

4.5 Clearly $2^n \equiv 2 \bmod (2)$. Also, $2^9 = 512 \equiv 1 \bmod (73)$ and $n \equiv 1 \bmod (9)$, so $2^n \equiv 2^1 = 2 \bmod (73)$. Finally, $2^{10} = 1024 \equiv -79 \bmod (1103)$, so $2^{30} \equiv -79^3 \equiv 2$ and hence $2^{29} \equiv 1 \bmod (1103)$; now $n = 29.5553 + 1 \equiv 1 \bmod (29)$, so $2^n \equiv 2^1 = 2 \bmod (1103)$. Thus $2, 73, 1103$ divide $2^n - 2$, so $2^n \equiv 2 \bmod (n)$.

4.6 For M_p (p prime), Corollary 4.4 gives $2^p \equiv 2 \bmod (p)$, so imitate the last paragraph of the proof of Theorem 4.7, with p replacing n. For F_n ($n \geq 0$), $t + 1 | t^e - 1$ if e is even, so put $t = 2^{2^n}$ and $e = 2^{2^n - n}$ (even since $2^n > n$), giving

$$F_n = 2^{2^n} + 1 \mid (2^{2^n})^{2^{2^n - n}} - 1 = 2^{2^n . 2^{2^n - n}} - 1 = 2^{2^{n+2^n - n}} - 1 = 2^{2^{2^n}} - 1.$$

Thus $2^{2^{2^n}} \equiv 1 \bmod (F_n)$, so $2^{F_n} = 2^{2^{2^n} + 1} = 2^{2^{2^n}}.2 \equiv 2 \bmod (F_n)$.

4.7 $5^{10} \equiv 1 \bmod (11)$ and $341 \equiv 1 \bmod (10)$, so $5^{341} \equiv 5 \bmod (11)$. However, $5^3 = 125 \equiv 1 \bmod (31)$, and $341 \equiv 2 \bmod (3)$, so $5^{341} \equiv 5^2 \not\equiv 5 \bmod (31)$. Thus $5^{341} \not\equiv 5 \bmod (341)$, so 341 fails the base 5 test. $7^3 = 343 \equiv 2 \bmod (341)$, so $7^{341} = (7^3)^{113}.7^2 \equiv 2^{113}.7^2 \bmod (341)$. Now $2^{10} \equiv 1 \bmod (341)$, so $2^{113} = (2^{10})^{11}.2^3 \equiv 2^3 = 8 \bmod (341)$, giving $7^{341} \equiv 8.7^2 = 392 \equiv 51 \not\equiv 7 \bmod (341)$. Thus 341 fails the base 7 test.

4.8 Starting with 1, applying g, f, g, g, f, g, g with $a = 2$, and reducing mod (91) each time, we get the values $1, 2, 4, 32, -45, 23, -34, 37$. Thus $2^{91} \equiv 37 \not\equiv 2 \bmod (91)$, so 91 fails the base 2 test and is composite.

4.9 $133 = 2^7 + 2^2 + 2^0$, so the binary notation is 10000101. Start with 1, apply g, f, f, f, f, g, f, g with $a = 2$, and reduce mod (133), to get $1, 2, 4, 16, -10, -33, 50, -27, -5$. Thus $2^{133} \equiv -5 \not\equiv 2 \bmod (133)$, so 133 fails the base 2 test and is composite.

4.10 Obvious if $3|a$ or $11|a$ respectively. Otherwise, $a^2 \equiv 1 \bmod (3)$ and $561 \equiv 1 \bmod (2)$ imply $a^{561} \equiv a \bmod (3)$, while $a^{10} \equiv 1 \bmod (11)$ and $561 \equiv 1 \bmod (10)$ imply $a^{561} \equiv a \bmod (11)$.

4.11 We need to show that $a^n \equiv a \bmod (n)$ for all a; since n is a product of distinct primes p, it is sufficient to show $a^n \equiv a \bmod (p)$ for all such $p|n$. This is obvious if $p|a$; otherwise, $a^{p-1} \equiv 1 \bmod (p)$ by Theorem 4.3, so $a^{n-1} \equiv 1 \bmod (p)$ since $p - 1|n - 1$, giving $a^n \equiv a \bmod (p)$.

4.12 Use Lemma 4.8: $1729 = 7.13.19$, and $1728 = 2^6.3^3$ is divisible by $6, 12$ and 18. Similarly, $2821 = 7.13.31$, and $2820 = 2^2.3.5.47$ is divisible by $6, 12$ and 30.

4.13 Lemma 1.8 needs $n-1$ divisible by $6, 22$ and $p-1$. Now $n = 7.23.p \equiv 1 \bmod (6)$ iff $p \equiv 5 \bmod (6)$, while $n \equiv 1 \bmod (22)$ iff $p \equiv 8 \bmod (11)$, and $n \equiv 1 \bmod (p - 1)$ iff $161 \equiv 1 \bmod (p - 1)$, that is, $p - 1|160$. The prime $p = 41$ satisfies all these, so $n = 7.23.41 = 6601$ is a Carmichael number.

4.14 $q_{i+1} = (2x_{i+1} - 3)/5^{i+1} = (2(x_i + 2.5^i q_i) - 3)/5^{i+1} = ((2x_i - 3) + 4.5^i q_i)/5^{i+1} = (q_i + 4q_i)/5 = q_i$. Since $2x_1 - 3 = 5$ we have $q_1 = 1$ and hence $q_i = 1$ for all i, so $k_i \equiv 2 \bmod (5)$.

4.15 The roots of $f(x) \equiv 0 \bmod (5)$ are $x_1 = 1, -1$. Now $f(1) = 25 = 5q_1$ with $q_1 = 5$, and $f'(x) = 3x^2 + 8x + 19$ so $f'(1) = 30$; since $q_1 \equiv f'(1) \equiv 0 \bmod (5)$, each $k_1 \equiv 0, 1, \ldots, 4 \bmod (5)$ satisfies equation (4.3), so we get five roots $x_2 \equiv 1 + 5k_1 \equiv 1, 6, 11, 16, 21 \bmod (5^2)$.

Taking $x_1 = -1$ gives $f(-1) = -15$, so $q_1 = -3$, and $f'(-1) = 14$, so $-3+14k_1 \equiv 0 \bmod (5)$, giving $k_1 \equiv 2 \bmod (5)$ and $x_2 \equiv -1+5k_1 \equiv 9 \bmod (5^2)$.

4.16 The only solution of $f(x) \equiv 0 \bmod (5)$ is $x \equiv x_1 = 2$, with $f(x_1) = 5q_1$ where $q_1 = 1$. Now $f'(x) = 3x^2 - 1$, so $f'(x_1) = 11$; solving $q_1 + 11k_1 \equiv 0 \bmod (5)$ gives $k_1 \equiv -1 \bmod (5)$, so the general solution of $f(x) \equiv 0 \bmod (5^2)$ is $x \equiv x_2 = x_1 + 5k_1 \equiv -3 \bmod (5^2)$. Repeating this, $f(x_2) = 5^2q_2$ where $q_2 = -1$, and $f'(x_2) = 26$. Solving $q_2 + 26k_2 \equiv 0 \bmod (5)$ gives $k_2 \equiv 1 \bmod (5)$, so the general solution of $f(x) \equiv 0 \bmod (5^3)$ is $x \equiv x_3 = x_2 + 5^2k_2 \equiv 22 \bmod (5^3)$.

4.17 Using $x^p = x$, and reducing coefficients mod (p), we can assume that $g(x) = \sum_{i=0}^{p-1} a_i x^i$ with $0 \le a_i < p$ for all i; there are p choices for each a_i, and hence p^p such polynomials g. If two such polynomials g_1, g_2 induce the same function, then $g_1 - g_2$ (of degree $d < p$) has p roots in $\mathbb{Z}_p$, so $g_1 = g_2$ by Corollary 4.2; thus distinct polynomials g induce distinct functions, so there are p^p polynomial functions. If A, B are finite sets, there are $|B|^{|A|}$ functions $A \to B$ (since there are $|B|$ possible images of each $a \in A$); hence there are p^p functions $\mathbb{Z}_p \to \mathbb{Z}_p$, so each is a polynomial function.

4.18 Since $(x - 1)\Phi_q(x) = x^q - 1$, the roots of $\Phi_q(x)$ in $\mathbb{Z}_p$ satisfy $x^q = 1$; they also satisfy $x^{p-1} = 1$ by Theorem 4.3; if $p \not\equiv 1 \bmod (q)$ then $1 = \gcd(p - 1, q) = (p - 1)u + qv$ by Theorem 1.7, so $x^1 = 1$; however, $\Phi_q(1) = q$, so there are no roots if $p \ne q$, and 1 is the only root if $p = q$. If $p \equiv 1 \bmod (q)$ then $x^{p-1} - 1 = (x^q - 1)(x^{p-1-q} + x^{p-1-2q} + \cdots + 1)$; this has $p - 1$ roots in $\mathbb{Z}_p$ by Theorem 4.3, and its factors have at most q and $p - 1 - q$ roots by Theorem 4.1, so $x^q - 1$ has exactly q roots, and $\Phi_q(x)$ has $q - 1$ (excluding 1).

4.19 Mod (3), the equation becomes $2x^2 + 1 \equiv 0$, with roots $x \equiv \pm 1$. Mod (7), it becomes $x^6 + 4x^2 + 3x + 3 \equiv 0$, with roots $x \equiv -2, 3$. The Chinese Remainder Theorem gives the roots $x \equiv 5, 10, 17, 19 \bmod (21)$.

4.20 If $\gcd(a, p) = 1$ then there is a unique class b of solutions of $ab \equiv 1 \bmod (p)$. This pairs off classes $a, b \in \mathbb{Z}_p \setminus \{0\}$; distinct pairs a, b cancel in $(p - 1)! \bmod (p)$, leaving only the self-paired factors $a \equiv b \equiv 1$ and $a \equiv b \equiv p - 1$, so $(p - 1)! \equiv 1.(p - 1) = -1 \bmod (p)$.

4.21 Modify Wilson's Theorem, using $(p-1)(p-2)(p-3) \equiv (-1).(-2)$ $.(-3) \equiv -6 \bmod (p)$.

4.22 $10585 = 5.29.73$; now $10584 = 2^3.3^3.7^2$ is divisible by $4, 28$ and 72, so 10585 is a Carmichael number by Lemma 4.8.

4.23 If $n = 13.61.p$ then to apply Lemma 4.8 we need $n - 1$ divisible by $12, 60$ and $p - 1$, that is, $13.61.p \equiv 1 \bmod (60)$ and mod $(p - 1)$. Equivalently, $p \equiv 37 \bmod (60)$ and $p - 1 | 13.61 - 1 = 792$, satisfied by $p = 37$ and 397, so $n = 29341$ and 314821 are Carmichael numbers.

Chapter 5

5.1 The units in $\mathbb{Z}_{12}$ are $[1]$, $[5]$, $[7]$ and $[11]$, all self-inverse. The units in $\mathbb{Z}_{15}$ are $[1]$, $[2]$, $[4]$, $[7]$, $[8]$, $[11]$, $[13]$ and $[14]$, with inverses $[1]$, $[8]$, $[4]$, $[13]$, $[2]$, $[11]$, $[7]$ and $[14]$ respectively.

5.2 $[a][b] = [ab]$ and $[b][a] = [ba]$; since $ab = ba$ for all $a, b \in \mathbb{Z}$, we have $[a][b] = [b][a]$ for all $[a], [b] \in \mathbb{Z}_n$.

5.3 If $r \in R$ then ar is coprime to n since a and r are, so ar is a unit. If $r, r' \in R$ and $ar \equiv ar' \bmod (n)$, then multiplying by an inverse u of a we get $r \equiv uar \equiv uar' \equiv r'$, so $r = r'$. Thus $r \neq r'$ implies $ar \not\equiv ar'$, so the $\phi(n)$ elements of aR lie in the $\phi(n)$ different classes of units, one in each class.

5.4 $U_{14} = \{\pm[1], \pm[3], \pm[5]\}$, so $\phi(14) = 6$. Then $(\pm 1)^6 = 1$, $(\pm 3)^6 = 729 \equiv 1$ and $(\pm 5)^6 = 15625 \equiv 1$.

5.5 The four rows of five entries are **1**, **2**, **3**, 4, 5; 6, **7**, 8, **9**, 10; **11**, 12, **13**, 14, 15, and 16, **17**, 18, **19**, 20, with units in boldface. There are $\phi(20) = 8$ units, with $\phi(4) = 2$ in each of $\phi(5) = 4$ columns.

5.6 $42 = 2.3.7$, so $\phi(42) = 42(1 - \frac{1}{2})(1 - \frac{1}{3})(1 - \frac{1}{7}) = 12$; a reduced set of residues, with 12 elements, is given by the set $\{1, 5, 11, 13, 17, 19, 23, 25, 29, 31, 37, 41\}$.

5.7 $\phi(n)$ is odd iff $n \leq 2$. Clearly $\phi(1) = \phi(2) = 1$ is odd. Any $n > 2$ is divisible by an odd prime p or by 4; $\phi(n) = \prod_i p_i^{e_i-1}(p_i - 1)$ (Corollary 5.7) then gives $(p-1)|\phi(n)$ or $2|\phi(n)$ respectively, so $\phi(n)$ is even. $\phi(n) = 2, 4, 6, 8, 10, 12$ for $n = 3, 5, 7, 15, 11, 13$. If $\phi(n) = 14$ then $7|\phi(n)$, so either $7^2|n$ or a prime $p|n$ where $7|(p-1)$; in the first case $6|\phi(n)$, in the second case $p > 15$ so $14 < (p-1)|\phi(n)$, each contradicting $\phi(n) = 14$.

5.8 If a prime-power p^e divides n then $(p-1)p^{e-1}$ divides $\phi(n) = m$, so $p^e \leq mp/(p-1) \leq 2m$. There are only finitely many prime-powers $p^e \leq 2m$, and hence only finitely many products n of such prime-powers.

5.9 $\phi(n)/n = \prod_{p|n}(1 - \frac{1}{p})$; since $(1 - \frac{1}{2})(1 - \frac{1}{3})(1 - \frac{1}{5}) = \frac{4}{15} > \frac{1}{4}$, no n with at most three prime factors p has $\phi(n)/n < 1/4$; the smallest four primes achieving this are $2, 3, 5, 7$, with $\phi(n)/n = 8/35$, so the smallest n with $\phi(n)/n < 1/4$ is $n = 2.3.5.7 = 210$.

5.10 For each $d|n$, $\{1, 2, \ldots, n\}$ contains a set A_d of n/d multiples of d. Now $\gcd(i, n) > 1$ iff some prime p divides i and n, so $\phi(n) =$

$n - |\cup_{p|n} A_p|$. The Inclusion-Exclusion Principle gives $|\cup_{p|n} A_p| = \sum_{p|n} |A_p| - \sum_{p,q|n} |A_p \cap A_q| + \sum_{p,q,r|n} |A_p \cap A_q \cap A_r| - \cdots$, where the summations are over sets of distinct primes dividing n. Now $|A_p| = n/p$, $A_p \cap A_q = A_{pq}$ has n/pq elements, etc., so $\phi(n) = n - \sum_{p|n} n/p + \sum_{p,q|n} n/pq - \cdots$, which is the expression obtained by expanding $n \prod_{p|n}(1 - \frac{1}{p})$.

5.11 If $n = p^e$ (p prime) then $\phi(n)/n = 1 - 1/p$; choosing $p > 1/\varepsilon$ (possible by Theorem 2.6) we get $\phi(n)/n > 1 - \varepsilon$.

5.12 $\phi(1)+\phi(2)+\phi(3)+\phi(4)+\phi(6)+\phi(12) = 1+1+2+2+2+4 = 12$, with $S_1 = \{12\}, S_2 = \{6\}, S_3 = \{4,8\}, S_4 = \{3,9\}, S_6 = \{2,10\}, S_{12} = \{1,5,7,11\}$.

5.13 If $n = p^e$ then $d = p^f$ where $0 \leq f \leq e$; now $\phi(d) = p^f - p^{f-1}$ for $f \geq 1$, and $\phi(d) = 1$ for $f = 0$, so $\sum_{d|n} \phi(d) = \sum_{f=1}^{e}(p^f - p^{f-1}) + 1 = p^e$.

5.14 $\phi(1000) = 1000(1 - \frac{1}{2})(1 - \frac{1}{5}) = 400$, so $a^{400} \equiv 1 \bmod (1000)$; now $2001 \equiv 1 \bmod (400)$, so $a^{2001} \equiv a \bmod (1000)$ and the last three decimal digits agree.

5.15 EULER, with $k = 7$. (Apologies to Bribo and Karkx, if they exist.)

5.16 TDNZZ, FERMAT (the decoding transformation is $x \mapsto 15x + 7$ mod (26), the inverse of encoding).

5.17 Encoding is $x \mapsto x^5 \bmod (29)$; now $9^2 \equiv -6$, so $9^4 \equiv (-6)^2 \equiv 7$ and $9^5 \equiv 7.9 \equiv 5$. Decoding is $x \mapsto x^{17} \bmod (29)$; now $11^2 \equiv 5, 11^4 \equiv 5^2 \equiv -4, 11^8 \equiv (-4)^2 \equiv -13, 11^{16} \equiv (-13)^2 \equiv -5$, so $11^{17} \equiv -5.11 \equiv 3$.

5.18 $e = 25$, since $10^2 \equiv 13, 10^4 \equiv -5, 10^8 \equiv -4, 10^{16} \equiv -13, 10^{24} \equiv -4. -13 \equiv -6, 10^{25} \equiv -6.10 \equiv 27$.

5.19 $n = 10147 = 73.139$ (both factors prime), so $\phi(n) = 72.138 = 9936$; the inverse of $e = 119$ in U_{9936} is $f = 167$, since $ef = 19873 = 2.9936 + 1$, so the decoding transformation is $x \mapsto x^{167} \bmod (10147)$.

5.20 $\phi(m)/m = \prod_{p|m}(1 - \frac{1}{p})$ and $\phi(n)/n = \prod_{q|n}(1 - \frac{1}{q})$ (with p and q primes), so $\phi(m)\phi(n)/mn = \prod_{p,q}(1 - \frac{1}{p})(1 - \frac{1}{q})$. Similarly, $\phi(mn)/mn = \prod_{r|mn}(1 - \frac{1}{r}) \geq \prod_{p,q}(1 - \frac{1}{p})(1 - \frac{1}{q})$, since each prime r dividing mn divides m or n, with equality iff no $p = q$, that is, $\gcd(m,n) = 1$.

5.21 $\phi(d) = d \prod_{p|d}(1 - \frac{1}{p})$ and $\phi(n) = n \prod_{q|n}(1 - \frac{1}{q})$, so $\phi(n)/\phi(d) = (n/d) \prod_{r}(1 - \frac{1}{r})$, where r ranges over the primes dividing n but not

d. Thus $\phi(n)/\phi(d)$ is a multiple of $\prod_r r. \prod_r (1 - \frac{1}{r}) = \prod_r (r - 1) \in \mathbb{Z}$, and is therefore an integer.

5.22 By Corollary 5.7, if $n = \prod p^e$ then $\phi(n) = \prod p^{e-1}(p-1)$, so $\phi(n) \equiv 2$ mod (4) iff $n = 4, p^e$ or $2p^e$ for some prime $p \equiv 3$ mod (4).

5.23 If $p|n$ then $p - 1 = \phi(p)|\phi(n) = 16$, so $p = 2, 3, 5$ or 17. If $p^e|n$ ($e > 1$) then $p^{e-1}|\phi(p^e)|16$, so $p = 2$ and $e \leq 5$. Thus $n = 2^e 3^a 5^b 17^c$ with $e \leq 5$ and $a, b, c \leq 1$. Examining the different cases gives $n = 17, 32, 34, 40, 48$ or 60.

5.24 (a) Put $1/2 = \phi(n)/n = \prod_{p|n}(1 - \frac{1}{p})$; for $p = 2, 3, 5, \ldots$ the values of $1 - \frac{1}{p}$ are $\frac{1}{2}, \frac{2}{3}, \frac{4}{5}, \ldots$, and the only product of these equal to $1/2$ is given by $p = 2$, so $n = 2^e$ ($e \geq 1$).

(b) Similarly, for $\phi(n)/n = 1/3$ the only possible product is $\frac{1}{2}.\frac{2}{3}$ (otherwise the denominator is divisible by a prime $p > 3$), so $n = 2^e 3^f$ ($e, f \geq 1$).

Chapter 6

6.1 In U_9, the elements $1, 2, 4, 5, 7, 8$ have order $1, 6, 3, 6, 3, 2$ respectively. In U_{10}, the elements $1, 3, 7, 9$ have order $1, 4, 4, 2$ respectively.

6.2 Follow the proof of Lemma 6.2 until $k | \gcd(l, m) = h$. Then $2^h = 1$ in U_n (since $2^k = 1$ and $k | h$), so $n | 2^h - 1$.

6.3 U_{10} is cyclic, generated by 3, since 3, $3^2 = 9$, $3^3 = 7$ and $3^4 = 1$ are the elements of U_{10}; U_{12} is non-cyclic, since $1^2, 5^2, 7^2, 11^2 = 1$ in U_{12}, so no element has order $\phi(12) = 4$.

6.4 If $n = 18$ we can take $a = 5$: the integers $a = 2, 3$ and 4 are not units mod (18), but $a = 5$ is, and it has $\phi(18) = 6$ distinct powers, namely $5, 7, 17, 13, 11, 1$. Similarly, for $n = 23, 27, 31$ we can take $a = 5, 2, 3$.

6.5 If a is a primitive root, then U_n is cyclic of order $m = \phi(n)$, generated by a, and hence generated by a^i iff $\gcd(i, m) = 1$. The number of such primitive roots a^i is $\phi(m) = \phi(\phi(n))$.

6.6 Since $5^1 = 5$, $5^2 = 4$, $5^3 = 6$, $5^4 = 2$, $5^5 = 3, 5^6 = 1$ in U_7, every element of U_7 is a power of 5.

6.7 For $d = 1, 2, 5, 10$, the elements of order d form the sets $\{1\}$, $\{10\}$, $\{3, 4, 5, 9\}$ and $\{2, 6, 7, 8\}$. The elements 2, 6, 7, 8 are generators.

6.8 In U_{25}, the powers of 2 are $2, 4, 8, 16, 7, 14, 3, 6, 12, 24 = -1, -2 = 23, -4 = 21, -8 = 17, -16 = 9, 18, 11, 22 = -3, -6 = 19, -12 = 13, 1$, so 2 has order $20 = \phi(25)$ and is therefore a primitive root.

6.9 2 is a primitive root mod (3^2), since it has order $\phi(3^2) = 6$ in U_{3^2}; hence it is a primitive root mod (3^e) for all e.

6.10 3 is a primitive root mod (7), with $3^6 = 729 \not\equiv 1 \bmod (7^2)$, so 3 is a primitive root mod (7^e) for $e = 2$, and hence for all e.

6.11 The elements $1, 3, 5, 7, 9, 11, 13, 15$ have orders $1, 4, 4, 2, 2, 4, 4, 2$ respectively.

6.12 $2^{e-1} \pm 1$, $-1 \not\equiv 1 \bmod (2^e)$, while $(2^{e-1} \pm 1)^2 = 2^{2(e-1)} \pm 2.2^{e-1} + 1 \equiv 1$ and $(-1)^2 \equiv 1 \bmod (2^e)$, so they have order 2. Conversely, if a has order 2 then $a^2 \equiv 1$, so $2^e | a^2 - 1 = (a-1)(a+1)$; either $a - 1$ or $a + 1 \equiv 2 \bmod (4)$, so 2^{e-1} divides $a + 1$ or $a - 1$; thus $a \equiv \pm 1$ mod (2^{e-1}) and hence $a \equiv \pm 1$ or $2^{e-1} \pm 1 \bmod (2^e)$. The element 1 has order 1, the other three have order 2.

6.13 First adapt Lemma 6.9 to show that $2^{n+2} \| 3^{2^n} - 1$ for all $n \geq 1$ (though not for $n = 0$). Now show that 3 has order 2^{e-2} in U_{2^e}.

by imitating Theorem 6.10 for $e \geq 4$ (it is obvious for $e = 3$). The powers 3^i then give half the elements of U_{2^e}, represented by integers congruent to 1 or 3 mod (8), so the elements -3^i, congruent to -1 or -3, give the other half.

6.14 5 is a primitive root mod (3^e) for $e = 2$, and hence for all e; being odd, it is a primitive root mod (2.3^e) for all e. Similarly, 3 is a primitive root mod (7^2), and hence mod (7^e) and mod (2.7^e) for all e.

6.15 5 is a primitive root mod (23) by Exercise 6.4. Now $4 \equiv 5^4$ mod (23), so putting $x = 5^i$ we get $5^{6i} \equiv 5^4$ mod (23), that is, $6i \equiv 4$ mod (22), with solutions $i \equiv 8, 19$ mod (22), so $x \equiv 5^8, 5^{19} \equiv \pm 7$ mod (23).

6.16 Equivalently, $x^4 \equiv 4$ mod (9) and $x^4 \equiv 4$ mod (11), with solutions $x \equiv \pm 4$ mod (9) and $x \equiv \pm 3$ mod (11), so the Chinese Remainder Theorem gives four roots $x \equiv \pm 14, \pm 41$ mod (99).

6.17 Using Theorem 6.10 we put $7 = -5^2$ and $x = \pm 5^i$ in U_{32}, where $0 \leq i < 8$. Then $-5^2 = (\pm 5^i)^{11} = \pm 5^{11i}$, so $x = -5^i$ with $11i \equiv 2$ mod (8), that is, $i = 6$, giving $x = -5^6 = 23$ in U_{32}.

6.18 $520 = 2^3.5.13$, so $U_{520} \cong U_8 \times U_5 \times U_{13} \cong C_2 \times C_2 \times C_4 \times C_4 \times C_3$, giving $e(520) = 12$; $123 \equiv 3$ mod (12), so $11^{123} \equiv 11^3 = 1331 \equiv 291$ mod (520).

6.19 If G is cyclic, it has an element of order $|G|$, and every element has order dividing $|G|$, so $e(G) = |G|$. Conversely, let $e(G) = |G| = \prod p_i^{e_i}$, with each p_i prime; since $e(G)$ is the lcm of the orders of the elements, G contains some g_i of order $p_i^{e_i}$ for each i; these elements commute and have coprime orders, so $\prod g_i$ has order $\prod p_i^{e_i}$ and therefore generates G. Thus $e(n) = \phi(n)$ if and only if U_n is cyclic, that is, $n = 1, 2, 4, p^e$ or $2p^e$, where p is an odd prime, by Theorem 6.11.

6.20 If not, then Theorem 6.15 gives $n = pq$ for primes $p < q$, with $p - 1, q - 1$ dividing $n - 1 = pq - 1$. Hence $q - 1$ divides $(pq - 1) - p(q - 1) = p - 1$, so $q \leq p$, which is false.

6.21 The hypotheses imply that a has order $p - 1$ in U_p, so $|U_p| \geq p - 1$. Hence $U_p = \{1, 2, \ldots, p - 1\}$, so these are all coprime to p and hence p is prime. Since a has order $p - 1 = |U_p|$, it is a primitive root. Take $p = F_4 = 65537$ and $a = 3$, so $q = 2$ and $(p - 1)/q = 2^{15}$; by repeated squaring and reduction mod (p) (with a calculator), check that $3^{2^{15}} \equiv -1$ mod (p), so $3^{2^{16}} \equiv 1$ and p is prime.

6.22 Wilson's Theorem (Corollary 4.5) gives $(p-1)! = -1$ in U_p; multiplying by $(p-1)^{-1} = (-1)^{-1} = -1$ gives $(p-2)! = (-1)^2 = 1$ in U_p, so $(p-2)! \equiv 1 \bmod (p)$ in $\mathbb{Z}$. If p is odd, then multiplying again by $(p-2)^{-1} = (-2)^{-1} = (p-1)/2$ gives $(p-3)! = (p-1)/2$ in U_p.

6.23 Let there be just n primes p. The n corresponding Mersenne numbers M_p are mutually coprime, and hence divisible by disjoint sets of primes; since there are only n primes, each M_p is divisible by a unique prime and is therefore a prime-power. However, 11 is prime and M_{11} ($= 23 \times 89$) is not a prime-power.

6.24 For $p = F_0 = M_2 = 3$ and for $p = F_1 = 5$ only. If $p = F_n = 2^{2^n} + 1$ with $n \geq 2$, then 2 has order $2^{n+1} < 2^{2^n} = \phi(p)$ in U_p; if $p = M_l = 2^l - 1$ with $p \geq 3$, then 2 has order $l < 2^l - 2 = \phi(p)$ in U_p.

6.25 There are $\phi(\phi(n))$ primitive roots $a^i \in U_n$, where a is any primitive root and $\gcd(i, \phi(n)) = 1$. Taking $n = 18$ (so $\phi(n) = 6$) and $a = 5$, we get $\phi(6) = 2$ primitive roots $5^1 \equiv 5$ and $5^5 \equiv 11 \bmod (18)$. For $n = 27$ there are $\phi(18) = 6$ primitive roots $2, 2^5 \equiv 5, 2^7 \equiv 20, 2^{11} \equiv 23, 2^{13} \equiv 11$ and $2^{17} \equiv 14 \bmod (27)$.

6.26 (a) Modify the first part of the proof of Theorem 6.7, writing $h = g + rp$ $(r = 1, \ldots, p-1)$, so that $h^{p-1} \equiv 1 - rpg^{p-2} \bmod (p^2)$; since r is coprime to p the proof continues, so h is a primitive root mod (p^2). Thus for each of the $\phi(p-1)$ primitive roots g mod (p), there are either $p-1$ or p primitive roots h mod (p^2) of the form $g + rp$ $(r = 0, \ldots, p-1)$. By Exercise 6.5, the total number of primitive roots mod (p^2) is $\phi(\phi(p^2)) = \phi(p(p-1)) = \phi(p)\phi(p-1) = (p-1)\phi(p-1)$, so each g yields exactly $p-1$ primitive roots h.

(b) Use Lemma 6.4, with $\phi(25) = 20$ divisible by the primes $q = 2$ and $q = 5$: thus 2 is a primitive root mod (25) since $2^4, 2^{10} \not\equiv 1$ mod (25), but 7 is not since $7^4 \equiv 1 \bmod (25)$. Similarly $(-7)^4 \equiv 1 \bmod (25)$, so -7 is not a primitive root mod (25).

Chapter 7

7.1 $1, 2, 7, 11$, using the quadratic formula and the four square roots of $b^2 - 4ac = 1$ in $\mathbb{Z}_{15}$.

7.2 No square roots, so no solutions; $\pm 4, \pm 10$, solutions $1, -3, 4, -6$.

7.3 $Q_1 = \{1\}, Q_2 = \{1\}, Q_3 = \{1\}, Q_4 = \{1\}, Q_5 = \{1, 4\}, Q_6 = \{1\}, Q_7 = \{1, 2, 4\}, Q_8 = \{1\}, Q_9 = \{1, 4, 7\}, Q_{10} = \{1, 9\}, Q_{11} = \{1, 3, 4, 5, 9\}, Q_{12} = \{1\}$.

7.4 Each of the $\phi(n)$ units $s \in U_n$ has a square $a \in Q_n$, and each $a \in Q_n$ has N square roots $s \in U_n$, so $\phi(n) = |Q_n|N$.

7.5 $\phi(60) = 60.\frac{1}{2}.\frac{2}{3}.\frac{4}{5} = 16$ and $N = 2^3 = 8$, so $|Q_{60}| = 16/8 = 2$. Thus $Q_{60} = \{1^2 = 1, 7^2 = 49\}$. By Example 3.18, the square roots of 1 are $\pm 1, \pm 11, \pm 19, \pm 29$; multiplying these by 7 gives $\pm 7, \pm 17, \pm 13, \pm 23$ as the square roots of 49.

7.6 $g = 2$ is a primitive root in U_{25} by Example 6.7, so taking successive powers of $g^2 = 4 \bmod (25)$ gives $4, 16, 14, 6, 24, 21, 9, 11, 19, 1$ as the elements of Q_{25}.

7.7 $-1 \equiv 28 = 2^2.7$, so $\left(\frac{-1}{29}\right) = \left(\frac{2}{29}\right)^2.\left(\frac{7}{29}\right) = \left(\frac{7}{29}\right)$, since $\left(\frac{2}{29}\right) = \pm 1$; now $7 \equiv 36 = 6^2$, so $\left(\frac{-1}{29}\right) = 1$ and $-1 \in Q_{29}$.

7.8 $3^3 \equiv -2$, so $3^{14} \equiv (-2)^4.3^2 \equiv 144 \equiv -1$ and hence $3 \notin Q_{29}$; $5^2 \equiv -4$, so $5^6 \equiv -64 \equiv -6$ and hence $5^{14} \equiv (-6)^2.(-4) \equiv -144 \equiv 1$, giving $5 \in Q_{29}$.

7.9 $P = \{1, \ldots, 14\}$, so $3P = \{3, 6, 9, 12, -14, -11, -8, -5, -2, 1, 4, 7, 10, 13\}$ with $\mu = 5$ odd, giving $3 \notin Q_{29}$; $5P = \{5, 10, -14, -9, -4, 1, 6, 11, -13, -8, -3, 2, 7, 12\}$ with $\mu = 6$ even, so $5 \in Q_{29}$; $10P = \{10, -9, 1, 11, -8, 2, 12, -7, 3, 13, -6, 4, 14, -5\}$ with $\mu = 5$ odd, so $10 \notin Q_{29}$.

7.10 Theorem 7.5 gives $\left(\frac{-2}{p}\right) = \left(\frac{-1}{p}\right)\left(\frac{2}{p}\right)$, so $-2 \in Q_p$ if and only if $\left(\frac{-1}{p}\right) = \left(\frac{2}{p}\right) (= \pm 1)$; Corollaries 7.7 and 7.10 show that this is equivalent to $p \equiv 1$ or $3 \bmod (8)$.

7.11 383 is prime, but $219 = 3.73$, so $\left(\frac{219}{383}\right) = \left(\frac{3}{383}\right)\left(\frac{73}{383}\right)$; quadratic reciprocity gives $\left(\frac{3}{383}\right) = -\left(\frac{383}{3}\right) = -\left(\frac{2}{3}\right) = 1$, and $\left(\frac{73}{383}\right) = \left(\frac{383}{73}\right) = \left(\frac{18}{73}\right) = \left(\frac{2}{73}\right)\left(\frac{3}{73}\right)^2 = \left(\frac{2}{73}\right) = 1$ by Corollary 7.10, so $\left(\frac{219}{383}\right) = 1$ and $219 \in Q_{383}$.

7.12 $-3 \in Q_p \Leftrightarrow p = 2$ or $p \equiv 1 \bmod (6)$; $5 \in Q_p \Leftrightarrow p = 2$ or $p \equiv \pm 1 \bmod (5)$; $6 \in Q_p \Leftrightarrow p \equiv \pm 1$ or $\pm 5 \bmod (24)$; $7 \in Q_p \Leftrightarrow p = 2$

or $p \equiv \pm 1, \pm 3, \pm 9 \mod (28)$; $10 \in Q_p \Leftrightarrow p \equiv \pm 1, \pm 3, \pm 9$ or ± 13 $\mod (40)$; $169 = 13^2 \in Q_p \Leftrightarrow p \neq 13$.

7.13 By repeated squaring we have $3^{2^i} \equiv 9, 81, -121, -8, 64, -16, -1$ $\mod (257)$ for $i = 1, \ldots, 7$, so $3^{128} \equiv -1$ and $F_3 = 257$ is prime.

7.14 $16^2 = 6 + 5^3.2$; solving $2 + 32k \equiv 0 \mod (5)$ gives $k = -1$ and $s = 16 + 5^3.(-1)$, so the square roots are ± 109.

7.15 $-3 \equiv 2^2 \mod (7)$, so take $r = 2$; then $2^2 = -3 + 7.1$, so $q = 1$; solving $1 + 4k \equiv 0 \mod (7)$, take $k = -2$, so $s = 2 + 7.(-2) = -12$, giving ± 12 as the square roots of -3 in $\mathbb{Z}_{7^2}$. Repeat with $r = 12$, $12^2 = -3 + 7^2.3$, so $q = 3$; solving $3 + 24k \equiv 0 \mod (7)$, take $k = -1$, so $s = 12 + 7^2.(-1) = -37$, giving ± 37 as the square roots of -3 in $\mathbb{Z}_{7^3}$.

7.16 $41 \equiv 3^2 \mod (2^5)$, so take $r = 3$; then $3^2 = 41 + 2^5.(-1)$, so $q = -1$; taking $k = 1$ makes $q + rk = 2$ even, so $s = 3 + 2^4.1 = 19$ is a square root of $41 \mod (2^6)$. Multiplying this by ± 1 and by $2^5 \pm 1 = \pm 31$ we find that the square roots of $41 \mod (2^6)$ are ± 19 and ± 13.

7.17 $49 \equiv 1 \mod (2^4)$, with square roots $s \equiv \pm 1, \pm 7 \mod (2^4)$, and $49 \equiv 4 \mod (3^2)$, with square roots $s \equiv \pm 2 \mod (3^2)$. Hence $s \equiv \pm 7, \pm 25, \pm 47, \pm 65 \mod (144)$.

7.18 $168 = 2^3.3.7$. Now $25 \equiv 1 \mod (2^3)$, with square roots $s \equiv 1 \mod (2)$; $25 \equiv 1 \mod (3)$, with square roots $s \equiv \pm 1 \mod (3)$; $25 \equiv 4 \mod (7)$, with square roots $s = \pm 2 \mod (7)$. Hence $s \equiv \pm 5, \pm 19 \mod (42)$, that is, $s \equiv \pm 5, \pm 19, \pm 23, \pm 37, \pm 47, \pm 61, \pm 65, \pm 79 \mod (168)$.

7.19 There are no roots in $\mathbb{Z}$ since 5, 41 and 205 are not perfect squares. Now argue as in Example 7.16: $41 \equiv 1 \mod (8)$, so $41 \in Q_{2^e}$ for all e; $41 \equiv 1 \mod (5)$, so $41 \in Q_{5^e}$ for all e; $(\frac{5}{41}) = (\frac{41}{5}) = (\frac{1}{5}) = 1$, so $5 \in Q_{41^e}$ for all e; if $p \neq 2, 5, 41$ then $(\frac{205}{p}) = (\frac{5}{p})(\frac{41}{p})$, so at least one of $5, 41, 205 \in Q_{p^e}$ for all e.

7.20 If $p_1, \ldots, p_k$ are the only primes $p \equiv 1 \mod (2^r)$, define $a = 2p_1 \ldots p_k$ and $m = a^{2^{r-1}} + 1$, divisible by an odd prime p. Then a has order 2^r in U_p, so $2^r | p - 1$ by Lagrange's Theorem. Thus $p \equiv 1 \mod (2^r)$, so $p = p_i$ for some i and hence $p|a$. But $p|m$, so $p|m - a^{2^{r-1}} = 1$, which is impossible.

7.21 By Theorem 7.15, $-1 \in Q_n$ iff $-1 \in Q_{p^e}$ for all $p^e || n$. Theorem 7.14 gives $-1 \in Q_{2^e}$ iff $e = 0$ or 1. If $p > 2$ then Theorem 7.13 gives $-1 \in Q_{p^e}$ iff $-1 \in Q_p$, that is, iff $p \equiv 1 \mod (4)$ by Corollary 7.7.

Thus $-1 \in Q_n$ iff n is not divisible by 4 or by any prime $p \equiv 3$ mod (4).

7.22 Argue as in Example 7.16. The roots $\pm\sqrt{q}, \pm\sqrt{r}, \pm\sqrt{qr}$ of $h(x)$ are not integers. However, at least one of $q, r, qr \equiv 1$ mod (8), so it lies in Q_{2^e} for all e; $(\frac{q}{r}) = 1$, so $q \in Q_{r^e}$ for all e; quadratic reciprocity gives $(\frac{r}{q}) = (\frac{q}{r}) = 1$, so $r \in Q_{q^e}$ for all e; if $p \neq 2, q, r$ then $(\frac{qr}{p}) = (\frac{q}{p})(\frac{r}{p})$, so at least one of $q, r, qr \in Q_{p^e}$ for all e; hence $h(x) \equiv 0$ mod (n) has a solution for all n.

7.23 If $a \in Q_n$ then $a^i \in Q_n$ for all i; now Q_n is a proper subgroup of U_n for $n > 2$, so some $b \in U_n$ is not a power of a.

7.24 By Exercise 7.23, no quadratic residue can be a primitive root. Now $|Q_p| = |U_p|/2$, and if $p = F_n$ then by Exercise 6.5 the number of primitive roots is $\phi(\phi(p)) = \phi(2^{2^n}) = 2^{2^n-1} = |U_p|/2$, so the quadratic residues and the primitive roots account for all the elements of U_p. Conversely, if p has this property then there are $|U_p \setminus Q_p| = (p-1)/2$ primitive roots, so $(p-1)/2 = \phi(\phi(p)) = \phi(p-1)$; hence $p - 1$ is a power of 2 by Exercise 5.24(a), so $p = F_n$ for some n by Lemma 2.11.

7.25 $923 = 13.71$. Now $43 \equiv 2^2$ mod (13), so $43 \in Q_{13}$, and quadratic reciprocity (used twice) gives $(\frac{43}{71}) = -(\frac{71}{43}) = -(\frac{28}{43}) = -(\frac{7}{43}) = (\frac{43}{7}) = (\frac{1}{7}) = 1$, so $43 \in Q_{71}$. Hence $43 \in Q_{923}$.

7.26 $513 = 3^3.19$. The square roots of 7 mod (3^3) are ± 13, and mod (19) they are ± 8, so the Chinese Remainder Theorem gives the square roots $\pm 68, \pm 122$ mod (513).

7.27 There are $(p-1)/2$ summands $(\frac{a}{p}) = 1$, where $a \in Q_p$, and $(p-1)/2$ summands $(\frac{a}{p}) = -1$, where $a \in U_p \setminus Q_p$, so $\sum(\frac{a}{p}) = 0$. If g is a primitive root mod (p) then $\sum_{Q_p} a \equiv 1 + g^2 + g^4 + \cdots + g^{p-3} = (1 - g^{p-1})/(1 - g^2)$, with $g^{p-1} \equiv 1$ but $g^2 \not\equiv 1$ mod (p) for $p > 3$, so $\sum a \equiv 0$ mod (p).

Chapter 8

8.1 Theorem 7.15 gives $Q_n \cong Q_{n_1} \times \cdots \times Q_{n_k}$ if $n_1, \ldots, n_k$ are mutually coprime, so $|Q_n| = \prod_i |Q_{n_i}|$.

8.2 The proof of Lemma 8.1 shows that if $\gcd(m, n) = 1$ then the divisors d of mn correspond to pairs of divisors a, b of m and n, so $\tau(mn) = \tau(m)\tau(n)$. Similarly, $\sigma(mn) = \sum_{d|mn} d = \sum_{a|m} \sum_{b|n} ab = \sum_{a|m} a \cdot \sum_{b|n} b = \sigma(a)\sigma(b)$.

8.3 The function $N^k(n) = n^k$ is multiplicative, so $\sigma_k(n) = \sum_{d|n} N^k(n)$ is multiplicative by Lemma 8.1.

8.4 $\tau(n) = \prod_i (e_i + 1)$ is odd if and only if each e_i is even, that is, if and only if n is a perfect square.

8.5 $496 = 2^4.31$ has proper divisors $1, 2, 4, 8, 16, 31, 62, 124, 248$, with sum 496.

8.6 No: $M_{11} = 2^{11} - 1 = 23.89$ is not prime.

8.7 8128 and $33, 550, 336$, corresponding to $M_7 = 127$ and $M_{13} = 8191$ (see Exercise 2.14).

8.8 $n\sigma_{-1}(n) = n \sum_{d|n} d^{-1} = \sum_{d|n} n/d = \sum_{e|n} e = \sigma(n)$ (where $e = n/d$), so $\sigma(n) = 2n$ if and only if $\sigma_{-1}(n) = 2$.

8.9 $\mu(1) = 1 \in \mathbb{Z}$, and since $\mu(n) = -\sum_{d|n, d<n} \mu(d)$, strong induction gives $\mu(n) \in \mathbb{Z}$ for all $n \geq 1$.

8.10 $n = pq$ has proper divisors $d = 1, p, q$, with $\mu(1) = 1$, $\mu(p) = \mu(q) = -1$, so $\mu(pq) = -1 + 1 + 1 = 1$; $n = p^2$ has proper divisors $d = 1, p$, with $\mu(1) = 1$, $\mu(p) = -1$, so $\mu(p^2) = -1 + 1 = 0$.

8.11 The values of $\mu(n)$ for $n = 1, \ldots, 30$ (grouped in blocks of five) are given by 1, −1, −1, 0, −1; 1, −1, 0, 0, 1; −1, 0, −1, 1, 1; 0, −1, 0, −1, 0; 1, 1,−1, 0, 0; 1, 0, 0, −1, −1. This suggests that $\mu(n) = \pm 1$ if n is square-free, and $\mu(n) = 0$ otherwise (see Theorem 8.8).

8.12 Apply Theorem 8.6 to the equations $\sum_{d|n} 1 = \tau(n)$ and $\sum_{d|n} d = \sigma(n)$ defining τ and σ. For $n = 12$ we have $\sum_{d|12} \tau(d)\mu(n/d) = 1.0 + 2.1 + 2.0 + 3. - 1 + 4. - 1 + 6.1 = 1$, while $\sum_{d|12} \mu(d)\tau(n/d)$ is the same sum in reverse order; similarly, $\sum_{d|12} \sigma(d)\mu(n/d) = 1.0 + 3.1 + 4.0 + 7. - 1 + 12. - 1 + 28.1 = 12 = \sum_{d|12} \mu(d)\sigma(n/d)$.

8.13 $f(2) = 2^1\mu(2) + 2^2\mu(1) = -2 + 4 = 2$, blocks $WM \equiv MW$; $f(3) = 2^1\mu(3) + 2^3\mu(1) = -2 + 8 = 6$, blocks $WWM \equiv WMW \equiv MWW$ and $MMW \equiv MWM \equiv WMM$; $f(4) = 2^1\mu(4) + 2^2\mu(2) + 2^4\mu(1) =$

$0 - 4 + 16 = 12$, blocks $WWWM \equiv WWMW \equiv WMWW \equiv MWWW$ and $WWMM \equiv WMMW \equiv MMWW \equiv MWWM$ and $MMMW \equiv MMWM \equiv MWMM \equiv WMMM$.

8.14 With k sexes we have $\sum_{d|n} f(d) = k^n$, so $f(d) = \sum_{e|d} k^e \mu(d/e)$ by Theorem 8.6.

8.15 $|\mu(d)| = 1$ or 0 as d is square-free or not, so imitating the proof of Theorem 8.8 we get $\sum_{d|n} |\mu(d)| = \sum_{r=0}^{k} \binom{k}{r} = 2^k$, where n is divisible by k distinct primes. Alternatively, the function $f(n) = \sum_{d|n} |\mu(d)|$ is multiplicative (since $|\mu(n)|$ is), and $f(p^e) = 2$ if p is prime and $e \geq 1$.

8.16 $\tau(n) = \sum_{d|n} 1 = \sum_{d|n} u(d)u(n/d) = (u * u)(n)$, so $\tau = u * u$, and similarly $\sigma = N * u$.

8.17 $(f * I)(n) = \sum_{de=n} f(d)I(e) = f(n)$, since $I(e) = 1$ or 0 as $e = 1$ or $e > 1$, so $f*I = f$. Prove $I*f = f$ similarly, or use the commutativity of $*$.

8.18 $\tau = u * u$ implies $\tau * \mu = u$, and $\sigma = N * u$ implies $\sigma * \mu = N$.

8.19 Let m and n be coprime. Then $d|mn$ if and only if $d = ab$ where $a|m$ and $b|n$, so $f(mn) = \sum_{d|mn} g(d)h(mn/d) = \sum_{a|m} \sum_{b|n} g(ab) \times h(mn/ab)$. Now g is multiplicative and $\gcd(a,b) = 1$, so $g(ab) = g(a)g(b)$; similarly $h(mn/ab) = h(m/a)h(n/b)$, so

$$
\begin{aligned}
f(mn) &= \sum_{a|m} \sum_{b|n} g(a)g(b)h(m/a)h(n/b) \\
&= \sum_{a|m} g(a)h(m/a) . \sum_{b|n} g(b)h(n/b) = f(m)f(n).
\end{aligned}
$$

8.20 Suppose $f(n) \neq 0$ for some n. Since f is multiplicative, $f(n) = f(1.n) = f(1)f(n)$, so $f(1) = 1 \neq 0$ and $f \in G$. Let $g = f^{-1}$. If g is not multiplicative, choose the least mn such that $\gcd(m,n) = 1$ and $g(mn) \neq g(m)g(n)$. Lemma 8.11 gives $g(1) = 1/f(1) = 1$, so $mn > 1$. Then $(g * f)(mn) = I(mn) = 0$, so $0 = \sum_{d|mn} g(d)f(mn/d) = \sum_{a|m} \sum_{b|n} g(ab)f(mn/ab) = \sum_{ab<mn} g(a)g(b)f(m/a)f(n/b) + g(mn) = \sum_{a|m} g(a)f(m/a) . \sum_{b|n} g(b)f(n/b) - g(m)g(n) + g(mn) = I(m)I(n) - g(m)g(n) + g(mn) = g(mn) - g(m)g(n)$, so $g(mn) = g(m)g(n)$, against our choice of mn. Hence g is multiplicative and non-zero, so $g \in M$ and M is closed under Dirichlet inverses. If $f, g \in M$ then $f * g$ is multiplicative by Theorem 8.14, and non-zero since $(f * g)(1) = f(1)g(1) = 1$, so $f * g \in M$. Since M is non-empty, it is a subgroup of G.

8.21 (a) If m or n is even, $\chi(mn) = 0 = \chi(m)\chi(n)$; if m and n are odd, $\chi(mn) = 1 = \chi(m)\chi(n)$ or $\chi(mn) = -1 = \chi(m)\chi(n)$ as $m \equiv n$ or $m \not\equiv n$ mod (4).

(b) χ is multiplicative, and hence, by Lemma 8.1, so is $g = \chi * u$, given by $g(n) = \sum_{d|n} \chi(d) = \tau_1(n) - \tau_3(n)$. Now $g(2^e) = 1, g(p^e) = e + 1$, and $g(q^e) = 1$ or 0 as e is even or odd, where p, q are primes $\equiv 1, 3$ mod (4) respectively. Hence, if $n = 2^e \prod p_i^{e_i} \prod q_j^{f_j}$ then $g(n) = \prod(e_i + 1)$ if each f_j is even, and $g(n) = 0$ otherwise.

8.22 If $g(n)$ denotes the sum of the primitive n-th roots of 1 in $\mathbb{C}$, then $\sum_{d|n} g(d)$ is the sum of all the n-th roots of 1; this is $\sum_{j=1}^{n} \zeta^j$ where $\zeta = e^{2\pi i/n}$, equal to 1 or 0 as $n = 1$ or $n > 1$. Thus $g * u = I$ and hence $g = I * \mu = \mu$ by the Möbius Inversion Formula.

8.23 If $\gcd(m, n) = 1$, then $d^2 | mn$ iff $d = ab$ where $a^2 | m$ and $b^2 | n$; now imitate the proof of Lemma 8.1.

8.24 If $d | n = \prod_j p_j^{e_j}$ then $\Lambda(d) = 0$ unless $d = p_j^e$ where $0 < e \le e_j$, in which case $\Lambda(d) = \ln(p_j)$; hence $\sum_{d|n} \Lambda(d) = \sum_j e_j \ln(p_j) = \ln(n)$. Theorem 8.6 gives $\Lambda(n) = \sum_{d|n} \ln(d)\mu(n/d) = \sum_{d|n} \ln(n/d)\mu(d) = \sum_{d|n} \ln(n)\mu(d) - \sum_{d|n} \ln(d)\mu(d) = -\sum_{d|n} \ln(d)\mu(d)$ since $\ln(n) = 0$ or $\sum_{d|n} \mu(d) = 0$ as $n = 1$ or $n > 1$.

8.25 Use induction on k. The case $k = 1$ is trivial, and the case $k = 2$ follows from the definition of $*$. If $k > 2$ then $(f_1 * \cdots * f_k)(n) = (f * f_k)(n) = \sum_{dd_k=n} f(d)f_k(d_k)$, where $f = f_1 * \cdots * f_{k-1}$; by the induction hypothesis, $f(d) = \sum f_1(d_1) * \cdots * f_{k-1}(d_{k-1})$, summing over all products $d_1 \ldots d_{k-1} = d$, so substituting for $f(d)$ gives the result for $f_1 * \cdots * f_k$.

8.26 The rows of A are generators of the subgroup, with $\det(A)$ equal to its index. For each $d | n$ there is a unique a $(= n/d)$ and there are d possible values of b $(= 0, 1, \ldots, d-1)$, so the number of matrices (and hence subgroups) is $\sum_{d|n} d = \sigma(n)$. Similarly, subgroups of index n in $\mathbb{Z}^k$ correspond to matrices

$$\begin{pmatrix} d_1 & b_{12} & b_{13} & \cdots & b_{1k} \\ 0 & d_2 & b_{23} & \cdots & b_{2k} \\ 0 & 0 & d_3 & \cdots & b_{3k} \\ \vdots & \vdots & \vdots & \ddots & \vdots \\ 0 & 0 & 0 & \cdots & d_k \end{pmatrix}$$

where $d_1 \ldots d_k = n$ and $0 \le b_{ij} < d_j$ for all i, j, so by Exercise 8.25 the number of such subgroups is

$$\sum_{d_1 \ldots d_k = n} d_2 d_3^2 \ldots d_k^{k-1} = (u * N * N^2 * \cdots * N^{k-1})(n).$$

Chapter 9

9.1 The general term in the expansion of the product is $(-1)^k p_1^{-s} \ldots p_k^{-s}$, where $p_1, \ldots, p_k$ are distinct primes; this is $\mu(n)/n^s$ where $n = p_1 \ldots p_k$. Each square-free $n \in \mathbb{N}$ arises once, and every other n has $\mu(n) = 0$, so the product is $\sum \mu(n)/n^s$. The inverse of the product is $\prod_p (1 - p^{-s})^{-1} = \zeta(s)$, by equation (9.2), so the product is $1/\zeta(s)$.

9.2 $\zeta(s) = 1 + 2^{-s} + (3^{-s} + 4^{-s}) + (5^{-s} + \cdots + 8^{-s}) + (9^{-s} + \cdots + 16^{-s}) + \cdots \geq 1 + 2^{-s} + (4^{-s} + 4^{-s}) + (8^{-s} + \cdots + 8^{-s}) + (16^{-s} + \cdots + 16^{-s}) + \cdots = 1 + 2^{-s} + 2.4^{-s} + 4.8^{-s} + 8.16^{-s} + \cdots = 1 + 2^{-s} + 2^{1-2s} + 2^{2-3s} + 2^{3-4s} + \cdots = (1 + 1 + 2^{1-s} + (2^{1-s})^2 + (2^{1-s})^3 + \cdots)/2 = (1 + f(s))/2$. As $s \to 1+$, $2^{1-s} \to 1-$, so $f(s) \to +\infty$ and hence $\zeta(s) \to +\infty$.

9.3 $P_k(1) = \sum_{n \in A_k} 1/n \geq \sum_{n \leq p_k} 1/n$, since $n \leq p_k$ implies $n \in A_k$; also $\sum_{n \leq p_k} 1/n \to +\infty$ as $k \to \infty$ since the harmonic series diverges, so $P_k(1) \to +\infty$. If $n = p_1^{e_1} \ldots p_k^{e_k}$, with each $e_i \geq 1$, then Corollary 5.7 gives $\phi(n)/n = 1/P_k(1)$, so $\phi(n)/n \to 0$ as $k \to \infty$.

9.4 In the expansion of $Q_k(s) = \prod_{i=1}^{k} (1 - p_i^{-s})$, each $n \in A_k$ contributes $\mu(n)/n^s$, and no other n contributes, so $Q_k(s) = \sum_{n \in A_k} \mu(n)/n^s$. Now $\sum_{n=1}^{\infty} \mu(n)/n^s$ converges for $s > 1$ (since $\sum |\mu(n)/n^s|$ converges by comparison with $\sum 1/n^s$), so imitating the proof of Theorem 9.3 we have

$$\left| Q_k(s) - \sum_{n=1}^{\infty} \frac{\mu(n)}{n^s} \right| = \left| \sum_{n \notin A_k} \frac{\mu(n)}{n^s} \right| \leq \sum_{n \notin A_k} \left| \frac{\mu(n)}{n^s} \right| \leq \sum_{n > p_k} \left| \frac{\mu(n)}{n^s} \right|$$
$$= \sum_{n=1}^{\infty} \left| \frac{\mu(n)}{n^s} \right| - \sum_{n \leq p_k} \left| \frac{\mu(n)}{n^s} \right|.$$

This approaches 0 as $k \to \infty$, so $Q_k(s) \to \sum_{n=1}^{\infty} \mu(n)/n^s$, giving the first equation. The second equation follows immediately from Theorem 9.3.

9.5 If $\gcd(x, y) = d > 1$ then the integer point $(x/d, y/d)$ lies strictly between O and $A = (x, y)$, so the latter is invisible. Conversely, the integer points on OA are the multiples $q(x', y') = (qx', qy')$ ($q = 0, 1, 2, \ldots$) of the closest such point (x', y') to O (other than O itself), so $x = qx'$ and $y = qy'$ for some $q | \gcd(x, y)$; if A is invisible then $q > 1$, so $\gcd(x, y) > 1$.

9.6 Imitate methods A, B, C, which deal with $s = 2$. In A, use $\Pr(\gcd(x_1, \ldots, x_s) = n) = P(s)/n^s$ to show $1 = P(s)\zeta(s)$. In B, $\gcd(x_1, \ldots, x_s) = 1$ iff, for each prime p, some $x_i \not\equiv 0 \bmod (p)$; these independent events have probabilities $1-p^{-s}$, so $P(s) = \prod_p(1-p^{-s})$. In C, $\gcd(x_1,\ldots,x_s) > 1$ iff $x_1 \equiv \cdots \equiv x_s \equiv 0 \bmod (p)$ for some prime p; this has probability p^{-s}, so $S_1 = \sum_p p^{-s}, S_2 = \sum_{p<q}(pq)^{-s},\ldots$, giving $P(s) = \sum \mu(n)/n^s$.

9.7 Let $Q = \Pr(\mathrm{Sq}(x) = 1)$.

(A) $\Pr(\mathrm{Sq}(x) = n) = Q/n^2$, so $1 = Q\zeta(2)$.

(B) x is square-free iff $x \not\equiv 0 \bmod (p^2)$ for each prime p; these independent events have probabilities $1 - p^{-2}$, so $Q = \prod_p(1 - p^{-2})$.

(C) $\mathrm{Sq}(x) > 1$ iff $x \equiv 0 \bmod (p^2)$ for some prime p; this has probability p^{-2}, so use the Inclusion–Exclusion Principle, with $S_1 = \sum_p p^{-2}, S_2 = \sum_{p<q}(pq)^{-2},\ldots$, giving $Q = \sum \mu(n)/n^2$.

9.8 Imitate methods A, B, C in Exercises 9.6 and 9.7, replacing $\mathrm{Sq}(x)$ with $\mathrm{D}_s(x)$, the largest s-th power dividing x. Let $Q(s) = \Pr(\mathrm{D}_s(x) = 1)$.

(A) $\Pr(\mathrm{D}_s(x) = n) = Q(s)/n^s$, so $1 = Q(s)\zeta(s)$.

(B) x is s-th power-free iff $x \not\equiv 0 \bmod (p^s)$ for each prime p, so $Q(s) = \prod_p(1 - p^{-s})$.

(C) $\mathrm{D}_s(x) > 1$ iff $x \equiv 0 \bmod (p^s)$ for some prime p, so $S_1 = \sum_p p^{-s}, S_2 = \sum_{p<q}(pq)^{-s},\ldots$, giving $Q(s) = \sum \mu(n)/n^s$. Thus $Q(s) = P(s)$.

9.9 If π^{2k} is algebraic then $f(\pi^{2k}) = 0$ for some non-zero $f(x) \in \mathbb{Z}[x]$ (the set of polynomials with integer coefficients); then $g(\pi) = 0$ where $g(x) = f(x^{2k}) \in \mathbb{Z}[x]$ is non-zero, so π is algebraic, which is false. Hence π^{2k} is transcendental. Equation (9.10) gives $\zeta(2k) = q_k\pi^{2k}$ for some $q_k \in \mathbb{Q}$; if $f(\zeta(2k)) = 0$ for some non-zero $f(x) \in \mathbb{Z}[x]$, then $f(q_k\pi^{2k}) = 0$ and multiplying by a suitable power of the denominator of q_k we get $g(\pi^{2k}) = 0$ for some non-zero $g(x) \in \mathbb{Z}[x]$, which is impossible. Hence $\zeta(2k)$ is transcendental, and so by a similar argument is $P(2k) = 1/\zeta(2k)$. Any transcendental number is irrational, since a rational number a/b ($a, b \in \mathbb{Z}, b \neq 0$) is a root of $f(x) = bx - a \in \mathbb{Z}[x]$ and hence algebraic.

9.10 $\tau = u * u$, and $\sum u(n)/n^s = \zeta(s)$ converges absolutely for $s > 1$, so use Theorem 9.6.

9.11 $\sigma_k = N^k * u$, where $N^k(n) = n^k$ has Dirichlet series $\sum n^k/n^s = \zeta(s - k)$, absolutely convergent for $s > k + 1$. Hence Theorem 9.6 gives $\sum \sigma_k(n)/n^s = \zeta(s - k)\zeta(s)$ for $s > \max(k + 1, 1)$.

9.12 λ is multiplicative, and hence so is $\lambda * u = h$, given by $h(n) = \sum_{d|n} \lambda(d)$. If p is prime, then $h(p^e) = \sum_{i=0}^{e} \lambda(p^i) = 1-1+1-\cdots = 1$ or 0 as e is even or odd; hence if $n = p_1^{e_1} \ldots p_k^{e_k}$ then $h(n) = 1$ or 0 as every e_i is even or not, that is, as n is a square or not. The Dirichlet series $\sum \lambda(n)/n^s$ converges absolutely for $s > 1$, by comparison with $\zeta(s)$, so Theorem 9.6 gives $\zeta(s) \sum_n \lambda(n)/n^s = \sum_n h(n)/n^s = \sum_m 1/(m^2)^s = \sum_m 1/m^{2s} = \zeta(2s)$ for $s > 1$, where we put $n = m^2$ if n is a square.

9.13 Define $f(n) = 1$ or 0 as n is prime or not. Then $\nu(n) = \sum_{d|n} f(d)$, so $\nu = f * u$. Now Theorem 9.6 gives the result, valid for $s > 1$ (where the Dirichlet series $\zeta(s)$ and $\sum_p 1/p^s$ for u and f converge absolutely).

9.14 The function $|\mu|$ is multiplicative, and is bounded above by 1, so Corollary 9.8 applies for $s > 1$. If p is prime then $|\mu(p^e)| = 1$ or 0 as $e \le 1$ or $e > 1$, so $\sum |\mu(n)|/n^s = \prod_p (1 + p^{-s}) = \prod_p ((1 - p^{-2s})/(1 - p^{-s})) = \prod_p (1 - p^{-s})^{-1} / \prod_p (1 - p^{-2s})^{-1} = \zeta(s)/\zeta(2s)$. By Exercise 9.12, λ has Dirichlet series $\zeta(2s)/\zeta(s)$, so $\lambda * |\mu| = I$ by Theorem 9.6, giving $\sum_{d|n} \lambda(d)|\mu(n/d)| = 0$ for all $n > 1$. Now $|\mu(n/d)| = 1$ or 0 as n/d is square-free or not, giving the result.

9.15 Imitate the proofs of Theorems 9.3 (with $s = 1$) and 9.7 (with $f(n) = 1/n$): $\prod_{i=1}^{k} (1 - p_i^{-1})^{-1} = \prod_{i=1}^{k} (1 + p_i^{-1} + p_i^{-2} + \cdots) = \sum_{n \in A_k} n^{-1} \ge \sum_{n < p_{k+1}} n^{-1} \to +\infty$ as $k \to \infty$ (because $\sum_{n=1}^{\infty} n^{-1}$ diverges), so $\prod_{i=1}^{k} (1 - p_i^{-1}) \to 0$ as $k \to \infty$ and hence $\prod_{p \le x} (1 - p^{-1}) \to 0$ as $x \to +\infty$.

9.16 If $f(n) = 0$ for all n, then $F(s)$ converges absolutely for all s, so $\sigma_a = -\infty$. If $f(n) = 2^n$ for all n, then $|f(n)/n^s| \to \infty$ as $n \to \infty$ for each s, so $F(s)$ converges absolutely for no s, and $\sigma_a = +\infty$.

9.17 Theorem 9.10 gives $\zeta'(s) = -\sum \ln(n)/n^s$. Exercise 8.24 gives $u * \Lambda = \ln$, so Theorem 9.6 gives $\zeta(s) \sum \Lambda(n)/n^s = \sum \ln(n)/n^s = -\zeta'(s)$.

9.18 $\tau_k = u * \cdots * u$ (k terms) by Exercise 8.25, so repeated use of Theorem 9.6 gives the result.

9.19 $\zeta(s)^4/\zeta(2s) = \prod_p (1 - p^{-2s})/(1 - p^{-s})^4 = \prod_p (1 + p^{-s})/(1 - p^{-s})^3$. Now $(1 + t)/(1 - t)^3 = \sum_{e=0}^{\infty} (e + 1)^2 t^e$, so $\zeta(s)^4/\zeta(2s) = \prod_p \sum_e (e + 1)^2 p^{-es}$. In the expansion of this, each integer $n = \prod p_i^{e_i}$ gives a term n^{-s} with coefficient $\prod (e_i + 1)^2 = \tau(n)^2$, so $\zeta(s)^4/\zeta(2s) = \sum_n \tau(n)^2/n^s$.

9.20 For each prime p there are at most x/p^2 multiples of p^2 between 1 and x, so $q(x) \ge x - \sum_p x/p^2 \ge x(1 - \sum_{n \ge 2} 1/n^2) = x(2 - (\pi^2/6))$. Each square-free integer $m \leq x$ is a product of distinct primes $p \leq x$; there are $\pi(x)$ such primes, and hence $2^{\pi(x)}$ sets of such primes, so $q(x) \leq 2^{\pi(x)}$. Thus $2^{\pi(x)} \geq x\big(2-(\pi^2/6)\big)$, so $\pi(x) \geq \log_2 x + \log_2\big(2-(\pi^2/6)\big)$.

9.21 Exercise 8.26 shows that $f_k = u * N * N^2 * \cdots * N^{k-1}$; the functions $u, N, \ldots, N^{k-1}$ have Dirichlet series $\zeta(s), \zeta(s-1), \ldots, \zeta(s-k+1)$, so repeated use of Theorem 9.6 shows that f_k has Dirichlet series $\zeta(s)\zeta(s-1)\ldots\zeta(s-k+1)$. This is valid provided $\sigma, \sigma-1, \ldots, \sigma-k+1 > 1$, that is, $\sigma > k$.

Chapter 10

10.1 The primes $p < 100$ in S_2 are $2 = 1^2 + 1^2, 5 = 2^2 + 1^2, 13 = 3^2 + 2^2, 17 = 4^2 + 1^2, 29 = 5^2 + 2^2, 37 = 6^2 + 1^2, 41 = 5^2 + 4^2, 53 = 7^2 + 2^2, 61 = 6^2 + 5^2, 73 = 8^2 + 3^2, 89 = 8^2 + 5^2, 97 = 9^2 + 4^2$; these are the primes $p = 2$ or $p \equiv 1 \bmod (4)$ in this range.

10.2 $130 = 11^2 + 3^2$, $260 = 8^2 + 14^2$, $847 = 7.11^2 \notin S_2$, $980 = 28^2 + 14^2$, $1073 = 28^2 + 17^2$. (These are not unique, e.g. $130 = 9^2 + 7^2$.)

10.3 If $x^2 + y^2 = 50$ then $|x|, |y| \leq \sqrt{50}$; by inspection, there are twelve pairs: $(\pm 1, \pm 7)$, $(\pm 7, \pm 1)$, $(\pm 5, \pm 5)$.

10.4 (a) $\Rightarrow$ (b): $uv = 1$ in $\mathbb{Z}[i]$ gives $d(u)d(v) = 1$ in $\mathbb{Z}$, so $d(u) = 1$ since $d(u) \geq 0$.

(b) $\Rightarrow$ (c): if $u = x + yi$, then $1 = d(u) = x^2 + y^2$ implies $x = \pm 1$ and $y = 0$, or $x = 0$ and $y = \pm 1$, that is, $u = \pm 1$ or $\pm i$.

(c) $\Rightarrow$ (a): ± 1 and $\pm i$ have inverses ± 1 and $\mp i$ in $\mathbb{Z}[i]$.

10.5 If $a, b \neq 0$, then $d(ab) = d(a)d(b) \geq d(b)$ since $d(a), d(b) \geq 1$. Then $d(ab) = d(b)$ iff $d(a) = 1$, that is, a is a unit, by Exercise 10.4.

10.6 $221 = 13.17$ with $13 \equiv 17 \equiv 1 \bmod (4)$, so $r(221) = 4.2.2 = 16$. Factors $z = u(3 \pm 2i)(4 \pm i) = u(10 \pm 11i)$ or $u(14 \pm 5i)$ give 16 representations of 221, each equivalent to $10^2 + 11^2$ or $14^2 + 5^2$.

10.7 $16660 = 2^2.5.17.7^2$, so $r(16660) = 4.2.2 = 16$. Factors $z = 14v(2 \pm i)(4 \pm i) = 14v(7 \pm 6i)$ or $14v(9 \pm 2i)$ give 16 representations of 16660, each equivalent to $98^2 + 84^2$ or $126^2 + 28^2$.

10.8 Exercise 8.21 shows that, if $n = 2^e \prod p_i^{e_i} \prod q_j^{f_j}$ where p_i and q_j are primes congruent to 1 or 3 mod (4), then $\tau_1(n) - \tau_3(n) = \prod(e_i + 1)$ if each f_j is even, and $\tau_1(n) - \tau_3(n) = 0$ otherwise. Compare this with the similar formula for $r(n)$ obtained in Section 10.2.

10.9 $x^2 \equiv 0, 1$ or $4 \bmod (8)$ for all $x \in \mathbb{Z}$, so $x_1^2 + x_2^2 + x_3^2 \not\equiv 7 \bmod (8)$.

10.10 $x^2 \equiv 0$ or $1 \bmod (4)$ for all $x \in \mathbb{Z}$, so if $n = x_1^2 + x_2^2 + x_3^2 \equiv 0 \bmod (4)$ then each $x_i^2 \equiv 0$ and hence x_i is even. Thus $n/4 = (x_1/2)^2 + (x_2/2)^2 + (x_3/2)^2 \in S_3$.

10.11 Suppose that $n = 4^e(8k + 7) \in S_3$. Applying Exercise 10.10 repeatedly gives $8k + 7 \in S_3$, contradicting Exercise 10.9.

10.12 There are $3!.2^3 = 48$ representations of 14, obtained from $1^2 + 2^2 + 3^2$ by permuting the terms $1, 2, 3$ and multiplying them by ± 1. Similarly, $11 = 1^2 + 1^2 + 3^2$ gives $3.2^3 = 24$ representations (there are only three different permutations of $1, 1, 3$).

10.13　$247 = 13.19 = (3^2 + 2^2 + 0^2 + 0^2)(4^2 + 1^2 + 1^2 + 1^2)$, so identity (10.3) gives $247 = (12-2)^2 + (3+8)^2 + (3-2)^2 + (3+2)^2 = 10^2 + 11^2 + 1^2 + 5^2$. (This is not unique: we could have used $13 = 2^2 + 2^2 + 2^2 + 1^2$, for instance, or identity (10.4) instead of (10.3).) Similarly, $308 = 2^2.7.11$ with $7 = 2^2 + 1^2 + 1^2 + 1^2$ and $11 = 3^2 + 1^2 + 1^2 + 0^2$, so identity (10.3) gives $77 = (6-1-1)^2 + (2+3-1)^2 + (2+3+1)^2 + (1-1+3)^2 = 4^2 + 4^2 + 6^2 + 3^2$ and hence $308 = 8^2 + 8^2 + 12^2 + 6^2$. Finally $465 = 31.15 = (3^2 + 3^2 + 3^2 + 2^2)(3^2 + 2^2 + 1^2 + 1^2) = (9-6-3-2)^2 + (6+9+3-2)^2 + (3-3+9+4)^2 + (3+3-6+6)^2 = 2^2 + 16^2 + 13^2 + 6^2$.

10.14　192, since $5^2 + 1^2 + 1^2 + 1^2$, $4^2 + 2^2 + 2^2 + 2^2$ and $3^2 + 3^2 + 3^2 + 1^2$ each give $4.2^4 = 64$ representations.

10.15　If $q = a + bi + cj + dk$ then $q\bar{q} = (a + bi + cj + dk)(a - bi - cj - dk) = (a^2 + b^2 + c^2 + d^2) + 0i + 0j + 0k = |q|^2$. The second part follows immediately from formula (10.8).

10.16　$|q_1 q_2|^2 = (q_1 q_2).\overline{(q_1 q_2)} = (q_1 q_2).(\overline{q_2}.\overline{q_1}) = q_1.(q_2 \overline{q_2}).\overline{q_1} = q_1.|q_2|^2.\overline{q_1} = q_1 \overline{q_1}.|q_2|^2 = |q_1|^2.|q_2|^2$. Now put $q_i = a_i + b_i i + c_i j + d_i k$ for $i = 1, 2$, and use Exercise 10.15 and formula (10.8).

10.17　The intersection of $B_n(1)$ with a hyperplane $x_n = x$ $(-1 \leq x \leq 1)$ is an $(n - 1)$-dimensional ball of radius $r = \sqrt{1 - x^2}$; this has $(n - 1)$-dimensional volume $V_{n-1}r^{n-1}$, so $V_n = \int_{-1}^{1} V_{n-1}(1 - x^2)^{(n-1)/2}dx$. Putting $x = \cos\theta$ gives $V_n = -\int_\pi^0 V_{n-1} \sin^n \theta \, d\theta = 2V_{n-1} \int_0^{\pi/2} \sin^n \theta \, d\theta = 2V_{n-1}I_n$, and hence $V_n = (V_n/V_{n-1})(V_{n-1}/V_{n-2})\ldots(V_2/V_1)V_1 = 2I_n.2I_{n-1}\ldots 2I_2.2 = 2^n I_n I_{n-1} \ldots I_2$. Integrating by parts, $I_n = \int_0^{\pi/2} \sin^{n-1}\theta. \sin\theta \, d\theta = (n - 1) \times \int_0^{\pi/2} \sin^{n-2}\theta \cos^2\theta \, d\theta = (n - 1) \int_0^{\pi/2} \sin^{n-2}\theta(1 - \sin^2\theta) \, d\theta = (n - 1)I_{n-2} - (n - 1)I_n$, so $I_n = (n - 1)I_{n-2}/n$ for $n \geq 2$. Since $I_0 = \int_0^{\pi/2} 1 \, d\theta = \pi/2$ and $I_1 = \int_0^{\pi/2} \sin\theta \, d\theta = 1$, we have $I_n = (\frac{n-1}{n})(\frac{n-3}{n-2})\ldots(\frac{1}{2})(\frac{\pi}{2})$ or $(\frac{n-1}{n})(\frac{n-3}{n-2})\ldots(\frac{2}{3})$ as n is even or odd. Substituting for each factor in $2^n I_n I_{n-1} \ldots I_2$, then cancelling and collecting terms, we get the formula for V_n.

10.18　$\text{vol}(B_n(r)) = V_n r^n$, and Exercise 10.17 gives $V_n = 2, \pi, 4\pi/3, \pi^2/2$ for $n = 1, 2, 3, 4$.

10.19　A regular octagon, inscribed in the unit disc, is divided by the radii through its vertices into eight isosceles triangles, each of area $\frac{1}{2}\sin\frac{\pi}{4} = 1/2\sqrt{2}$. The disc has area π, so $\pi > 8/2\sqrt{2} = 2\sqrt{2}$.

10.20 Putting $x_1 = x/a, x_2 = y/b$ transforms X into the interior $x_1^2 + x_2^2 <$
1 of the unit circle, a disc of area π; this transformation multiplies
areas by $(dx_1/dx)(dx_2/dy) = 1/ab$, so X has area πab.

10.21 Let $v = (x_i)$ and $w = (y_i)$ be in $B_n(r)$, so $\sum x_i^2, \sum y_i^2 < r^2$; since
$\sum (x_i - y_i)^2 \geq 0$ we also have $2\sum x_i y_i \leq \sum x_i^2 + \sum y_i^2 < 2r^2$.
Now $tv + (1 - t)w = (tx_i + (1 - t)y_i)$, with $\sum (tx_i + (1 - t)y_i)^2 =$
$t^2 \sum x_i^2 + 2t(1 - t) \sum x_i y_i + (1 - t)^2 \sum y_i^2 < (t + (1 - t))^2 r^2 = r^2$
whenever $0 \leq t \leq 1$, so $tv + (1 - t)w \in B_n(r)$.

10.22 If $a > 2^n$ the set $X = \{\sum \alpha_i v_i \mid 0 < \alpha_1 < a, 0 < \alpha_i < 1$ for
$i = 2, \ldots, n\}$ is convex, but not centrally symmetric, with $\text{vol}(X) =$
$a\,\text{vol}(F) > 2^n\text{vol}(F)$, and it contains no lattice points; the set $X' =$
$X \cup (-X)$ is centrally symmetric, but not convex, with $\text{vol}(X') =$
$2\,\text{vol}(X) > 2^n\text{vol}(F)$, and it contains no lattice points.

10.23 Check closure to show that Λ is a subgroup of $\mathbb{R}^4$. Clearly each
$v_i \in \Lambda$. If $(x, y, z, t) \in \Lambda$ then $z = ux + vy + kp$ and $t = vx - uy + lp$
for some k, l so $(x, y, z, t) = xv_1 + yv_2 + kv_3 + lv_4$. Thus $v_1, \ldots, v_4$
generate Λ, and since they are linearly independent, Λ is a lattice.

10.24 If $p = 2x^2 + y^2$ in $\mathbb{Z}$ then $x, y \not\equiv 0 \bmod (p)$, so $-2 = (y/x)^2$ in
U_p, giving $-2 \in Q_p$. For the converse, let $-2 \in Q_p$, say $u^2 \equiv -2$
$\bmod (p)$, and imitate the proof of Theorem 10.2, with $X = \{(x, y) \in$
$\mathbb{R}^2 \mid 2x^2 + y^2 < 2p\}$, the interior of an ellipse, of area $\sqrt{2}\pi p$ (Exercise
10.20), and $\Lambda = \{(x, y) \in \mathbb{Z}^2 \mid y \equiv ux \bmod (p)\}$, a lattice with
$\text{vol}(F) = p$; now $\pi > 2\sqrt{2}$ (Exercise 10.19), so $\text{vol}(X) > 2^2\text{vol}(F)$;
Minkowski's Theorem gives a non-zero $(x, y) \in X \cap \Lambda$, so $p = 2x^2 + y^2$.
Exercise 7.10 gives $-2 \in Q_p$ if and only if $p \equiv 1$ or 3 mod (8).

10.25 If $p = x^2 + xy + y^2$ then $x^3 - y^3 = (x - y)(x^2 + xy + y^2) \equiv 0 \bmod (p)$;
now $x, y \not\equiv 0 \bmod (p)$, so $(x/y)^3 = 1$ in U_p; if $x \equiv y \bmod (p)$ then
$p|3x^2$, so $p = 3$ since $x \not\equiv 0 \bmod (p)$; if $x \not\equiv y \bmod (p)$ then x/y
has order 3 in U_p, so 3 divides $|U_p| = p - 1$ by Lagrange's Theorem.
Conversely, $p = 3$ clearly has the required form, so let $p \equiv 1 \bmod (3)$;
then U_p contains an element u of order 3 by Theorem 6.5, with
$1 + u + u^2 = (u^3 - 1)/(u - 1) = 0$ in $\mathbb{Z}_p$; now imitate the proof of
Theorem 10.2, with $X = \{(x, y) \in \mathbb{R}^2 \mid x^2 + xy + y^2 < 2p\}$, the
interior of an ellipse, and $\Lambda = \{(x, y) \in \mathbb{Z}^2 \mid y \equiv ux \bmod (p)\}$; the
ellipse has semiaxes $a = 2\sqrt{p/3}$ and $b = 2\sqrt{p}$ (along the lines $y = x$
and $y = -x$), so X has area $\pi ab = 4\pi p/\sqrt{3} > 4p = 2^2\text{vol}(F)$, and
Minkowski's Theorem gives the required $(x, y) \in X \cap \Lambda$.

10.26 The basic idea is that the number of integer lattice-points in a large
disc is given approximately by its area. For each $(x, y) \in \mathbb{Z}^2$, let

$S(x, y) = \{(u, v) \in \mathbb{R}^2 \mid |u - x|, |v - y| \leq 1/2\}$, a square of side-length 1, centred at (x, y). Each $(u, v) \in S(x, y)$ is within distance $1/\sqrt{2}$ of (x, y); thus if $x^2 + y^2 = m \leq n$ then $\sqrt{u^2 + v^2} \leq \sqrt{x^2 + y^2} + 1/\sqrt{2} \leq \sqrt{n} + 1/\sqrt{2} \; (= a$, say), so $S(x, y)$ lies in the disc $D(a)$ of radius a centred at $(0, 0)$. The squares have area 1, and meet only at their edges, while $D(a)$ has area πa^2, so at most πa^2 squares $S(x, y)$ can be inside $D(a)$. Thus πa^2 bounds the number of $(x, y) \in \mathbb{Z}^2$ with $x^2 + y^2 \leq n$, so $\sum_{m=0}^{n} r(n) \leq \pi a^2 = \pi(n + \sqrt{2n} + 1/2)$. Similarly, if $\sqrt{u^2 + v^2} \leq b = \sqrt{n} - 1/\sqrt{2}$ then $(u, v) \in S(x, y)$ for some (x, y) with $x^2 + y^2 \leq n$, so these squares $S(x, y)$ contain the disc $D(b)$ and hence $\sum_{m=0}^{n} r(n) \geq \pi b^2 = \pi(n - \sqrt{2n} + 1/2)$. Both $\pi(n \pm \sqrt{2n} + 1/2)/n \to \pi$ as $n \to \infty$, so $\frac{1}{n} \sum_{m=0}^{n} r(m) \to \pi$ and hence $\frac{1}{n} \sum_{m=1}^{n} r(m) \to \pi$. If $r_k(n)$ denotes the number of representations of n as a sum of k squares, then a similar argument in $\mathbb{R}^k$ shows that $\sum_{m=1}^{n} r_k(n)/n^{k/2} \to V_k$ as $n \to \infty$, so $\frac{1}{n} \sum_{m=1}^{n} r_k(n) \sim n^{(k-2)/2} V_k$.

10.27 In the notation of Exercises 8.21 and 10.8 we have $r = 4(\tau_1 - \tau_3) = 4\chi * u$, so apply Theorem 9.6, with $L(s)$ and $\zeta(s)$ the Dirichlet series for χ and u.

Chapter 11

11.1 4961 and 6480 are coprime (since $6480 = 2^4.3^4.5$, and $2, 3, 5$ do not divide 4961) and $4961^2 + 6480^2 = 8161^2$ (e.g. by calculator).

11.2 Since 1^2 and 2^2 are not sums of two positive squares, a Pythagorean triple must have $c \geq 3$. Since $a^2 < a^2 + 1^2 < (a+1)^2$, $a^2 + 1^2$ is not a square, so $b \geq 2$, and similarly $a \geq 2$. If $b = 2$ then $a^2 < a^2 + b^2 = a^2 + 4 < (a+1)^2$ since $a \geq 2$; thus $a^2 + b^2$ is not a square, so $b \neq 2$, and similarly $a \neq 2$. Thus $a, b, c \geq 3$. Let $k \geq 3$; if k is odd, then $k^2 = 2l + 1$ with $l \geq 1$, and $l^2 + k^2 = (l+1)^2$; if k is even, then $k^2 = 4l$ with $l \geq 2$, and $(l-1)^2 + k^2 = (l+1)^2$.

11.3 If $c = k$ then $a, b \leq k - 1$, giving only finitely many possibilities for a and b. If $a = k$ then $k^2 = c^2 - b^2 \geq c^2 - (c-1)^2 = 2c - 1$, so $b < c \leq (k^2 + 1)/2$, giving only finitely many possibilities for b, c.

11.4 Argue as in Exercises 11.2 and 11.3: $k = 3$ and $k = 4$ yield only $(3, 4, 5)$, $k = 5$ yields this and $(5, 12, 13)$, $k = 6$ yields only $(6, 8, 10)$, and $k = 7$ yields only $(7, 24, 25)$.

11.5 Primitivity implies a and b are not both even; if both are odd, then $a^2 + b^2 \equiv 2 \bmod (4)$, whereas $c^2 \equiv 0$ or $1 \bmod (4)$; hence one is odd and one even. Similarly, if $a, b \not\equiv 0 \bmod (3)$ then $a^2 + b^2 \equiv 2 \bmod (3)$, whereas $c^2 \equiv 0$ or $1 \bmod (3)$. Primitivity implies at most one of a, b, c is divisible by 5; since squares are $\equiv 0$ or $\pm 1 \bmod (5)$, at least one must be divisible by 5; hence exactly one is divisible by 5.

11.6 If $c = 2b$ then $a^2 + b^2 = 4b^2$, so $a^2 = 3b^2$. Thus $3|a$, so $3|b$ and hence $3|c$, say $a = 3a_1, b = 3b_1, c = 3c_1$, giving a smaller Pythagorean triple (a_1, b_1, c_1) with $c_1 = 2b_1$. Iterating, we get a contradiction by descent. This shows that $\sqrt{3} \neq a/b$ for any $a, b \in \mathbb{Z}$, so $\sqrt{3} \notin \mathbb{Q}$.

11.7 If $a = 2b$ then $c^2 = 5b^2$, giving $5|c$, so $5|b$ and hence $5|a$. Now imitate Exercise 11.6.

11.8 If a prime p divides uv, it divides u or v, so it divides u^2 or v^2 respectively; if p also divides $u^2 + v^2$ then it divides both u^2 and v^2, so it divides both u and v, which is impossible since $\gcd(u, v) = 1$.

11.9 Use descent, showing that each solution (x, y, z) generates another with smaller x. We may assume x, y are coprime, so (y^2, z, x^2) is a primitive Pythagorean triple. If y is odd then $y^2 = u^2 - v^2, z = 2uv, x^2 = u^2 + v^2$ giving $u^4 - v^4 = (xy)^2$, a solution with $u < x$. If y is even then $y^2 = 2uv, z = u^2 - v^2, x^2 = u^2 + v^2$ with u, v coprime. If

v is odd then $2u$ and v, being coprime with product y^2, are squares, say $2u = (2a)^2, v = b^2$, so $x^2 = 4a^4 + b^4$ and $(2a^2, b^2, x)$ is a primitive Pythagorean triple; hence $2a^2 = 2de, b^2 = d^2 - e^2, x = d^2 + e^2$ with d, e coprime, so $a^2 = de$ implies $d = f^2, e = g^2$, giving a solution $f^4 - g^4 = b^2$ with $f < x$. If v is even, use a similar argument with $u = a^2, 2v = (2b)^2$.

11.10 Let the area $ab/2 = d^2$, so $2ab = 4d^2$; since $a^2 + b^2 = c^2$, we get $(a + b)^2 = c^2 + 4d^2$ and $(a - b)^2 = c^2 - 4d^2$. Multiplying these gives $(a^2 - b^2)^2 = c^4 - (2d)^4$, so Exercise 11.9 gives $a = b$, contradicting Theorem 11.2.

11.11 Expanding the RHS shows that $b^p + c^p = (b + c)(b^{p-1} - b^{p-2}c + b^{p-3}c^2 - \cdots + c^{p-1})$. Replacing 3 with p, the argument in Section 11.8 shows these two factors are mutually coprime, so both are p-th powers. Now follow the argument in the text, using conditions (1) and (2) of Theorem 11.8, with p and q replacing 3 and 7.

11.12 If p does not divide $q - 1$, then $1 = \gcd(p, q - 1) = pu + (q - 1)v$ for some u, v; if $x \in U_q$ then $x^{q-1} = 1$, so $x = x^{pu+(q-1)v} = (x^u)^p$ is a p-th power; clearly $0 = 0^p$ is also a p-th power. If $q = kp + 1$ and g is a primitive root mod (q), then the p-th powers in U_q are the elements $(g^i)^p = g^{pi}$; we have $g^{pi} = g^{pj}$ if and only if $q - 1 | p(i - j)$, that is, $i \equiv j \mod (k)$, so the k classes $[i] \in \mathbb{Z}_k$ give k distinct p-th powers in U_q.

11.13 $0^p \equiv 0, 1^p \equiv 1, (-1)^p \equiv -1 \mod (q)$. If $a \not\equiv 0 \mod (q)$ then $(a^p)^2 = a^{q-1} \equiv 1$, so $a^p \equiv \pm 1$. Since $p \not\equiv 0$ or $\pm 1 \mod (q)$, this proves (2). For (1), $x^p, y^p, z^p \equiv 0$ or $\pm 1 \mod (q)$, so if $x^p + y^p + z^p \equiv 0$ then x^p, y^p or $z^p \equiv 0$ since $q > 3$, so x, y or $z \equiv 0$.

11.14 $p = 3, 5, 11, 23, 29, 41, 53, 83, 89$.

11.15 The 7th powers in $\mathbb{Z}_{29}$ are $[0]$ and (by Corollary 6.6) the 4th roots of $[1]$, namely $\pm[1], \pm[12]$, so conditions (1) and (2) follow by inspection. For $p = 13$ take $q = 53$; the 13th powers in $\mathbb{Z}_{53}$ are $[0], \pm[1], \pm[23]$.

11.16 $(-\zeta^r)^p = -\zeta^{rp} = -1$ for $0 \le r \le p - 1$, and these terms $-\zeta^r$ are all distinct, so $x^p + 1 = \prod_{r=0}^{p-1}(x + \zeta^r)$. Now put $x = a/b$ and multiply by b^p.

11.17 $1 + \zeta + \cdots + \zeta^{p-1} = (1 - \zeta^p)/(1 - \zeta) = 0$ since $\zeta^p = 1$. Hence if $z = a_0 + a_1\zeta + \cdots + a_{p-1}\zeta^{p-1}$ then substituting $-1 - \zeta - \cdots - \zeta^{p-2}$ for ζ^{p-1} gives $z = b_0 + b_1\zeta + \cdots + b_{p-2}\zeta^{p-2}$ with $b_r = a_r - a_{p-1}$. Subtracting two such representations of z would give $f(\zeta) = 0$ for some non-zero polynomial $f(x)$ of degree at most $p - 2$, with integer coefficients; among such polynomials, that of least degree must divide $\Phi_p(x)$ (otherwise a remainder of smaller degree vanishes at ζ); this contradicts the irreducibility of $\Phi_p(x)$.

11.18 The recurrence relation (9.12) in Chapter 9 implies that $B_n \in \mathbb{Q}$ for all n, by induction on n. Now $B_0 = 1, B_1 = -1/2, B_2 = 1/6, B_3 = 0, B_4 = -1/30, B_5 = 0, B_6 = 1/42, B_7 = 0, B_8 = -1/30, B_9 = 0, B_{10} = 5/66$, so odd primes $p \leq 13$ do not divide the numerators of $B_2, B_4, \ldots, B_{p-3}$, and hence are regular.

Bibliography

A. D. Aczel, *Fermat's Last Theorem: Unlocking the Secret of an Ancient Mathematical Problem*, Four Walls Eight Windows, New York, 1996.

T. M. Apostol, *Introduction to Analytic Number Theory*, Springer-Verlag, New York, 1976.

T. M. Apostol, A proof that Euler missed: evaluating $\zeta(2)$ the easy way, *Math. Intelligencer* 5 (1983), 59–60.

R. V. Churchill, *Fourier Series and Boundary Value Problems* (2nd ed.), McGraw-Hill, New York, 1963.

D. Cox, Introduction to Fermat's Last Theorem, *Amer. Math. Monthly* 101 (1994), 3–14.

H.-D. Ebbinghaus *et al.*, *Numbers*, Springer-Verlag, New York, 1991.

H. M. Edwards, *Fermat's Last Theorem. A Genetic Introduction to Algebraic Number Theory*, Springer-Verlag, New York, 1977.

F. Q. Gouvêa, "A marvellous proof", *Amer. Math. Monthly* 101 (1994), 203–222.

R. L. Graham, D. E. Knuth and O. Patashnik, *Concrete Mathematics*, Addison-Wesley, Reading, MA, 1989.

G. H. Hardy, *A Mathematician's Apology*, Cambridge University Press, Cambridge, 1940.

G. H. Hardy and E. M. Wright, *Introduction to the Theory of Numbers* (5th ed.), Oxford University Press, Oxford, 1979.

G. A. Jones and D. Singerman, *Complex Functions, an Algebraic and Geometric Viewpoint*, Cambridge University Press, Cambridge, 1987.

D. E. Knuth, *Art of Computer Programming, Vol. 1: Fundamental Algorithms*, Addison-Wesley, Reading, MA, 1968.

N. Koblitz, *Course in Number Theory and Cryptography* (2nd ed.), Springer-Verlag, New York, 1994.

E. Kranakis, *Primality and Cryptography*, Wiley, Chichester, 1986.

B. Mazur, Number theory as gadfly, *Amer. Math. Monthly* 98 (1991), 593–610.

P. Ribenboim, *13 Lectures on Fermat's Last Theorem*, Springer-Verlag, New York, 1979.

K. A. Ribet, Wiles proves Taniyama's conjecture; Fermat's Last Theorem follows, *Notices Amer. Math. Soc.* 40 (1993) 575–576.

R. L. Rivest, A. Shamir and L. M. Adleman, A method for obtaining digital signatures and public key cryptosystems, *Comm. ACM.* 21 (1978), 120–126.

H. E. Rose, *Course in Number Theory*, Oxford University Press, Oxford, 1988.

J.-P. Serre, *Course in Arithmetic*, Springer-Verlag, New York, 1973.

S. Singh, *Fermat's Last Theorem: The Story of a Riddle that Confounded the World's Greatest Minds*, Fourth Estate, London, 1997.

J. Stein, Computational problems associated with Racah algebra, *J. Comp. Physics* 1 (1967), 397–405.

R. Taylor and A. Wiles, Ring-theoretic properties of certain Hecke algebras, *Ann. Math.* 141 (1995), 553–572.

E. C. Titchmarsh, *Theory of the Riemann Zeta Function*, Clarendon Press, Oxford, 1951.

A. van der Poorten, *Notes on Fermat's Last Theorem*, Wiley, New York, 1996.

A. Wiles, Modular elliptic curves and Fermat's Last Theorem, *Ann. Math.* 141 (1995), 443–551.

Index of symbols

The symbol $\square$ is used in the text to mark the end of a proof. The following symbols, in regular mathematical use, are used without further comment:

$\mathbb{C}$ the set of complex numbers

$\mathbb{R}$ the set of real numbers

$\mathbb{Q}$ the set of rational numbers

$\mathbb{Z}$ the set of integers $\{0, \pm 1, \pm 2, \ldots\}$

$\mathbb{N}$ the set of natural numbers $\{1, 2, \ldots\}$

S^k the set of ordered k-tuples from a set S

$n!$ factorial n $(= 1.2.3 \ldots n)$

$\binom{n}{r}$ the binomial coefficient $(= n!/r!\,(n-r)!)$

$|x|$ the modulus of x

$\approx$ is approximately equal to

$\log_r a$ the logarithm of a to the base r

$\sqrt{n}$ the square root of n

∞ infinity

$\rightarrow$ tends towards, or approaches

$f'(x)$ the derivative of the function $f(x)$

$\wedge, \vee$ the logical connectives 'and' and 'or'

$\cap, \cup$ intersection and union

$\sum$ sum

$\prod$ product

$\times$ direct product

C_n cyclic group of order n

$\cong$ isomorphism of groups or rings (see Appendix B)

$\det(A)$ determinant of a matrix A

The following symbols are defined in the text, on the pages indicated, and are then used without comment.

$\lfloor x \rfloor$ the greatest integer $i \le x$ 2

$\lceil x \rceil$ the least integer $i \ge x$ 3

$b | a$ b divides a 3

$b \nmid a$ b does not divide a 3

gcd the greatest common divisor   5

lcm the least common multiple   12

f_n the n-th Fibonacci number   16

$\{x\}$ the closest integer to x 17

$\Phi_p(x)$ the p-th cyclotomic polynomial   21

min minimum   23

max maximum   23

$p^e \| n$ p^e is the highest power of a prime p dividing n 23

$\pi(x)$ the number of primes $p \le x$ 27

$\lg x$ $\log_2 x$, the logarithm of x to the base 2   27

$\operatorname{li} x$ $\int_2^x (\ln t)^{-1} dt$ 27

$\ln x$ $\log_e x$, the natural logarithm of x to the base e   27

F_n the n-th Fermat number $2^{2^n} + 1$ 30

M_p the Mersenne number $2^p - 1$ 31

$a \equiv b$ a is congruent to b modulo (n), also denoted by $a \equiv b \bmod (n)$ or $a \equiv_n b$ 38

$a \not\equiv b$ a is not congruent to b modulo (n) 39

$[a]$ the congruence class of a mod (n), also denoted by $[a]_n$ 40

$\mathbb{Z}_n$ the set of congruence classes mod (n) 40

$\mathbb{Q}_p$ the field of p-adic numbers   81

U_n the group of units mod (n) 85

$\phi(n)$ Euler's function $|U_n|$ 85

F^* the group of non-zero elements of a field F 103

$e(G)$ the exponent of a group G 116

$e(n)$ the universal exponent of n 116

Q_n the set of quadratic residues mod (n) 120

$\left(\frac{a}{p}\right)$ the Legendre symbol of a mod (p) 123

$\tau(n)$ the number of divisors of n, also denoted by $d(n)$ 144

$\sigma(n)$ the sum of the divisors of n 144

$\sigma_k(n)$ the sum of the k-th powers of the divisors of n 144

$u(n)$ the unit function, equal to 1 for all n 145

$N(n)$ the function equal to n for all n 145

$I(n)$ the identity function, equal to $\lfloor \frac{1}{n} \rfloor$ for all n 149

$\mu(n)$ the Möbius function   149

$f * g$ the Dirichlet product of f and g 157

$\chi(n)$ the function equal to $0, 1$ or -1 as n is even or $n \equiv 1$ or 3 mod (4) 162

$\tau_1(n)$ number of divisors d of n such that $d \equiv 1$ mod (4) 162

$\tau_3(n)$ number of divisors d of n such that $d \equiv 3$ mod (4) 162

$\Lambda(n)$ the Mangoldt function   162

$\zeta(s)$ the Riemann zeta function   164

B_n the n-th Bernoulli number   177

$\lambda(n)$ Liouville's function   182

σ_a the abscissa of absolute convergence   186

σ_c the abscissa of convergence   186

S_k the set of all sums of k squares   191

Index of names

Index